Diseases of Field and Horticultural Crops and Their Management

DISEASES OF FIELD AND HORTICULTURAL CROPS AND THEIR MANAGEMENT

Volume II

(As per ICAR Fifth Dean's Committee Recommendations)

BHUPENDRA SINGH KHARAYAT
Subject Matter Specialist (Plant Protection)
KVK Champawat
G.B. Pant University of Agriculture and Technology
Pantnagar, Uttarakhand (India)

PHI Learning Private Limited
Delhi-110092
2026

In fond memory of ***Shri Asoke K. Ghosh*** *(October 1942 – February 2024), Founder Chairman and Managing Director of PHI Learning, whose vision endlessly inspires.*

The Legacy Continues....

Published by Pushpita Ghosh, PHI Learning Private Limited, Rimjhim House, 111, Patparganj Industrial Estate, Delhi-110092 and Printed by Syndicate Binders, A-20, Hosiery Complex, Noida, Phase-II Extension, Noida-201305 (N.C.R. Delhi).

₹695.00

Diseases of Field and Horticultural Crops and Their Management—Volume II
Bhupendra Singh Kharayat

ISBN-978-81-19364-65-7 (Print Book)
ISBN-978-81-19364-63-3 (e-Book)

For

B.Sc. (Hons) Agriculture Students

Contents

Preface

The book is based on the latest syllabus of ICAR. The Fifth Dean's Committee of ICAR has revised and recommended a new syllabus for B.Sc. (Hons) Agriculture in 2016. *Diseases of Field and Horticultural Crops and their Management,* Volume II, is a core course with 3(2+1) credit hours and is taught in the sixth semester. There has been a long-felt need of a lucid, concise, updated, and to-the-point book on crop diseases and their management for the students and teachers The book covers symptoms, etiology, disease cycle and epidemiology, and management of important diseases of the following crops: Field Crops, such as Wheat, Sugarcane, Sunflower, Mustard, Chickpea, Lentil, Cotton, Pea, and Horticultural Crops, such as Mango, Citrus, Grapevine, Apple, Peach, Strawberry, Potato, Cucurbits, Onion and Garlic, Chillies, Turmeric, Coriander, Marigold and Rose. Some additional diseases are included, such as "Huanglongbing (HLB) of Citrus", "Botrytis Grey Mold of Strawberry", "Wart Disease of Potato," and "Gummy Stem Blight of Cucurbits" in this book considering their destructive nature and economic importance, particularly in North India. The book will definitely be useful for the readers to clear the concept and enhance the understanding of plant diseases and their management as the special emphasis is given in the writing of disease diagnosis, morphology and biology of pathogens, disease cycle and epidemiology, and crop disease management. The revised and latest names of the pathogens and latest recommended practices for plant disease management have been included in the book. Additional lists of important crop diseases are also given in the book. This volume also includes exercise such as MCQs and Match the Following questions with answers which are highly useful for ICAR-JRF and other examinations such as ICAR-SRF, ICAR-NET, IARI, BHU, AO, AFO, Bank PO, IFS, CSE-UPSC exams, etc. Healthy critics and valuable suggestions from the students, teachers, and readers will be highly welcomed for the continuous improvement of the book and these will be incorporated in subsequent editions. The readers can send their feedback at phi@phindia.com or bhupendrakharayat@gmail.com. I hope the book will be highly useful for students, teachers, scientists, plant protection specialists, horticulture or agriculture offices, and farmers. In the end, I would like to suggest that the students, teachers, and readers must watch the published videos on different crop diseases on my YouTube channel “e-Plant Health”.

Bhupendra Singh Kharayat

Acknowledgements

I wish to express my deep sense of gratitude and indebtedness to all those who helped me directly or indirectly during the preparation of this book. First and foremost, I am extremely thankful to all my family members for their constant encouragement and support shown throughout the preparation and publication of Volume I & II of the book. I am deeply indebted and highly thankful to all researchers who permitted me to use their work. I am also thankful to my guide Dr. Yogendra Singh (Professor, Plant Pathology, G.B.P.U.A. & T., Pantnagar). I am also thankful to my colleague Dr. Rajani Pant, Dr. S.C. Bhatt, and my friend Mahesh Pokharia for their continuous support and for being a continuous source of energy during the preparation of this manuscript. I am also thankful to all my students for their feedback and encouraging words while writing these books. My sincere and special thanks to all team members of PHI Learning Private Limited, for extending their support in processing the manuscript, and early publication of the book.

Bhupendra Singh Kharayat

1

Wheat (*Triticum aestivum* L.)

1.1 RUSTS

On 25th April each year, the Romans had a festival called the "Robigalia" during which a procession was marched out to the sacred trees where priests prayed to Robigus to spare the crops. In addition to prayer, the wine was poured on the altar, and the sacrifice of sheep, reddish dogs, foxes, and cows were done in an attempt to appease the Robigus in the belief that he would not send the rusts to destroy their crops. Nobody knew at the time that barberries (*Berberis vulgaris)* were an alternate host in the stem rust life cycle (*Puccinia graminis*), it was only in 1865 that a famous mycologist, Heinrich Anton de Bary discovered the complete life cycle of the stem rust and demonstrated that *P. graminis* required two different hosts (Wheat and Barberry bush) during the different stages of its development and thus discovered the heteroecious nature of wheat stem rust pathogen. Subsequently, Johan Hubert Craigie (1927), a Canadian pathologist, successfully demonstrated that pycnia are the sexual structure of rust fungi. He designated two mating types (+) and (–) for haploid pycnia (spermatia and receptive hyphae) in wheat black rust (*Puccinia graminis* f. sp. *tritici*). The systematic investigations on cereal rusts in India were initiated by Rai Bahadur and Karam Chand Mehta of Agra College in 1922–23. A historic account of wheat rust epidemics in India has been given by Nagarajan and Joshi (1975). Both black and yellow rusts of wheat can cause 100% losses whereas brown rust can cause 50% yield loss. Yellow and stem rusts are the most widely studied diseases worldwide. On wheat, leaf, stripe, and stem rusts are distinguished from each other based on the color, size, and arrangement of the pustules on the plant surface and the plant part typically affected.

Almost wherever wheat is grown, production is significantly constrained by one or more of the three rust diseases. These include leaf rust (*Puccinia triticina*); stem rust (*P. graminis* f. sp. *tritici*); and stripe rust (*P. striiformis* f. sp. *tritici*). *Puccinia* is an obligate biotroph, which can infect wheat during its growing period. Among the three wheat rust pathogens, leaf rust causes less damage as compared to stripe and stem rust, however, due to its recurrent global occurrence, it would be anticipated that total annual losses incurred would be greater than any other rust. In India, stem rust and stripe rust are geographically constrained. Both stem rust and stripe rust can cause up to 100% crop loss, whereas leaf rust can cause 45%–50% crop loss, under favorable climatic conditions. In India stripe rust of wheat (caused by *Puccinia striiformis* f. sp. *tritici* is a threat in 10 million hectares of Northern India, whereas stem rust (caused by *P. graminis* f. sp. *tritici* threatens about 7 million hectares of Central and Peninsular India. In contrast, leaf rust (caused by *P. triticina*) is prevalent wherever wheat is grown.

1.1.1 Stem or Black Rust

Black stem rust of wheat is a major concern for the world's food security, causing crop losses of up to 70%. Black stem rust of wheat caused serious disease outbreaks in North America during the first half of the 20th century, with wheat yield losses estimated at 19%–28% in some states. Anton deBary first demonstrated the heteroecious life cycle of a rust fungus with *Puccinia graminis*, the causal agent of stem rust. Anton deBary, in 1865, first recognized the nature of the heteroecious life cycle. In 1884, Eriksson discovered host–specific subspecies or «special forms» of the fungus. Each special form is designated in Latin as a *forma specialis* or "f. sp." All of the *formae speciales* have an identical appearance but vary in host range. About 1916, E.C. Stakman and others determined that within *P. graminis* f.sp. *tritici* are further genetic subdivisions called races. Later, races were found within other *formae speciales* as well. Nagarajan and Singh gave "*Indian Stem Rust Rules*" in 1975.

Symptoms

Stem rust causes blister–like lesions or pustules or sori that are large, oval to spindle–shaped or elliptical or elongated and full of orange–red to dark–brick red powdery spores (urediniospores) and is most commonly observed on stems, both sides of leaves, and leaf sheaths of infected plants (Figure 1.1). However, pustules can also be present on awns, glumes, spikelets, and seeds. The uredial pustules erupt irregularly through the host epidermis and are surrounded by

Figure 1.1 Stem Rust of Wheat. (Source: USDA ARS).

tattered host tissues. This is one of the most characteristic features of stem rust and helps to distinguish stem rust from leaf rust. Numerous urediniospores are release from young pustules. Later uredinia convert to telia by producing dark–colored, overwintering, non–infectious, thick walled, teliospores, hence the name black rust. Sometimes telia may develop independently of uredia. Severe infections weaken stems, leading to lodging and shriveled, prematurely ripened grain. On barberry, the alternate host, pycnia appear usually on the upper leaf surfaces, often as in small clusters, and exude pycniospores in sticky honeydew. Later raised yellow–orange aecial cups produce on the underside of leaves which produce aeciospores.

Causal Organism and Etiology

Stem rust or black rust or black stem rust of wheat caused by the fungus *Puccinia graminis* f. sp. *tritici*. It belongs to the kingdom Fungi, phylum Basidiomycota, class Urediniomycetes, order Uredinales, family Pucciniaceae, and genus *Puccinia*. The hyphae are intercellular and send haustoria into the host cells. Pycnia are flask–shaped and consist of pycniosporophores and receptive hyphae with a pycnial chamber in the center. Aecia are cup–shaped and are enclosed by a peridium. The aeciospores are echinulate, have six germ pores, and arise in chains from the base of the aecial cup. Aeciospores are capable of infecting the grass host, that is, wheat. The uredospores are stalked, oval, brown, dikaryotic, and single–celled with spiny thick walls. Teliospores are firmly fixed to their stalk, diploid, two–celled, with a thick and smooth wall. Teliospore germinates to produce basidiospores which infect the alternate hosts such as common barberry. Basidiospores are colorless, small, ellipsoid, elongate, unicellular, uninucleate, haploid, and thin–walled.

Disease Cycle and Epidemiology

Puccinia graminis is typically a macrocyclic (produce all five types of spores), heteroecious (requires two host i.e. a cereal grain and barberry, to complete its life cycle) and pleomorphic rust fungus (i.e., produce more than one type of spores). The wheat plant on which the fungus passes its dikaryophase is called the primary host and the barberry is called the secondary or alternate host in which the haplophase is seen. *Puccinia graminis* with a complete life cycle have five different spore forms, numbered 0 to IV, in addition to their specific names (Table 1.1).

Table 1.1 Different spore forms in the life cycle of Puccinia graminis.

Stages	Sequence of Spore Producing Structures and Spores	Nuclear Condition
Stage 0	Pycnium, bearing spermatia and receptive hyphae.	Homokaryotic, monokaryotic, haploid (N)
Stage I	Aecium, bearing aeciospores.	Homokaryotic, monokaryotic, haploid (N)
Stage II	Uredinium, bearing uridiniospores.	Hetrerokaryotic, dikaryotic (N + N)
Stage III	Telium, bearing teliospores.	Hetrerokaryotic, dikaryotic (N + N)
Stage IV	Basidia, bearing basidiospores.	Hetrerokaryotic, dikaryotic (N + N)

Note:—Spore forms 0 (pycnia) and I (aeciospores) occur on barberry and spore forms II (urediniospores), III (teliospores), and IV (basidiospores) occur on wheat.

Rusts often have alternate hosts and this cereal pathogen infects barberry, *Berberis vulgaris*, and *Berberis canadensis*. The basidiospores of *Puccinia graminis* are unable to infect the cereal host but will infect the alternate host, barberry. Aceiospores produced on *Barberis*, infect wheat leaves through stomata, and from dikaryotic mycelium, which forms the uredinia and, later, the telia. But in the plain of India, the primary infection is caused late in the wheat seasons by urdiniospores, blown–down from hills. Urediniospores spread the infection, while the teliospores, after the dormant phase, germinate and form basidiospores that infect the alternate host (e.g. *Barberis*) and form the pycnia and aecia. Sexual reproduction is achieved by the transfer of gametes, called spermatia (singular spermatium), which are transferred between pustules on barberry leaves by insects. Black stem rust is favored by hot days (25°C–30°C), mild nights (15°C–20°C), and wet leaves from rain or dew. Both aeciospores and urediniospores, which infect wheat, require free water for germination. A single uredinium may contain approximately 50,000 to 4,00,000 spores and there may be 4–5 generations of urediniospores in the growing season of wheat. Urediniospores serve as "repeating spores" or conidia.

Ug99: A New Race of Wheat Stem Rust

The Ug99 strain (race TTKSK) of the wheat stem rust pathogen *Puccinia graminis* f. sp. *tritici* (Pgt) presents a significant threat to global wheat production. Races of Pgt are defined by the combination of host resistance genes for which a given Pgt isolate displays either virulence or avirulence. In 1998, a new race of stem rust fungus with a notably unique virulence to resistance gene *Sr31* located on translocation 1BL, was found in Uganda and subsequently named Ug99. Since then it has spread through Africa and the Middle East causing devastating epidemics. Genes *Sr22*, *Sr25*, *Sr26*, *Sr33*, *Sr35*, *Sr45*, and *Sr50* are, at present, possibly the most useful race–specific resistance genes, used in combinations.

1.1.2 Leaf or Brown Rust

Wheat leaf or brown rust is one of the most damaging diseases. Leaf rust, is the most common and widely distributed of the three wheat rust diseases. Leaf rust damages are lesser than stem and yellow rust, but its harmful effects cause more annual damage, due to its prevalent occurrence. It alone can cause 30%–50% yield losses under favorable environmental conditions and by growing the susceptible cultivar. In China, leaf rust develops on approximately 15 million hectares of wheat, causing a yield loss of approximately three million tons annually.

Symptoms

Leaf rust pustules are dusty, lighter brown to orange, small, circular to oval, erumpent (breaking through the host surface) and mostly develop only on the upper surfaces of leaves and rarely on leaf sheaths, glumes, and awns. As the disease progresses, pustules (uredinia) can cover nearly the entire upper leaf surface (Figure 1.2). Leaf rust lesions do not cause conspicuous tearing of epidermal tissues like stem rust. Later, the pustules turn from orange (uredinia) to black (telia). Telia develop beneath the epidermis but it does not rupture and telia do not expose. They produce brown–black teliospores. Often multiple spore stages of the fungus can be seen on a single leaf. Telia are not always formed especially if rust infections occur late in the growing season.

Figure 1.2 Leaf Rust of Wheat. (Courtesy: Dr. Erick D. De Wolf, Kansas State University, Throckmorton, Manhattan, KS, USA).

Causal Organism and Etiology

Wheat leaf rust is caused by an obligate biotroph, basidiomycete fungus *Puccinia triticina*. Leaf rust is characterized by the uredinial stage. Uredinia are orange to brown, up to 1.5 mm in diameter, erumpent, round to ovoid, with uredinia that are scattered on both the upper and the lower leaf surfaces of the primary host. Urediniospores produced by uredinia are orange–brown, sub–globoid, average 20 μm in diameter, and are, with 7–10 germ pores, scattered in thick, echinulate walls. The teliospores (16 μm in width) are similar to those of *Puccinia striiformis*, thick smooth walled, dikaryotic, brown–black and are 2–3 celled. Teliospores allow the rust to survive the hot and dry summers. Teliospores germinate and produce basidiospores.

Disease Cycle and Epidemiology

The fungus *Puccinia triticina* is also a macrocyclic with five spore stages and heteroecious rust, requires a telial/uredinial host (usually wheat) and an alternative (pycnial/aecial) host (*Thalictrum speciosissimum* or *Isopyrum fumaroides* or *Anchusa* or *Clematis*) to complete life cycle. But the pycnial and aecial stages are not reported from India. The alternate host currently provides little direct inoculum of *P. triticina* to wheat. The leaf rust pathogen survives the period between wheat crops in many areas on a green–bridge of volunteer (self–sown) wheat. Urediniospores are wind–disseminated to very long distances (thousands of kilometers) from their source plant and infect host plants. Wind–blown uredospores from hills cause the primary infection. They germinate and penetrate through stomata on a lower surface and form dikaryotic mycelium. Urediniospores germinate in 30 minutes after coming in contact with free water at temperatures of 15°C–25°C. At optimum and constant temperature, spore germination to sporulation can occur within 7–10 days. At low temperatures (10°C–15°C) or diurnal fluctuations, longer periods are required. Urediniospores are produced on wheat hosts and re–infect (secondary infections) the telial host with free water on the leaf surface and temperatures of 10°C–25°C. As the host plant matures and the uredinial infections develop teliospores. Teliospores can germinate shortly after development, and basidiospore infection can occur throughout the wheat growing cycle. The leaf rust development is favored by mild days, 20°C–25°C, and nights, 15°C–20°C, with adequate moisture for night–time dew development.

1.1.3 Stripe or Yellow Rust

Stripe rust, also called yellow rust, is an old and devastating disease of wheat (*Triticum aestivum* L.) caused by the biotrophic fungus *Puccinia striiformis* f. sp. *tritici* (*Pst*). Yellow rust can cause significant crop yield reduction in highly susceptible wheat varieties with average losses from 10%–70%. The pathogen reduces grain yield and quality. Stripe rust is the current major rust disease affecting winter cereal production across the world.

Symptoms

Initial symptoms of stripe rust appear as chlorotic patches or yellow spots or flecks or yellow streaks (pre–pustules), followed by small, oval, or elongated, bright yellow pustules (containing masses of yellow–colored urediniospores), arranged in a distinct linear patterns along the leaf veins, hence the name, stripe rust (Figure 1.3). Pustules can also form on leaf sheaths, heads, stalks, awns, and glumes and occasionally immature grains can be infected. Mature pustules erupt and release yellow–orange masses of urediniospores. Under favorable conditions, the entire leaf surface may be covered with pustules. Later, dark brown to black telial pustules develop under the epidermis and form streaks on the leaves and leaf sheaths. The telia are sub–epidermal and remain as flat black crusts. As the disease progresses, tissues around the pustules turn brown and dry, resulting in a scorched appearance. Stripes or necrotic spots vary in size depending on cultivar resistance and environmental conditions. Yellow rust can be distinguished from leaf rust by its yellow pustules which are arranged in stripe–like rows following the leaf veins. However, the oval, darker–colored pustules are produced by leaf rust.

Causal Organism and Etiology

Yellow rust is caused by the biotrophic fungus *Puccinia striiformis* f. sp. *tritici* (*Pst*). Urediniospores are yellow to orange, binucleate, unicellular, broadly ellipsoidal to broadly obovoid, echinulate, and with 6–18 scattered germ pores. The spore wall is colorless. At higher temperatures or during the later grow the stages of the host, black telia are often produced, which are pulvinate to oblong. The teliospores are dark brown, predominantly two–celled, with thick walls, mostly oblong–clavate, and rounded or often flattened at the tip.

Figure 1.3 Yellow or Stripe Rust of Wheat. (A) Yellow pustules on leaf, (B) Yellow rust symptoms on wheat glumes, (C) Pycnia (spermogonia) on upper surface of barberry leaf, and (D) Aecium on lower surface of barberry leaf.

Disease Cycle and Epidemiology

Puccinia striiformis f. sp. *tritici*, is also like the other two rust, a macrocyclic, heteroecious fungus that requires both primary (wheat or grasses) and alternate (*Berberis* or *Mahonia* spp.) host plants to complete its life cycle. Every year, the inoculum (uredospores) causes the annual reoccurrence starting from hills to plains. In India, the survival of the fungus is through its uredial stage on grass hosts at high altitudes (at 7,000 feet and above) on hills. The grass hosts include *Bromus japonicus* found in the plains also. Urediniospores have the capacity for wind dispersal over long distances (thousands of kilometers) from the initial infection sites. Pathogen survives in the cool temperatures of hills (Himalayas) and the primary infection takes place by mid of December in the foot–hills and sub–mountainous parts of north–western India. Urediniospores are the only known source of inoculum for wheat, and they germinate at low temperatures and cause infection on the wheat. On germination of urediniospores, the germ tube forms a small fragile appressorium over a stoma of the leaf. The infection peg from the appressorium enters through the stomatal opening and forms a large, thick–walled cylindrical sub–stomatal vesicle. From this, an infection hypa arises. The short branching hyphae form club–shaped haustoria. Due to its polycyclic nature, stripe rust epidemics can reach high infection rates in stripe rust–susceptible wheat cultivars, especially when temperature and relative humidity are favorable for disease development. Combined, temperature and relative humidity regulate several critical stages of *Puccinia striiformis* f. sp. *tritici* lifecycles such as spore germination, infection, latent period, sporulation, spore survival, and host resistance, all of which influence epidemic onset. Minimum temperature in the range of 7°C–13°C coupled with 85%–100% relative humidity during the night and maximum temperature in the range of 15°C–24°C during the day are congenial for infection, development, and spread of disease. Frequent rains along with wind during February–March may further enhance the disease severity and spread. Under natural conditions, it is common to find wheat co–infected with stripe rust, stem rust, and leaf rust. The difference in stem rust, leaf rust, and stripe rust of wheat is given in Table 1.2

Table 1.2 The difference in stem rust, leaf rust, and stripe rust of wheat.

Features	Stem rust/Black rust	Leaf/Orange/Brown rust	Stripe/Yellow rust
Causal organisms	*Puccinia graminis* f. sp. *tritici*	*Puccinia triticina*	*Puccinia striiformis* f. sp. *tritici*
Symptoms	Random, oblong pustules with torn margins, larger than those of leaf rust and stripe rust.	Random, circular to oval pustules	Small closely packed circular pustules during the vegetative stage, become stripes along the leaves of older plants
Pustule color	Orange–red to dark–red	Orange–brown	Orange–yellow
Pustule shape and size	Oval–shaped or elongated; small to large	Round or slightly elongated; small to medium	Round, blister–like; small
Pustule arrangement	Single and random	Single and random	Stripes
Plant part affected	Both sides of leaves, leaf sheaths, mainly stem, outside of head	Upper surface of leaf and leaf sheath	Upper surface of leaf, leaf sheath, awns, and inside glumes
Favorable environmental conditions	18°C–30°C, free moisture (dew/rain/irrigation)	15°C–20°C, free moisture (dew/rain/irrigation)	10°C–15°C, cool and wet conditions
Alternate host	*Barberries* spp., *Mahonia* sp.	*Thalictrum* sp., *Isopyrum* sp.	*Barberries* spp.
Races	240 races; only 14 are reported from India	100 races; only 14 are reported from India	14 races; only 10 are reported from India

Management

Cultural Methods

- Use balanced dose of fertilizers as per the local recommendation and avoid the excess use of nitrogenous fertilizers.
- Avoid water logging in the field.
- Collect and destroy infested crop debris and maintain field sanitation.

Host Resistance

- Grow yellow rust–resistant varieties like Unnat, PBW 644, PBW 723, PBW 725, PBW 550, PBW 752, WHD 943, PDW 291, PBW 660, HD–3043, HD–3059, HD3083, WH–1105, H–3086, etc.
- The variety HS 542 (Pusa Kiran) is resistant to stripe and leaf rust.
- The variety HW 1098 (Nilgiri Khapli) has high degree of resistance to stem, leaf, and yellow rust.

Chemical Control

- If yellow rust is noticed in the wheat crop, it is advised to spray the crop with Propiconazole (Tilt) 25 EC @ 0.1% (1 ml/litre) using the power sprayer or tractor mounted boom sprayer @ 0.5 litre per hectare of the fungicide shall be sufficient to effectively cover the wheat crop.

1.2 LOOSE SMUT

Loose smut is a common disease throughout the wheat–growing regions of the world. This seedborne disease is commonly present in wheat fields.

Symptoms

Entire ear, except the rachis and awns, are smutted. The grains are replaced by black charcoal like powder of teliospores of the fungus (Figure 1.4). Initially, the spore mass is covered by a white, silvary membrane of host origin, but later the membrane ruptures exposing the dry

Figure 1.4 Loose Smut of Wheat. The grains are replaced by black charcoal like powder.

olive brown to black, spore mass that are blown away by wind. The smutted head disintegrate easily and, only the rachis is left behind. The disease is called loose smut because the spores produced on smutted heads are easily removed by wind and rain. The infected plants appear normal until ear formation. Initially, the diseased plants show normal growth, slightly taller plants, and ears emergence little earlier than surrounding healthy plants. Before ear emergence, the diseased plants may have characteristic dark green flag leaves with chlorotic streaks, later turning yellow and necrotic.

Causal Organism and Etiology

Loose smut of wheat is caused by the basidiomycete fungus *Ustilago tritici*. *Ustilago* is now classified in class: Ustlilaginomycetes, subclass: Ustilaginomycetidae, order: Ustilaginales and family: Ustilaginaceae. The fungal mycelium is septate, binucleate (diplophase), irregularly lobed, hyaline to brown in color. *Ustilago tritici* produces a very large number of teliospores which are pale olive, spherical to oval, have minute echinulate walls. Mature spores are diploid and thick–walled and called smut spores or teliospores or chlamydospores.

Disease Cycle and Epidemiology

The disease is internally seedborne and monocyclic. The fungus overwinters as mycelium in the infected seed. Infected seeds are symptomless and appear normal. When the infected seed germinates, the fungus becomes active and grows systemically into the young shoots to the growing point of the plants; eventually, such systemically infected plants bear smutted ears. The hyphae grow intercellularly, deriving nutrition from the host cells through haustoria. The hyphae collect in the ovaries and the hyphal cells round off, transform into thick–walled teliospores. The teliospores are covered by the ovary wall and when it ruptures the spores get free and dispersed. Wind, rain, and insects carry the spores from a smutted head to the flowers of a healthy head. The spores germinate quickly under moist conditions on the feathery style of the ovary and form a septate promyclium. Each of the promycelium contains a haploid nucleus, formed after karyogamy and meiosis. A thin, hypha–like "infection thread" originates from each cell, which represents the basidiospores. Two infection threads of opposite mating types fuse to form a dikaryotic hypha, which penetrates the ovary wall, ovule, and finally infects the embryo. The fungus remains dormant as mycelium in the embryo and becomes internally seedborne. Such internally infected seeds give rise to systemically–infected plants. There is no secondary infection and the disease is monocyclic. The spores germinate quickly under moist conditions and grow down the stigma and pistil (female parts of the flower) to infect the young embryo (seed). Infections occur only during flowering which is favored by a cool and moist weather. Teliospores germination and disease development are favored by rainfall or wet weather and cool to moderate temperatures (16°C to 25°C) and high relative humidity (95%).

Management

Cultural Methods

- Use pathogen–free seeds in the healthy field.
- For seed production, disease–free field/areas should be identified for having crops without considerable inoculum load.

- Rouge out suspected diseased ears at the time of ear emergence and destroy them by burning or burying them under the ground to reduce the infestation of healthy seeds at a later stage. In general the plants showing yellowing of the boot leaf tip give smutted ear heads on emergence.

Physical Control

- **Hot water treatment**: Presoaked seed in normal water for 4 hours then dip in hot water at 52°C for 10 minutes then cool and dry the seeds in shade.
- **Solar heat treatment**: In India, Luthra and Sattar (1954) evolved this method against loose smut of wheat. This method is practiced in the hot months of May to June. In this method, the seeds are soaked in normal water for 4 hrs (8 am to 12 noon) on a bright hot summer day. After this pre–soaked seeds are spread in a thin layer and dry in the sun for 4 hrs (12 noon to 4 pm). The dormant mycelium becomes active on soaking the seeds in the water and dies when the seeds are dried in the hot sun and seeds become disinfected.
- **Anaerobic treatment**: Soak the seeds in water for 6 hours, then drain and keep them in airtight container at 28°C for 30 hours.

Chemical Control

- Treat the seeds with fungicides such as Benomyl 50% WP @ 2 g/kg seeds, or Carbendazim 50% WP @ 2 g/kg seeds, or Carboxin 75% WP @ 2–2.5 g/kg seeds, or Tebuconazole 2% DS @ 0.2 kg/10 kg seed, or Carboxin 37.5% + Thiram 37.5% DS @ 3.0 g/kg.

1.3 KARNAL BUNT

Karnal bunt is one of five bunt and smut diseases that affect wheat throughout the world. Karnal bunt was first identified by Manoranjan Mitra in 1931, in an infested experimental field at Botanical Experimental Station at Karnal, India. Karnal bunt is also known as new bunt or kernel smut or partial bunt. The green revolution mega cultivar "Sonalika" was the first one to fall prey to Karnal bunt, only to be followed by the other popular varieties viz. "HD2009," "WL711," and "UP262". Earlier it was a minor disease found only in Northwestern India. Karnal bunt is a limiting factor in wheat export because the pathogen is regulated by most countries as a quarantine pest. Seedborne nature of this disease threatens wheat trade between countries. The minimum seed certification standards prescribed for this disease is 0.05% and 0.25% for foundation and certified seed, respectively. For export purposes, it is 0%. The disease reduces seed quality, changes the chemical composition of infected grains, and makes seed inedible. Wheat containing 3% bunted grains is unfit for human consumption. In India, the disease is endemic to northwestern regions, foothills of the north with sporadic occurrence in Bihar, West Bengal, Gujarat, and Madhya Pradesh. The disease is of little agronomic significance, but it is the subject of strict quarantine control by several wheat–importing countries, and therefore a major biosecurity concern for wheat–exporting countries that are free of the disease.

Symptoms

The symptoms become evident only when grains fully developed. Not all the grains in a spike are affected, only a few grains and generally, 5–6 grains, rarely whole, are infected per spike. Hence, it is named partial bunt. The infected spikes can be detected by the shiny silvery black spikelets with glumes spread apart and swollen ovaries. The pathogen converts the infected ovary into fungal sorus where a mass of dark brown teliospores are produced which is covered by the pericarp (Figure 1.5). The pericarp often ruptures during harvest exposing the teliospores as a black, dusty mass. The embryo is not infected but gets shriveled under very severe infection. Partially smutted kernels may retain viability and produce healthy plants after germination. The disease is characterized by a rotten fishy odor, which is due to the presence of trimethylamine secreted by black powder or teliospores.

Figure 1.5 Karnal Bunt of Wheat.

Causal Organism and Etiology

Karnal bunt disease of wheat is caused by the hemibiotrophic fungus *Tilletia indica* (syn. *Neovossia indica*). It belongs to the phylum Basidiomycota, sub–phylum Ustilaginomycotina, and order Ustilaginales in the family Ustilaginaceae, which is a family of smut fungi. *Tilletia indica* Mitra is a hemibiotrophic and partially systemic pathogen of wheat, durum wheat, and triticale. It is a heterothallic (producing opposite mating types for sexual reproduction and characterized by the presence of individual self–sterility) fungus with bipolar mating (governed by a single allelic mating locus) system. The pathogen undergoes sexual reproduction after teliospore germination. Primary and secondary sporidia or hyphae as compatible mating types must fuse to form a dikaryon which enhances the chances of variation due to heterozygosity that plays a significant role in the production of new variants.

Disease Cycle and Epidemiology

Karnal bunt is a seed–, soil– and airborne disease. Teliospores are quite resistant propagules and survive for over 3 years in the soil and serve as a main source of inoculum. The teliospores get mixed into the soil at the time of harvesting, threshing or winnowing and sowing of infected and teliospore–contaminated seed. Infection occurs through airborne propagules, the allantoid sporidia. It is a monocyclic disease in which teliospores germinate in the presence of moisture on soil surface during the flowering stage of the crop to produce the large number of monokaryotic haploid, primary sporidia. The primary sporidia either produce secondary sporidia or germinate to develop hyphal mass which in turn generates allantoid (banana–shaped) and filiform secondary sporidia. The sporidia become airborne and land on plant surfaces where they may germinate and produce additional generations of allantoids–shaped sporidia. The sporidia germinate and produce infection hyphae which penetrate the stigma, ovary wall, or glume directly and grow into the young kernel. Disease remains localized in the kernel and does not develop systemically. Individual cells of the fungal hyphae are converted to teliospores as the kernel matures and are released when the sorus ruptures at harvest. The teliospores can be disseminated within and between fields and also through farm machinery, humans, animals, birds, feces of animals fed with bunted grains, straw contaminated with teliospores, irrigation water, or wind currents. Environmental conditions during flowering and the susceptibility of the host determine the severity of Karnal bunt. There is no secondary infection and the disease is monocyclic. Temperatures range of 18°C–22°C (minimum of 8°C–10°C) and free moisture are optimal for germination of teliospores, growth of the fungus, multiplication of allantoid sporidia, and infection. Relative humidity greater than 80% and cloudy weather with frequent rainfall during anthesis stages are most favorable for disease development.

Management

Cultural Methods

- Use healthy seeds for sowing.
- Delay sowing and increase row spacing during sowing.
- Decrease the seed rate during sowing.
- Intercropping and crop rotation with the non–host crops for two consecutive years can also reduce the disease by affecting the soil–borne primary inoculum.
- Changing the rice–wheat–rice cropping system to Maize–wheat cropping system reduces the teliospores populations in soil.
- Burn or destroy the infected crop residues after harvesting.
- Solarize the soil by mulching with a polyethylene tarp to raise soil temperature that reduces viability and germination of teliospores buried up to 10 cm deep in the soil.
- Effective land leveling and drainage can reduce disease incidence. In low–lying areas of the field accumulate water and are more prone to karnal bunt.
- Avoid over–fertilization.
- Avoid plant lodging using a balance dose of nitrogen and potash.

- Avid overhead irrigation.
- Avoiding irrigation during the period of awn emergence and end of flowering may hinder disease development.

Physical Control

- Treat the seeds with hot water at 60°C for 10 minutes.
- Treat the seeds with dry heat at 120°C–140°C for 10 minutes.

Chemical Control

- Treat the seeds with Vitavax @ 2–2.5 g/kg of seed to eliminate the seedborne infection.
- Treat the seeds with Formaldehyde, Ethanol, and Sodium hypochlorite.
- Propiconazole (Tilt) 25% EC or Bitertanol 25% WP @ 200 ml in 300 litres of water/acre.
- The foliar application of fungicides such as Mancozeb @ 0.25% or Carbendazim @ 0.1% or Triadimefon @ 0.2% or Cyproconazole (SAN 619) 0.1% or Diniconazole (S308) 0.1% or Folicur @ 0.2% or Contaf (0.1%) or Propiconazole (Tilt) @ 0.1% at the anthesis stage is also the most effective option.

1.4 POWDERY MILDEW

Powdery mildew is one of the most damaging diseases and the fourth most important disease of wheat after three rusts. Typically it can reduce wheat yield by 10%–35% and 50%–100% in severe cases. In India, it is mainly found in lower hills and some plain regions of Rajasthan.

Symptoms

The symptoms of powdery mildew are characterized by a powdery white to gray, cottony fungal growth on leaves, leaf sheaths, stems, floral bracts, heads, and awns (Figure 1.6). The fluffy white pustules begin as small white circular patches of fungal mycelium often surrounded by chlorosis, mostly visible on the upper leaf surface. Yellowish spots below the fungal growth can often be observed on the underside of leaves. The lower leaves may be completely covered with white fluffy fungal growth. Areas in and around fungal pustules remain green while the rest of the leaf appear chlorotic. These green areas are known as "green–islands" and are sites of photosynthetically active host tissue. Under favorable conditions, pustules often merge together and form large masses of fungal growth on the upper surface of leaves. As the diseased area expands, tissue around older pustules dies and turns grey–brown. In severe infection, the entire leaf turns yellow, deform, twist, and eventually collapse and die. The pathogen reduces photosynthesis and increases respiration and transpiration rates in host leaves due to severe chlorosis. Heavily colonized plants lose vigor, impairing heading and grain filling. If the infection occurs at the milking stage, the grains are dried and shriveled. The white powdery pustules produce large amounts of asexual conidia (also called oidia). As plants mature, small globose black dot–like sexual fruiting bodies called chasmothecia (formerly cleistothecia), develop within the fluffy growth of older lesions on infected plant parts.

Figure 1.6 Powdery Mildew of Wheat. (A) Powdery white to grey, cottony fungal growth on stem, (B) Powdery mildew symptoms of wheat ear, (C) Small globose black dot (chasmothecia) on glume and awns.

Causal Organism and Etiology

Powdery mildew of wheat is caused by the obligate biotrophic ascomycetous fungus *Blumeria graminis* f. sp. *tritici*. Oidia are single–celled, oval and colorless, produced in long chains. Chasmothecia immersed in the mycelial weft, globose to strongly depressed, dark brown to black, bearing rudimentary, myceloid, sub–hyline to pale brown appendages. Each chasmothecia contains 8–25 asci. Asci are ovate to cylindric, more or less sessile, each with 4–8 elliptical, and sub–hyaline to pale brown ascospores.

Disease Cycle and Epidemiology

Blumeria graminis f. sp. *tritici* is an obligate biotrophic parasite that completes its life cycles only on living hosts. It reproduces both asexually (conidia) and sexually (ascospores). These two types of spores play a vital role in its survival and rapid dispersal and serve as primary inoculum. In absence of a living host, the fungus may remain dormant and oversummers on wheat stubble in the form of the chasmothecia. Ascospores released from the chasmothecia serve as primary inoculum for epidemic initiation, although the chasmothecia often fail to mature and/or release ascospores, possibly due to the unavailability of nutrients for fungal development, shortness of the epidemic, or weather conditions. Germinated ascospores like oidia penetrate the epidermal cell of wheat leaves by producing haustoria. The haustoria have characteristic finger–like projections. After initial infections, under favorable conditions, the white powdery pustules produce large quantities of oidia by multiple cycles of asexual reproduction, which are easily dispersed by wind or rain, cause secondary infection, and spread the disease rapidly. In India, the complete disease cycle is still unknown. The fungus cannot survive under the hot summer months in planis. The chasmothecia has been shown as a major source of primary inoculum in temperate regions like Lahul Spiti in Himachal Pradesh, whereas in other parts of north India, oidia are the main source of primary inoculum blown from hills. In Rajasthan, the primary inoculum comes from Mount Abu and causes infections in Jodhpur region. Powdery mildew epidemics are favored by cool, humid conditions and are characterized by polycyclic development in which spore loads can build up to extremely high levels. Powdery mildew infection on wheat can take place in a range of 5°C–30°C but high relative humidity (more than 95%) and optimum temperature (15°C–20°C), are the most favorable for infection and rapid spread of the fungus, which can complete a repeating cycle in 7–10 days, resulting in the development of new virulent powdery mildew races. High relative humidity also favors spore germination, but free water inhibits spore germination. The fungus needs high humidity but not leaf wetness or rain or dew to infect the plant. Heavy rain washes spores off the leaf surface and temporarily slow disease progress. High seed rates or heavy application of nitrogen fertilizer promotes dense crops that stay humid over extended periods of time.

Management

Cultural Methods

- Follow crop rotation as the pathogen can survive in soil and it is highly specific. Powdery mildew fungus from one cereal (e.g. barley) does not infect other cereals (e.g. wheat).
- Sowing should be done with the appropriate seed rate, i.e.100 kg/ha.
- Avoid more application of nitrogen fertilizer but instead, potassium fertilizer must be recommended.
- Remove the crop debris of volunteer plants from the field.
- Regular irrigation.
- In regions where chasmothecia are formed, management of crop stubble is highly recommended.

Host Resistance

- Avoid the cultivation of susceptible varieties.
- Most of the present varieties do not have adequate resistance.

Biological Control

- AQ10 biofungicide a commercial formulation of pycnidial fungus *Ampelomyces quisqualis* can be used for foliar spray against cereal powdery mildew fungus *Blumeria graminis*, as it is a host for mycoparasite *Ampelomyces*.

Chemical Control

- Treat the seed with Bayleton @ 0.1%–0.2%.
- Seed dressing and foliar spray of Triazole fungicides such as Bayton, Bayleton, and Vigil @ 0.1%–0.2% is highly effective.
- Dusting wheat with Sulphur dust is the best option, but 1% Sodium thiosulphate
- Triadimefon 25% WP @ 200 g in 300 litres of water/acre.

1.5 ALTERNARIA LEAF BLIGHT

Alternaria leaf blight causes significant yield losses in wheat on the Indian subcontinent, whence it originates and has spread throughout the world. Alternaria leaf blight was first reported from Maharashtra in 1924 (Kulkarni, 1924).

Symptoms

Initial Alternaria leaf blight symptoms consist of small (less than 1 mm), oval elliptical shaped, yellow lesions, irregularly scattered on the leaves (Figure 1.7). As the lesions enlarge, they appear irregular and dark brown to grey, surrounded by a bright yellow margin. At a later stage, several lesions coalesce, covering large areas of the leaf, resulting in the death of the entire leaf and sometimes causing plant death. In some cases, the leaf starts drying up from the tip, prematurely, when lesions appear. Under moist conditions, lesions are covered with a black powdery mass of conidia. Infection usually starts on the lower leaves, but in severe cases, symptoms can be found on all plant parts including leaves, stem, leaf sheaths, and as well as on glumes, awns and grain if spikes are infected at the pre–anthesis stage. If the spikes are infected early, seeds do not form. Heavily infected fields present a burnt appearance and can be identified from a distance. The disease appears in the field along with spot blotch and is difficult to identify by seeing the symptoms.

Figure 1.7 Alternaria Blight of Wheat.

Causal Organism and Etiology

Alternaria leaf blight of wheat is caused by *Alternaria triticina.* A three–dimensional sporulation pattern (length of primary conidiophores, branching types and origin of branching, conidial shapes, and ornamentations) was introduced by Simmons and Roberts (1993) to facilitate the classification and identification of small–spored *Alternaria* isolates by microscopy. Mycelium is initially hyaline, becoming olive–buff to deep olive–buff, branched, septate. Conidiophores are similar to mycelium in color, septate, unbranched or occasionally branched, erect, broader towards the distal end, on the host single or fasciculate, emerging through stomata, amphigenous, geniculate. Conidia acrogenous, borne singly or in chains of 2–4, smooth, irregularly ovoid, both ends rounded, or ellipsoid, or conical–ellipsoid, gradually tapering into a beak; beak (a secondary conidiophore) concolorous with the main conidial body, straight.

Disease Cycle and Epidemiology

Alternaria triticina is both soilborne and seedborne. However, the soilborne inoculum is susceptible to extremes of high temperature and under the conditions of a hot summer may not survive until the next crop. Therefore, seedborne inoculum plays a major role in the perpetuation of the pathogen. The initial inoculum carried with the seed may be able to multiply in the soil; plants are then infected when they reach the susceptible three–leaf stage. Younger seedlings are not infected. The leaves in contact with the soil may be infected directly or spores may be splashed by rain on the lowermost leaves, from which the pathogen gradually spreads to the upper leaves. The spores germinate on the leaf surface producing germ tubes and appressoria. Penetration of the leaf occurs through both the upper and lower surfaces. Germ tubes usually penetrate directly through the epidermis and seldom enter the leaf through stomata. The mycelium is both intercellular and intracellular. The outer epidermal cells collapse first, followed by the mesophyll cells which collapse slightly in advance of the mycelium. The fungus does not invade the conducting vessels in the xylem and phloem. Disease development is maximum on 10–week–old plants (at the boot–leaf stage) when the relative humidity is high and the average maximum temperature does not exceed 25°C. Successful infection by the pathogen requires at least 48 hours in a saturated atmosphere. Maximum lesion development occurs at 25°C. Increases in levels of nitrogen and phosphorus fertilizer are reported to significantly increase or significantly decrease the severity of the disease. Conidia develop in the lesions under humid conditions and are dispersed by wind. The fungus can also survive between seasons in crop debris and may produce inoculum in the soil.

Management

Cultural Methods

- Use only certified seed.
- Timely sow the seed.
- Plant tall crops as a barrier for airborne inoculum.

Physical Control

- Presoaking of seeds in water for 4 hours followed by hot water treatment at 52°C for 10 minutes.

Host Resistance

- Grow resistance varieties

Chemical Control

- Spray Zineb 75% WP @ 6–8 kg in 300–400 litres of water/acre or Mancozeb 75% WP @ 6–8 kg in 300 litres of water per acre.

1.6 EAR COCKLE

Seedgall nematode has the distinction of being the first plant parasitic nematode recorded in 1743 by John Turbevill Needham. The present name of seedgall nematode *Anguina tritici* was given by Filipjev in 1936. In India, the nematode was first time reported by Milne reported from Punjab in 1919. The nematode alone causes ear cockle disease of wheat. The disease is locally known as "*gegla*", "*sehun*" or "*mamni*". *Anguina tritici* is often associated with a bacterium *Rathayibactor tritici* causing another disease of wheat "yellow ear rot" which is locally known as "tundu". *A. tritici* also acts as a vector for *Rathayibacter tritici.* Losses up to 100% have been reported in wheat while 65% losses in the rye. Crop losses occur frequently in areas with primitive agricultural practices, because seed flotation, mechanical separation, and crop rotation are not practiced. Tundu disease is more damaging than ear cockle. Apart from the loss in yield, the disease is known to cause toxicity due to Corynetotoxin in humans and cattles when large quantities of galls are consumed with flour or seeds.

Symptoms

The symptoms are visible on leaves, stems, and floral organs. Infected seedlings show crinkling, curling, and twisting of the leaves preventing normal emergence of ears, basal swelling of stem and stunted plant growth with increased tillering. Other leaf symptoms include wrinkling, twisting, curling of the margins towards the midrib, distortion, buckling, swelling, and bulging. A tight spiral coil evolves and dwarfing, loss of color, or development of a mottled yellowed appearance, and stem bending may also occur. Seed galls, which are green in the beginning, later turn hard, dark brown to black. The ear or inflorescence may be absent or, when present, generally wider and shorter with very short or no awns on glumes (Figure 1.8). Glumes may be loosely arranged, and seeds are replaced by galls. Seed galls in cereals are known as ear cockles. Galls or ear cockles are smaller, dark brown or black, and irregular in shape compared to healthy seeds (Figure 1.9). Galls contain a white mass consisting of dried nematodes in an anhydrobiotic state and become powdery if crushed. Infected seeds may be confused easily with bunt disease (*Tilletia tritici*). However, bunt balls give off a foetid or fishy odor when crushed, whereas a crushed seed gall reveals a white mass of nematode bodies.

Causal Organism and Etiology

Ear–cockle disease of wheat is caused by *Anguina tritici.* General morphology: annules are very fine, usually visible only in the esophageal region; lateral fields with four or more fine incisures, in adults visible only on young specimens. Lip region is low and flattened, slightly offset; lips are visible as six raised, radial ridges. Procorpus of the esophagus is swollen but constricted at the junction with the median bulb. Esophageal glands form a roughly pyriform

Figure 1.8 Ear Cockle of Wheat. Healthy (left) and infected ears of wheat (right). (Courtesy: Dr. Jonathan D. Eisenback, Department of Plant Pathology, Virginia Polytechnic Institute and State University, Blacksburg, Virginia, USA.

Figure 1.9 Ear Cockle of Wheat. *Anguina tritici* infected seed galls (right) and healthy wheat seeds (left). (Source: Dr. Ulrich Zunke, Department of Cell Biology, University of Hamburg, Hamburg, Germany, (Bugwood.org)).

bulb, but vary in shape, and sometimes, show irregular lobes, not overlapping intestine; cardia small. Tail is conoid, tapered to an obtuse or rounded tip. Chromosomes: 2n=38. Female: Body obese, spirally coiled ventrally when relaxed by heat. Isthmus of esophagus is sometimes swollen posteriorly and offset from glandular region by a deep constriction. Anterior branch of the genital tract is greatly developed; ovary usually with two or more flexures, with many oocytes arranged about a rachis and ending distally in a cap cell. Spermatheca pyriform, its broader end separated from the oviduct by a sphincter, its narrower end merging into the uterus. Posterior genital branch is simple post–vulval sac. Vulval lips are prominent. Several eggs may be present in uterus at one time. Male: Body is sometimes curved dorsally when heat relaxed, i.e. ventral surface outermost. Testis with one or two flexures; spermatocytes arranged about a rachis, ending distally in a cap cell. Vas deferens about 200 μm long, separated from testis by a constriction. Spicules paired, stout, arcuate, each having two ventral ridges running from tip to widest part: the head rolled or folded ventrally. Gubernaculum is simple, troughlike. Bursa arises just anterior to spicules and ends just short of tail–tip.

Disease Cycle and Epidemiology

The main source of survival and dispersal of *A. tritici* are infected seeds. Seedgalls mixed with healthy seeds at harvest and reach to the soil during sowing. Each seed gall may contain 1000 to 32,000 or more nematode juveniles (J_2) in a quiescent state and produce viable juveniles even upto 28–32 years. When these cockles come in contact with moist soil, absorb moisture, become soft, swell, rupture and release active infective second–stage juveniles (J_2). Meantime the wheat seed germinates and the active juveniles invade the growing seedlings just before their emergence from the soil. A water film must be present on the invaded plant to allow nematode movement. The infective J2 infects and feed ectoparasitically the stem growing point and leaf sheath until it is able to enter the inflorescence where it feeds as an endoparasite. As the embryo of seeds develops, *A. tritici* invades ovules where they induce galls, mate (amphimictic), lay thousands of eggs and reside as second–stage juveniles (J_2). One female can deposit up to 2000 eggs. Adults die following oviposition and the eggs hatch into second stage juveniles. Juveniles enter a state of anhydrobiosis in dry storage. The J_2 can remain as anhydrobiotes and viable within dried seed galls for as many years. *Anguina tritici* completes one generation per crop season and the total life period duration is 106–113 days. The major determinants in symptom expression are temperature, humidity, depth of sowing, and gall source. Lower the temperature at sowing time more is the disease incidence of ear cockle. Loamy light soils, wet and cool weather favors nematode growth.

Management

Cultural Methods

- Use healthy seeds from healthy crops.
- **Crop rotation**: Grow non–host crops for 2–3 years to eliminate the nematodes from field after harvest. Gram, mustard, cumin, carrots, onion, oat, maize and sorghum are considered non–hosts of *Anguina*, however, these crops cannot completely control the disease.
- **Rouging**: Remove and destroy infected plants from fields in the early stages.
- **Sowing time**: Early sowing of the crop usually escapes the infection.
- **Drainage**: Maintain good drainage in the field.

Physical Control

- **Cleaning of seeds**: Use of clean seeds free from nematode galls is the best way to manage the ear cockle disease of wheat. Since seed galls are the only source for the perpetuation of ear cockle and yellow ear rot, their removal can completely eliminate both diseases. Galls are lighter in weight than healthy wheat seeds and can be removed easily by winnowing or by floatation methods such as water and salt floatation.

- **Water floatation**: Putting contaminated seeds in plain water, followed by stirring, results in the floatation of galls on the water surface. Galls can be skimmed off with a sieve and buried or destroyed. With this method, about 95% of galls can be removed.
- **Salt floatation**: Use of 20% salt (brine) solution is very common for floatation and removing seedgalls upto 100%. Wheat seed should be washed with water several times, after brine treatment to remove adhering salt particles, otherwise, seed germination is impaired.
- **Hot–water treatment**: Presoak wheat seed in cold water for 4–6 hours then treat with hot water at 54°C for 10 minutes to kill the nematodes effectively.
- **Winnowing or fanning**: It is an effective method to remove the galls from seed lots. This method is also helpful in removing the galls; however, 100% of galls are not removed by this method.
- **Dry cleaning or sieving/screening**: It is a common practice and the most successful method for the eradication of seed gall nematodes, though complete removal of gall is not possible with this method because some large–sized galls remain on the sieves. Most of the galls can be separated from the contaminated seed using a course sieve (3 mesh).

Host Resistance

- Grow resistant/tolerant varieties. Varieties like Sonora, Lerma Rojo, NP 908, and S 227 had shown a certain degree of tolerance to the disease.

Chemical Control

- Seed treatment and soil application of Aldicarb sulfone @ 2 kg a.i. per hectare, gives good control of ear cockle and yellow ear rot diseases.
- Treat the soil with Nemaphos @ 10 kg a.i. per hectare which destroys active larvae.
- The soil application of Carbofuran 3% CG @ 2.60 kg/acre.

2 Sugarcane (*Saccharum officinarum* L.)

2.1 RED ROT

Red rot disease aptly called the "Cancer of sugarcane," and is considered the main constraint for sugarcane production in India. Being one of the oldest diseases of sugarcane, red rot is distributed worldwide. As early as 8th–7th century BC, this disease was mentioned in Vedic text as Aitereya Aryanaka. Red rot is also the oldest mentioned disease of sugarcane dating back to the times of Buddha. Red rot of sugarcane was first reported by Went in 1893 from Java (Indonesia), by the name "Root snot" or "Red smut", which was locally known as "*Sereh*". In India, the first documented epidemic of red rot occurred in 1895–1901. Epidemics of the disease have been very common ever since its occurrence in India. Red rot disease causes the most significant economic loss (5%–50%) in sugarcane production. About 10%–15% of nation's sugar produced is lost due to the red rot disease.

Symptoms

Yellowing, loss of color, shriveling, drooping, and drying up of margins of third and fourth leaf from the crown is the earliest symptoms. Then the entire top withers. In later stage, the cane shriveled, the rind shrink and becomes longitudinally wrinkled. Such canes are lighter in weight and easily broken. If diseased canes are split open longitudinally, especially when the withering of leaves starts, the pith appears red (Figure 2.1 A). Characteristic bands of clear white areas are seen running transversally across the full breadth of the reddened pith. Cavities filled with grayish or white mycelium are found in pith. The juice often gives a bad odor and does not set well on boiling due to the conversion of sucrose into glucose and alcohol by the enzymatic action of the pathogen. Grayish red or red discoloration occurs throughout the stalk length, especially at nodes. The rind becomes shrunken at the time of harvest with the longitudinal cavities filled with profuse cottony gray fungal mycelium, conidia, and drops of dark brown liquid. Numerous velvety, dark black, minute, dot-like acervuli with prominent setae (hair-like) develop on the rind of the dried stalk. Elongated red patches with ash-colored centers develop on the midrib which sometimes extends the entire length of the leaf (Figure 2.1 B). Black powdery masses of conidia and abundant acervuli are present on these patches. In the later stage, the red rot-infected canes show dull red to brown shades on the rind or nodal regions and during severe infection pinkish sporulation can be seen on rind and leaf scars.

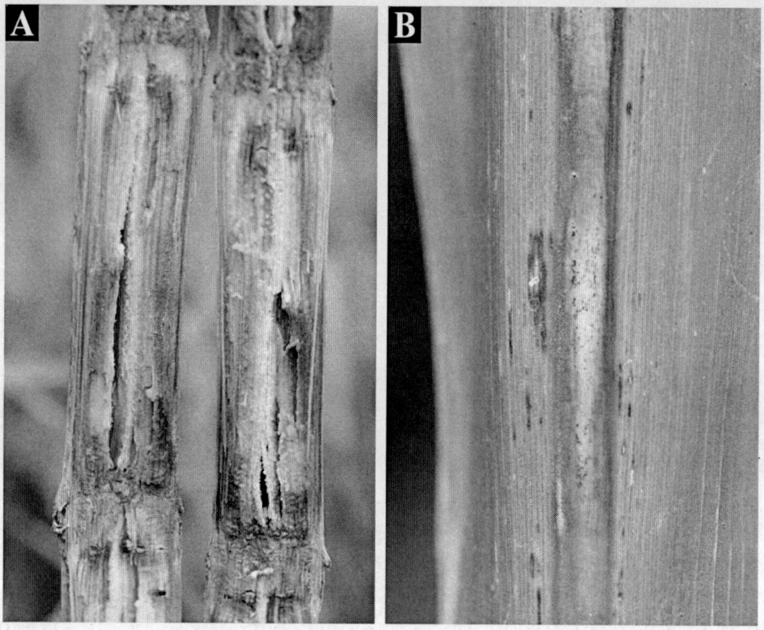

Figure 2.1 Red Rot of Sugarcane. (A) Typical red rot symptoms on cane and (B) Leaf symptom.

Causal Organism and Etiology

Red rot of sugarcane is caused by a facultative parasite *Colletotrichum falcatum* (teleomorph: *Glomerella tucumanensis*) belonging to the family Glomerellaceae of phylum Ascomycota. Mycelium is septate and grows in intracellular and intercellular space. Acervuli (asexual fruiting bodies) are minute, velvety and formed on the surface of the host part), often with setae (dark-pigmented, unbranched, thick-walled sterile hyphae usually pointed at the tip). Acervuli look salmon-pink in mass. Conidiophores are hyaline, linear, or club-shaped present on acervuli. Conidia are sickle-shaped (falcate), single-celled, thin-walled uninucleate, colorless, fusiform, tapering gradually to the base and more abruptly at the apex, and posses a large oil globule at the center. Chlamydospores are thick-walled and greenish-black. In the perfect stage, perithecia of *Glomerella tucumanensis* are globose, ostiolate, and superficial with the bottom embedded in the host tissue. Asci are numreous, hyaline, clavate, and slightly thickend at the apex. Each ascus contains 8 ascospores which are single-celled, hyline, and elliptical. Appressoria are thick-walled swellings at the end of a hypha or germ tube useful for attaching the fungus to the host surface before penetration of the tissue.

Disease Cycle and Epidemiology

Colletotrichum falcatum can grow in soil and produce acervuli. The infective propagules in the soil may be present as conidia, acervuli, setae, thick-walled hyphae, chlamydospores, asci and ascospores, and appressoria with infected crop debris and play important roles in disease development and spread. The perithecial stage of *C. falcatum* is not common in India. The pathogen has a number of collateral hosts, e.g. *Sorghum vulgare*, *S. halepense*, *Saccharum spontaneum*, *Leptochloa filiformis*, and *Miscanthus* sp. Generally, the infected setts or stubble fragments are the main source of primary inoculum and survival of *C. falcatum*. At the time of planting, most of the seed setts are internally infected or have dormant mycelium in bud scales, leaf scars, and nodal and intermodal tissues. When such setts are planted they invariably produce infected shoots. If such shoots survive in the field, the secondary infection is caused by conidia, produced in acervuli, on such shoots and disseminated by insects, wind flood water, rain splashes and irrigation water. Ratoon crops also serve as sources of perennation and multiplication of the fungus. The germ tube from conidia and infection threads form appressoria, which after melanization, mainly penetrates through the tissue of leaf scar, bud scales, leaves, leaf sheath, growth ring, and root primordia. The pathogen can also enter the stalk through rootlets, growth cracks, and cut ends of the setts. Red rot disease of sugarcane is favored by an optimum temperature range of 29.4°C–31°C and high humidity (90%). Other factors responsible for disease development are susceptible varieties, water-logging, soil pH 5–6, and drought conditions during the initial growth phase.

Management

Cultural Methods

- Healthy setts should be used for planting.
- Avoid cultivating the same varieties for several years.
- Discard any sett showing reddening at the cut ends, or at the nodal region.
- Remove the affected clump along with the root and burn.
- After harvesting burn diseased fallen leaves in the field.
- Avoid ratooning red rot-infected fields (more than 10% infection).
- Follow crop rotation with rice and other crops for 2–3 years in red rot-affected fields.
- Avoid planting of sugarcane in previously affected fields.
- Do not allow the movement of rain or floodwater from diseased to healthy fields blocks it by making a bund.
- Incorporate green manure in the field to suppress the inoculum at least for one year.
- Adopt wider row planting with compulsory heavy earthing up before the onset of monsoon to avoid the secondary spread of red rot pathogen through water flow.
- Fields with high red rot incidence should be harvested at the earliest.

Physical Control

- Treat the sugarcane setts with hot water at 52°C for 18 minutes.

- Heat treatment of seed canes in moist hot air at 54°C for 2.5 hours at 99% humidity.
- Hot air treatment of sugarcane setts at 54°C for 8 hours.

Host Resistance

- Grow recommended resistant/tolerant varieties such as Co 0238, Co 86032, Co 86249, CoSi 95071, CoG 93076, CoC 22, CoSi 6 CoG, BO 128, Co 89029, CoH 2201, CoSe 92423 and CoSe 95422.

Biological Control

- Sett treatment with *Trichoderma viride* (tv1) @ 4 g/L or *Pseudomonas fluorescens* (Pf1) @ 10 g/L of water for 10 minutes.

Chemical Control

- A one-hour sett treatment with Bavistin @ 0.5% suspension could reduce red rot incidence from infected setts.
- Sett treatment with 125 g of Carbendazim 50 WP in 250 litres of water along with 2.5 kg of urea for 5 minutes to manage red rot and sett rot.
- Sett treatment with Bavistin or Thiram @ 0.1% is also beneficial.
- Spray Nativo 75 WDG or Cabrio 60 WDG @ 500 ppm in red rot-infected fields during June–July.
- After harvesting, plough the field after broadcasting of bleaching powder (@ 10 kg mixed in 25 kg sand or soil/acre).

2.2 SMUT

The disease was first reported from Natal, South Africa in 1877. From India, the disease was first recorded by Sydow and Butler in 1906. The disease is also referred as "culmicolus smut" or "whip smut" or "the death of sugarcane". Sugarcane smut is one of the most serious diseases of sugarcane. Affected cane is severely stunted and production losses of 30%–100% are common in susceptible varieties.

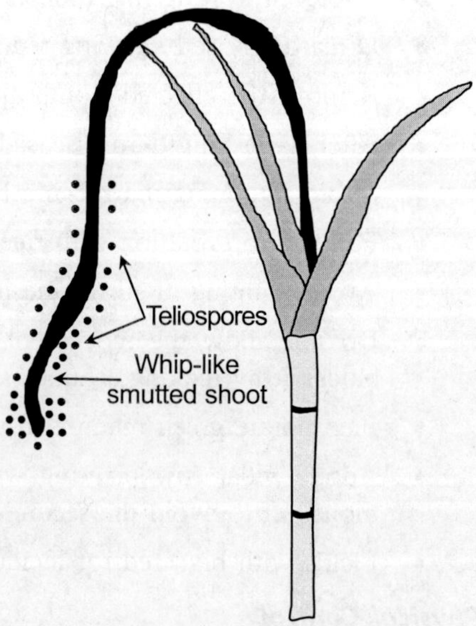

Figure 2.2 Smut of Sugarcane.

Symptoms

The most characteristic symptom of this disease is the production of often, very long, curved, pencil-thick, unbranched, whip-like structure (sorus) at the apex of the affected stalk containing plant tissues, fungal mycelium and black powdery mass of smut spores covered with thin silvery-white, a papery membrane made up of host epidermis (Figure 2.2). Soon the membrane ruptures exposing a dense black mass of smut spores (teliospores). The whip morphology differs from short to long, twisted, multiple whips, etc. Affected plants tiller profusely with spindly and more erect shoots with small narrow leaves (grass-like). Others symptoms are leaf and stem galls, bud proliferation, adventitious leaves, stalk distortion, plant stunting and thin cane formation. Before whip formation, shortening, and crinkling of the youngest leaves occurs.

Causal Organism and Etiology

The smut disease of sugarcane is caused by a fungus, *Ustilago scitaminea* (syn: *Sporisorium scitaminea*). Smut fungi are facultative biotrophs since they grow saprotrophically as yeast-like cells on culture media but require biotrophic infection of host cells to complete their life cycle. The mycelium is intercellular and sends haustoria into the host cells. The spores are spherical, light brown, with a puncate wall. They germinate forming 3- or 4-celled promycelium. From each cell, a sporidium is produced. The elongated, one-celled sporidia germinate producing an infection thread.

Disease Cycle and Epidemiology

Sporisorium scitaminea present in internally-infected "seedcane sets" and soil as teliospores, serve as a major source of primary inoculum. Dormant-infected buds on the seed setts also give rise to infected plants. The teliospores, on germination, produce a septate, four-celled promycelium that forms globular, four, sporidia (basidiospores) at the tip. Under favorable soil conditions (good humus and moisture), the sporidia may bud off more sporidia in a short chain. The wind disseminated sporidia, after dikryotization, infect auxiliary buds and cause shoot infection. The dikaryotic phase is established through fusion (anastomosis) between cells of promycelium or between cells of hyphae developing from the germinating teliospores, to give rise to the dikaryotic secondary mycelium. Once the dikrayotization is over, the dikaryotic infection hypha invades the host tissue through the buds at each node and becomes systemic. In the host tissue, these fungi grow both inter- and intracellularly, developing a knob-like haustorum, and systemically colonize the apical meristem, affecting the reproductive development of infected plants. When the whip fully emerges, the membrane ruptures releasing about one billion teliospores per whip per day in the air which causes secondary infection by infecting the buds of standing crops. Infective buds in mature plants are either symptomatic as a whip at the apex of the stalk or remain asymptomatic hidden in buds up to the next season. The pathogen survives and is spread by planting setts of smutted canes containing fungal propagules, spores borne on buds, infection of buds on standing canes, and ratooning the smutted canes. Generally, dry and hot weather favors smut disease development. Optimum temperature (25°C–30°C) and high relative humidity (100%) is essential for spore germination. High rainfall reduces the severity of smut development. Plants grown under stress conditions are more prone to infection. Ratoon crops are more prone to smut.

Management

Cultural Methods

- Avoid planting setts from smutted canes.
- Collect smut-infected clumps/whips carefully in a polythene bag and burn. This reduces secondary infections of buds.

- Avoid ratooning of the diseased crop.
- Adopt crop rotation in diseased fields.
- Disinfect the setts before planting using physical or chemical methods.

Physical Control

- Treat the setts with dry or moist hot air (at 54°C for 8 hours) or aerated steam therapy (AST) (at 50°C for one hour) has been found effective against smut, red rot, and ratoon stunting disease.
- Dip the setts in hot water at 55°C–60°C for 10 minutes before planting.
- Smut spread mainly through diseased seeds therefore, the seeds should be bought from reliable sources.
- Adopt three tier seed nursery program following the heat therapy (hot water, MHAT, AST) process of seed production. The seed cane is subjected to moist hot air treatment (MHAT) at 54°C and 95% RH for 4 hours. While for hot water treatment temperature should be 50°C for 2.0–2.5 hours.

Host Resistance

- Grow resistant/tolerant varieties such as Co 8371, Co 85004, Co 86032, Co 86249, Co 87025, Co 87044, Co 87263, Co 87268, Co 91010, CoM 88121, Co Pant, etc.

Chemical Control

- Dip the setts in suspension of fungicides such as Bavistin (@ 0.5%) or Vitvax (0.3%) or Dithane M-45 (0.3%) or Pyroquilon (@ 4 g a.i./L of water) or Carbendazim + Maneb @ 4.5 g a.i./L of water).
- Spray the Propiconazole 25EC @ 0.1% in clear weather. Need-based second spray is advised after 15 days.

2.3 WILT

Wilt of sugarcane was recorded almost 100 years ago in India and is one of the major fungal diseases after red rot affecting cane production and productivity. Although first reported by Butler (1906) from Bihar, detailed studies were made a few years later by Butler and Khan (1913). Wilt either alone or in combination with red rot causes significant damage to cane yield and quality. Many commercial varieties of sugarcane have been withdrawn from cultivation due to their susceptibility to this disease in India.

Symptoms

Symptoms are usually noticed when canes are half-grown until harvest time. The first noticeable symptom is the wilting of a single cane or an entire clump. Initially, the top leaves become yellow and/or wither, then dry up. When the diseased canes are split open, the pith has light purple or reddish streaks and becomes light and hollow, making it worthless for milling. When canes are split open in the early stage of disease development, the tissues, particularly of the lowest internode, have a brick red color with individual vascular strands a dark red. There are no white transverse bands as seen in the red rot disease. The reddening may be confined to a few joints or extend to the whole length of the cane. Brown patching on the rind is formed where the underlying tissues are dead. In the reddened pith, the fungus hyphae ramify throughout the cells in all directions. Gradually the pith becomes hollow and usually covered with a copious grayish-white, fluffy mycelial fungal growth and conidia. A foul odor occurs after splitting a diseased stalk, and a cottony mycelial growth with abundant microconidia is present in pith cavities. A diseased seed piece usually does not produce eyes and roots. Occasionally, an eye may swell or grow slightly before it dies. The interior of such a seed piece has a brown discoloration.

Causal Organism and Etiology

Wilt of sugarcane is caused by *Fusarium sacchari* (syn. *Cephalosporium sacchari*). The fungal mycelium is hyaline, septate, and thin-walled. The conidiophores are simple or branched and produce single-celled, hyaline, oval to elliptical microconidia. Macroconidia are straight with 3 to 5 septae measuring 27–73 × 3.4–5.2 mm. Blastoconidia are either straight, sickle shape, or pike shaped with 2–3 septae.

Disease Cycle and Epidemiology

Fusarium sacchari persists as saprophytic mycelia in infested residue. Abundant microconidia are produced on mycelia and disseminated by the wind to plant hosts. *Fusarium sacchari* is also disseminated by infected seed cane and by any means that transport residue. Infection commonly occurs through insect wounds. The sugarcane wilt pathogens are primarily soil-inhabitants. *C. sacchari* can survive for 2–3 years in soil. Perennial nature of the crop permits uninterrupted survival of the pathogens in sugarcane fields on crop debris and also through seed setts. The infected seed setts permit the growth of the pathogen in the shoots. Infection of healthy seed setts after planting mostly occurs through the cut ends when the pathogen grows into the cane parenchyma. Fungi could also enter setts via root primordial and leaf scars. Wounds at underground parts of the plants such as those caused during intercultural operations and those caused by insects and nematodes facilitate entry of the pathogen. After infecting the plants the fungus proliferates in tissues, particularly the vascular bundles where it can live for long periods without causing the death of the plant or producing visible symptoms. Generally, the symptoms are expressed when the plant are under stress such as drought. While infected seed setts are the chief means of transmission of the pathogen, the soilborne inoculum can also be disseminated by irrigation water, wind, and insects. High wilt incidence is favored by soil conditions such as high soil moisture, pH of 7.0–8.0, and high C:N ratio, etc. However, acidic soil fertilized with heavy dose nitrogen also favors the wilt.

Management

Cultural Methods

- Use healthy seed setts from disease-free areas for planting.
- Before planting, examine seed pieces at both the cut ends for any discoloration.
- Destroy infected crop residues.
- Avoid ratooning diseased crops.

- Follow crop rotation for 3 years if the soil is sick for wilt disease.
- Manage root borer.
- Maintain soil moisture and fertility properly to reduce disease incidence.
- Avoid prolonged drought and water-logging conditions.

Host Resistance

- Grow resistant varieties like Co 617 and BP17.

Chemical Control

- Dip the setts in a solution (@ 40 ppm) of boron or manganese or spray the plants with either of these minor elements. This helps in reducing disease incidence.
- Treat the setts with a fungicide like Bavistin @ 0.1% before planting.
- Apply Carbendazim @ 2 gm/L of water at the root zone area and the same as follow at 15 days interval.

2.4 POKKAH BOENG

"Pokkah boeng", is a Javanese term denoting a malformed or distorted top of affected sugarcane. It was originally described by Wakker and Went (1896) in Java, but no causal agent was established. Bolle (1927) was the first to demonstrate that the disease is caused by a fungus of the genus *Fusarium,* identified as *F. moniliforme*. Pokkah boeng is a reemerging disease of sugarcane that has been found recently to cause major yield losses in most sugarcane-producing regions, including South Africa, Malaysia, India, and China. Among fungal diseases, pokkah boeng has become a serious problem for sugarcane-growing countries, which can lead to severe yield losses.

Symptoms

The pokkah boeng disease manifests three phases viz. chlorotic phase, top-rot phase, and knife-cut phase.

Chlorotic phase: The earliest symptom of pokkah boeng disease appears as chlorosis at the base of the young leaves, pronounced wrinkling, twisting and shortening, malformation or distortion of the young leaves. Base of affected leaf becomes often narrow.

Acute phase or Top-rot phase: This most advanced and serious stage of pokkah boeng disease which occurs when the fungus penetrates through the growth point and causes death and rotting of the top of the young spindles. Red specks and stripes are also developed.

Knife-cut phase: This stage is associated with the top rot phase of the disease. When infection progressess from leaves to stems, internal and external ladder-like (knife-cut) lesions develop. These are characterized by one or more transverse cuts in the rind of the stalk/stem in such a uniform manner as if; the tissues are removed with a sharp knife. This is an exaggerated stage of a typical ladder lesion of a pokkah boeng disease.

Causal Organism and Etiology

Pokkahboeng disease of sugarcane is caused by *Fusarium fujikuroi* species complex (FFSC). Most frequent *Fusarium* species causing this disease are *Fusarium proliferatum*, *F. moniliformae*, *F. verticillioides*, *F. sacchari*, and *F. andiyazi*. *Fusarium* moniliforme showed different colour for its mycilium pale white, pink and purple mycilium and for pigmentation of metabolites also during the growth on artificial medium and its conidial growth is affected by glucose concentration.

Disease Cycle and Epidemiology

Basically, pokkah boeng disease of sugarcane is an airborne disease and primarily transmitted through the air and secondary infection takes place through the infected setts, irrigation water, splashed rains, and soil. The pathogen can survive for 12 months in the plant debris under natural conditions. These *Fusarium* spceies are also associated with a wide range of hosts such as banana, maize, cotton, mango, sugarcane, and other important crops. The conidia of *Fusarium* spp. disperse by air, deposited on plants, and then washed by rain into infection sites such as partially unfolded leaves and roots of the spindle during rapid and vigorous growth of the canes. The plants are most susceptible to the disease when these are 3–7 months-old. Then, the conidia or spores or ascospores germinate and infect the young tissues of the spindle. After the infection, the mycelium gets established in bud scales, root primordial or leaf scars, and later within the plant tissues. Pathogens also enter the host tissue through wounds caused by insects or borers or natural growth points. The disease is common during rainy months in the field. The disease is favored by warm, moist growing conditions. Pathogens growth, abundant sporulation, and disease development are favored by temperature range 20°C–32°C, and average relative humidity more than 70%–80% with a cloudy weather and drizzling rains. The pathogen could grow in a wider range of acidic to slightly alkaline pH from 6.5–7.5 of the soils. Cool and dry conditions favor the survival of a fungus in plant debris.

Management

Cultural Methods

- Use healthy seeds (setts) for planting.
- Rouge out canes showing 'top rot' or 'knife cut' symptoms from the fields.
- Destroy infected crop residues.
- Avoid ratooning diseased crops.
- Follow crop rotation for 3 years.
- Manage root borer.
- Maintain soil moisture and fertility properly.

Physical Control

- Similar as recommended for ratoon stunting disease.

Host Resistance

- Grow disease-resistant/tolerant varieties.

Chemical Control

- Spray fungicide such as Bavistin @ 1 g/L of water) or Copper oxychloride or Dithane M-45 @ 3 g/L of water. Give 2–3 sprays after every 15 days intervals.

2.5 GRASSY SHOOT

Grassy shoot disease (GSD) or sugarcane grassy shoot (SCGS) was first reported in 1949 in India, and then it has been recorded in most of the sugarcane-growing states of India.

Symptoms

Symptoms of sugarcane grassy shoot vary at different stages of plant growth and thus, the disease has been variously described as "new chlorotic disease", "albino disease", "yellow disease", "bunchy disease" or "leaf tuft". The disease is characterized by profuse tillering, dwarfing and grassy appearance of the stools, hence named "grassy shoot disease." The leaves are narrow, reduced in size, and have a soft texture, like grass leaves; cane with short-internodes, profuse growth gives rise to a dense or crowded bunch or a bunchy or grassy appearance, to clump. The leaves appear pale yellow or chlorotic and sometime, entirely devoid of any pigment (white leaf or albinism). Affected plants do not produce millable canes.

Causal Organism and Etiology

Grassy shoot of sugarcane is caused by Phytoplasma (16SrXI group phytoplasmas). The GSD phytoplasma is present in the sieve tubes of the phloem as ovoid, spherical, or irregular shaped bodies. The size of ovoid or spherical bodies is 300–400 nm in diameter and that of protruding filaments 30–50 nm in diameter. These cells are devoid of the true cell wall and bounded by a single triple-layered membrane. If stained with Diene's stain, the infected phloem becomes dark blue. The cells are physically small (less than 1 µm) and therefore they are difficult to detect with a conventional microscope.

Disease Cycle and Epidemiology

Sugarcane grassy shoot disease is caused by Phytoplasma which is transmitted by leafhoppers and aphids (e.g. green corn aphid, *Rhopalosiphum maidis* and sugarcane aphid, *Melanaphis sacchari* forma *indosacchari*) in a non-persistent manner in the field. The GSD phytoplasma survives and spread through infected cane used as seed setts. It may also be sap transmitted mechanically through knives. Transmission by dodder (*Cuscuta campestris*) is also reported. Ratooning of the diseased crops is also an important means of survival of GSD phytoplasma. Sorghum is also a natural host of GSD phytoplasma.

Management

Cultural Methods

- Phytosanitary control measures.
- Avoid ratooning if the incidence of grassy shoot disease is more than 15% in the crop.
- If disease symptoms appear within 14 days after planting, replaced such plants with healthy plants.

- Uprooted infected plants and burn.
- Rogue out infected plants in the secondary and commercial seed nursery.

Physical Control

- Treat the setts with aerated steam at 50°C for 1 hour to control primary infection.
- Inactivate the phytoplasma by dipping setts in hot water at 50°C for 2–3 hours or 54°C for 40 minutes or 55°C for 30 minutes.
- Treat the setts with hot air at 54°C for 8 hours.

Host Resistance

- Growing resistant varieties such as Co 86249, CoG 93076, and CoC 22.

Chemical Control

- Spray Dimethoate @ 1 ml/L of water or Methyl-demeton @ 2 ml/L of water for managing insect vectors.

2.6 RATOON STUNTING

Ratoon stunting (also known as ratoon stunting disease or RSD) is one of the most devastating diseases of sugarcane in the world causing losses upto 30% per year. Ratoon stunting disease was first observed by Steindl from Queensland, Australia in 1944–1945. He also coined the term "ratoon stunting" as the ratoon crops were more severely infected. In India, the disease was first observed by Prof. S.J.P. Chilton from Golagokarnath, (U.P) in 1956.

Symptoms

External symptoms: The stunting and unthrifty growth of the plants are the only external symptoms of the disease, which can occur not only due to RSD but due to several other factors like nutritional deficiency, lack of moisture, poor agronomic practices, etc. Diseased fields often have a characteristic 'up-and-down' appearance due to differing levels of stunting in adjacent stools. The stunting is not uniform in all the plants even if all the stools in a field are infected. There is up and down appearance, which means that all the plants are not of uniform height.

Internal symptoms: In the longitudinally splitted diseased stalks, the discoloration appears as red-orange, dots, commas, or short straight or bent lines of 2–3 mm length of nodal tissue. The discoloration is due to the plugging of many large xylem vessels with a colored gummy substance. However, similar discoloration is also associated with other sugarcane diseases. For a reliable diagnosis, all the nodes in the fully developed stalks should show discoloration. No symptoms can be seen in internodes.

Causal Organism and Etiology

Ratoon stunting disease is caused by the bacterium *Leifsonia xyli* subsp. *xyli* (formerly *Clavibacter xyli* subsp. *xyli*). The bacterium is gram-positive, non-acid fast, non-endospore forming, pleomorphic, fastidious, xylem inhabiting bacterium, and non-culturable. The bacterium is small, smooth cell-walled, rod-shaped often appear as coryneform (club-shaped), like matchstick and joined to form V-shaped structures, and filamentous. The cell measure 0.3–0.5 × 1–4 μm, but the length of the filaments may be 10 μm or even more. The bacterium differs from other xylem inhabiting bacterium in morphology, a very narrow host range (sugarcane only being the natural host) and in being only transmitted mechanically during the cutting of canes for seed setts.

Disease Cycle and Epidemiology

Leifsonia xyli subsp. *xyli* perpetuates in the diseased canes and infected stubbles and the latter give rise to infected ratoon crops. RSD bacterium is sap transmitted but no insect vector is known. During harvesting, field dispersal takes place through cutting knives and the use of seed setts from diseased cane as planting material. The bacterium penetrates through wounds and then it enters the xylem vessels. After penetration, it colonizes the xylem vessels of roots, stalks, and leaves. It seems that the bacterium generally multiplies and spreads in dysfunctional vessels. The enzymatic activity of the bacterium dissolves pit membranes to facilitate its movement through xylem vessels. The high concentration of the bacteria in the vascular bundle in susceptible varieties causes a blockage that ultimately leads to the poor and stunted growth of the sugarcane plant. Low rain (drought) or excessive rain (waterlogging) can contribute to yield losses due to the disease manifestation. The effect on sugarcane growth with RSD tends to increase when crops are stressed; losses are greatest when the precipitation is erratic under rainfed conditions or when irrigation management is poor.

Management

Cultural Methods

- The use of pathogen-free seed setts for planting.
- Strict quarantine measures should be followed to check the spread of the pathogen to disease-free areas.
- Give proper irrigation to the crop to minimize the losses.
- Apply Ammonium sulphate as fertilizer in sugarcane which reduces RSD incidence.
- Disinfect tools and implements such as cutting knives, cultivators, and harvester blades to prevent the spread of the pathogen. Remove all soil and plant material with water and detergent under high pressure. The commonly used disinfectants include 5%–15% Lysol (neutralized Cresylic acid), 1% Dettol, 50% Ethanol, or 10% Formalin, 0.1% Benzalkonium chloride (Cane Knife Steriliser) or Didecyldimethyl-ammonium chloride (SterimaX). The disinfectant should be left in contact with the implement for 5 minutes before use.

Physical Control

In the absence of a pathogen-free seed, the diseased canes/setts should be treated with heat therapy to eliminate the seedborne inoculum.

- **Hot water treatment**: In India, the hot water treatment is normally carried out at 50°C for 30 minutes. The same is done at 50°C for 3 hours in Queensland (Australia), Florida, and Hawaii (USA).

- **Hot air treatment**: In India, it is carried out at 54°C for 8 hours.
- **Moist-hot-air treatment**: The method was developed in India to check the loss of moisture in hot air treatment. Within one hour, the required temperature of 54°C is obtained, and it is maintained for 7 hours in a sealed chamber.
- **Aerated-steam method**: It is the most recent method in which an aerated-steam, a mixture of air and steam is used. In India, the recommended temperature is 54°C for 4 hours, while in the US the treatment is done at 51°C for 4 hours. In addition to RSD, the treatment is also effective for grassy shoot, leaf scald, and red rot of sugarcane.
- **Disadvantage of heat therapy**: Reduction in germination, high cost of equipment, and failure to eradicate the pathogen completely. The seed sett never becomes 100% disease-free.

Host Resistance

- Use of host resistance is the most economical method of disease control. No immune/resistant commercial varieties are known anywhere in the world.

3 Sunflower (*Helianthus annuus* L.)

3.1 SCLEROTINIA STEM ROT

The diseases caused by *S. sclerotiorum* on sunflowers, Sclerotinia root rot, Basal stalk rot, and wilt; Sclerotinia stem rot; and Sclerotinia head rot and midstalk rot, are responsible for high yield losses in the world. In commercial sunflower fields, yield losses of 10%–20% for Sclerotinia head rot and 5%–70% for Sclerotinia wilt/Basal stalk rot have been observed. In addition to yield, Sclerotinia head rot can also affect the quality of the seeds by reducing the oil content in the seed by 10%–15% and increasing the content of free fatty acid, leading to the rancidity of the oil. *S. sclerotiorum* causes three distinct diseases on sunflower: (1) Sclerotinia root rot, Basal stalk rot, and wilt; (2) Sclerotinia stem rot; and (3) Sclerotinia head rot and midstalk rot.

Symptoms

Sclerotinia root rot, basal stalk rot, and wilt: The first above-ground symptom of Sclerotinia root rot, basal stalk rot, and wilt are sudden wilting of the sunflower plants before or during flowering, often in clumps. Later, tan-colored, water-soaked lesions develop at the base of the plants, which can girdle the stem. As the disease progress, the stem appears bleached and shredded and has decayed pith. The infected plants can easily lodge in the field during high winds. Hard, black sclerotia (1–2 mm in diameter) are present inside, and sometimes on the outside, of the stem. The presence of these relatively large sclerotia can help to distinguish Sclerotinia Basal stalk rot and wilt from other stem diseases such as Verticillium wilt and charcoal rot.

Sclerotinia stem rot: Initially, small, water-soaked lesions develop on the plants near the soil line. As the disease progresses, secondary symptoms can appear as wilting, bleaching and shredding of the stem. Mycelial growth and abundant black sclerotia, are present inside (pith) and outside on the infected lesions of the stems. As the disease progresses, the stem break and shred with only the vascular elements visible. Symptoms of Sclerotinia stem rot can be confused with those of Phomopsis stem canker. While Phomopsis stem canker lesions are light to dark brown in color, the stalks of the sunflower plants affected by Sclerotinia stem rot have sclerotia forming inside and outside the stem.

Sclerotinia head rot and midstalk rot: Symptoms of Sclerotinia head rot include dark-colored, water-soaked lesions on the backside of the sunflower heads or the presence of white mycelial growth, which covers the developing seeds in the head. As the disease continues to develop, *S. sclerotiorum* rots the inside of the head and large sclerotia fill the head below the seed layer and around the seeds. Later the sunflower head disintegrates and shreds, leaving behind large sclerotia. Symptoms of Sclerotinia head rot can be confused with Rhizopus head rot. However, sunflower heads affected by Rhizopus head rot will exhibit gray-colored, threadlike mycelial strands with no sclerotia.

Causal Organism and Etiology

Sclerotinia sclerotiorum causes stem rot of sunflower. It is worldwide distributed plant pathogenic fungus which has the widest host range of any known plant pathogen. It nearly affects more than 400 species of 275 genera which includes cruciferous vegetables like cabbage and cauliflower, tomato, potato, sunflower, soybean, and lettuce. Apothecia consist of a stipe (20–80 mm long) and a flat to concave, tan-colored receptacle (2–10 mm diameter) that supports the asci-bearing tissue, the hymenium. The asci forcefully discharge ellipsoid-shaped ascospores, when triggered by the sudden decrease in relative humidity or from disturbance of the apothecium. Ascospores are single hyaline cells and have thin walls.

Disease Cycle and Epidemiology

Sclerotinia sclerotiorum overwinters in the soil or crop debris as sclerotia and can survive for 3 to 5 years. Depending on the environmental conditions, sclerotia can germinate carpogenically or myceliogenically. In carpogenic germination, the sclerotia germinate to form apothecia (cup-shaped) and infect foliar plant parts of the host by ascospores. The formation of apothecia is favored by air temperatures between 10°C–25°C, high relative humidity, and high soil moisture for one to two weeks. These spores require temperatures between 20°C–25°C, 48 to 72 hours of leaf wetness (e.g., dew, rain), and a nutrient source (e.g., dead or senescing plant tissues, or exudates from the nectaries) for germination. Ascospores survive for about two weeks after release from apothecia. The ascospores are ejected into the air, carried by the wind, and may land on sunflowers or other susceptible plants. Ascospores require a film of water and a food base such as dead or senescing plant tissue to germinate and grow before they infect the plant. They may also infect through wounds caused by hail, insects, etc. In myceliogenic germination, sclerotia of *Sclerotinia sclerotiorum* germinate to produce hyphae that can directly infect the roots of sunflower plants attack sunflower. Hyphal germination of sclerotia will occur when the soil is wet (water holding capacity of between 30 and 100%) and cool (temperatures of 12°C to 24°C) from rain or irrigation practices. Infection of plants following hyphal germination of sclerotia enables the fungus to penetrate the lateral roots, tap root, stem causing Sclerotinia root rot, basal stalk rot, and wilt. Prolonged wet weather during flowering is associated with head rot. Serious outbreaks of these diseases do occur in wet growing seasons or under irrigation.

Sclerotinia root rot, Basal stalk rot, and wilt: *S. sclerotiorum* can infect sunflower through the roots. In previously infected sunflower fields with white mold, the sclerotia can germinate to produce white cottony mycelium, upon contact with roots of sunflower plants. The fungus penetrates the lateral roots and then the taproot causing decay of the root system. The fungus grows through the roots into the stem and the plant wilts. As the disease progresses, sclerotia are produced in the decaying stem and roots which survive in the soil and serve as a primary source of inoculum for the next crop.

Sclerotinia stem rot: The ascospores produced by apothecia can originate from within the sunflower field or from the neighboring fields and disperse by the wind. These spores land on the leaves and petioles of sunflower plants. Upon availability of free moisture and nutrients, ascospores germinate and infect the leaves and petioles and progress to the stem, causing Sclerotinia stem rot.

Sclerotinia head rot and midstalk rot: Ascospores are the infective propagule and they require leaf wetness and nutrients to infect. Like stem rot, ascospores use dead florets and pollen as the nutrient base to germinate. The spores penetrate into the receptacle and decay the entire head. *S. sclerotiorum* can infect the sunflower seed, and exist as mycelia in the seed coat, but the infected seed is of less importance for disease development.

Management

Cultural Methods

- Deep ploughing help to reduce the number of sclerotia present on the soil surface.
- Follow crop rotation for 3–5 years with non-hosts crops such as wheat, sorghum, and corn.
- Do not plant highly susceptible crops more than once in four years, including dry edible beans, sunflowers, mustard, and canola.
- Complete field and crop sanitation.
- Soil amendment with FYM @ 5 tonnes per acre.
- Avoid moisture stress during high summer and water logging conditions.
- Collect and destroy plant debris.
- Avoid planting next to a field that had severe Sclerotinia in the past 4–5 years.
- Plant thoroughly cleaned seed.
- Avoid irrigating sunflower fields with high water holding capacity, especially during flowering, to prevent carpogenic germination of sclerotia.
- Maintain minimal plant population with wide row spacing.
- Avoid excessive rates of nitrogen in sunflower fields as excess nitrogen can promote dense canopies and provide the microclimate for disease development.
- Keep sunflower fields weed free. Remove weeds regularly such as broad-leaf weeds that may be present in a sunflower field can be hosts of *S. sclerotiorum and* can contribute to the survival of inoculum and dissemination and these include Jerusalem

artichoke (*H. tuberosus*), Canada thistle (*Cirsium arvense*), pigweed (*Amaranthus*), common lambsquarters (*Chenopodium album*), ragweed (*Ambrosia*), nightshade (*Solanaceae Juss.*), soybean, vetch (*Vicia*), velvetleaf (*Abutilon theophrasti*) and wild sunflower (*H. annuus*).

Host Resistance

- No major genes have been identified in cultivated sunflowers that can confer complete resistance to *S. sclerotiorum*. The genetics of resistance to diseases caused by *S. sclerotiorum* is quantitative, which complicates the efforts to breed for resistant varieties.

Biological Control

- Seed treatment with *Trichoderma viride* @ 6 g/kg seed.
- Seed treatment with *Pseudomonas fluorescens* or *P. putida* strains protect sunflower from Sclerotinia infection during the seedling stage.
- Apply *Trichoderma harzianum* in soil at the rate of 10 g/kg to reduce wilt.
- Apply and incorporate the fungus *Coniothyrium minitans* before sowing as it invades and destroys the pathogen in the soil. *Coniothryium minitans* can directly attack the sclerotia of *S. sclerotiorum.*
- Soil amendments like castor cake, neem cake, and oat straw reduce the disease incidence.

Chemical Control

- Seed treatment with Captan or Thiram at the rate of 3 g/kg of seed.
- Drenching the base of the plant with chestnut compound 3 g per litre of water.
- Seed treatment with Carbendazim at the rate of 0.2% followed by the addition of *Trichoderma harzianum* 10 g/kg soil and spraying Carbendazim @ 0.2% to 15 days old seedling.
- Other fungicides can also be used are Boscalid, Fluopyram, Penthiopyrad, Metconazole, Tebuconazole, Penthiopyrad, Azoxystrobin, and Pyraclostrobin at recommended rate.

3.2 ALTERNARIA LEAF SPOT OR BLIGHT

It is widely distributed wherever sunflower is grown and is considered the most destructive one. Due to this disease, seed yield (27%–80%), and oil yield (17%–35%) reduction has been reported. Further, a negative correlation between disease intensity and seed yield, and oil content has been observed. The parameters, viz., the number of seeds per head and seed yield per plant are mostly affected and also affect the quality of seeds by affecting germination and initial vigor of the seedlings. The disease reduces the seed yield (27%–80%) and oil yield (17%–33%).

Symptoms

Alternaria leaf spot symptom appears in all growth stages of sunflower plants and on leaves, petiole, stem, bracts, and flower heads. Initial symptoms appears in the form of circular to oval, dark brown to black, small lesions (0.2–0.5 mm in diameter), surrounded by distinct yellow halos with the grey white necrotic center. Later, the spots enlarge into the round to irregular spots with concentric rings and coalesces leading to necrosis (blighting) and withering of entire leaves. The spot appears first on the lower leaves and later appears on the middle and upper leaves. Stem lesions appear as elliptical to diamond-shaped sunken lesions which coalesce to form large blackened areas resulting in stem breakage which can lead to severe premature defoliation. In severe cases, the lesions appear on petioles, ray florets, and head as brown round spots with a slight depression in the center. Sometimes head rot also occurs.

Casual Organism and Etiology

Various species of fungus *Alternaria* viz. *A. helianthi, A. alternata*, and *Alternariaster helianthi* can cause Alternaria leaf spots of sunflower. *A. helianthi* is the most common pathogen responsible for the disease, but *A. zinniae* and *A. alternata* can occasionally attack sunflowers. The fungus *Alternaria helianthi* produces pale grey-yellow colored cylindrical conidiophores which are straight or curved, geniculate, simple or branched, septate, and bear single conidium (straight or slightly curved, cylindrical to long ellipsoid, multi septate, pale grey-yellow to pale brown).

Disease Cycle and Epidemiology

The fungus survives on seeds and infected plant debris and serves as primary inoculum. The fungus overwinters as mycelium on infected plant residues and in dry conditions survives for 20 weeks in soil. The disease occurs when conidia land on leaves or stems, germinate in the presence of free moisture, and directly penetrate and infect the plant. Plants are most susceptible to infection at flowering and continuing through maturity; plant stress also predisposes plants to the disease. The leaf spots resulting from this primary infection sporulate and produce a new crop of abundant conidia. The conidia readily disperse to the whole foliage of new plants, within and among the fields by splashing irrigation water, wind, and insects, and causes secondary infection. The fungus is seedborne with 22.9% seed transmissible nature. The disease is favored by warm (temperature 25°C–27°C), humid environments, frequent rain (free water), and 12 hours of wet foliage.

Management

Cultural Methods

- Early planting during Kharif escapes from the leaf spot. Altering the date of sowing in order to reduce disease pressure and sowing during August–September onwards is suggested.
- Mid-September planting of sunflowers remains free from most of the major diseases including *Alternaria.*
- Intercropping of sunflower with groundnut in the ratio of 6:2 reduces leaf spots.
- Spacing of 60 × 30 cm or 45 × 30 cm is optimum in reducing the buildup of leaf spot.
- Soil solarization in summer for 3–4 weeks greatly reduces the survival and increases the vulnerability of *Alternaria* to antagonists. This practice significantly reduces the incidence of Alternaria blight of sunflower and it can be adopted in areas where summer temperature goes above 43°C and irrigation facilities are available.

- Remove and burn the crop residue. This method is most helpful because the mycelium rests in the plant debris and soil.
- Burying infested crop residue to hasten decomposition helps minimize *Alternaria* infection.
- Summer deep ploughing reduces the inoculum level in the soil, which is present in plant debris as dormant mycelium. As the fungus is aerobic, deep ploughing summer of all undecayed organic matter and plant debris to at least 8–10 cm, greatly reduce Alternaria leaf blight incidence in sunflower.

Host Resistance

- Grow resistant/tolerant cultivars. At present, there is no cultivars are known to be resistant to Alternaria blight in sunflowers. But some tolerant sunflower varieties like BSH-1, GP-145 and AH-303 may be effective to Alternaria blight under field conditions. Some other resistance sources identified are EC-132846, EC-132847, EC-132361, and EC-126184.

Chemical Control

- Selection of healthy seeds and seed treatment with the fungicide namely Iprodione 25% + Carbendazim 25% or Mancozeb 63% + Carbendazim 12% @ 2 g per kg of seed during the sowing.
- Foliar application of Iprodione 25% + Carbendazim 25% @ 0.2% or Propiconazole 25% EC @ 0.1% or Mancozeb @ 2.5 g/L, two times at 30 and 45 days after sowing.

4 Mustard (*Brassica juncea* L.)

4.1 ALTERNARIA BLIGHT OR BLACK SPOTS

Cultivated crucifers including oilseed brassicas and vegetables such as cabbage, cauliflower, knol-khol, and radish are severely affected by several species of *Alternaria*. Yield losses may vary from 9%–28% depending on host species and variety and seed germination may reduce 23%–27%. In rapeseed (yellow and brown sarson) yield may reduce upto 70%. The yield is negatively correlated with the number and size of spots, sporulation of fungus, amount of defoliation, and resistance components of cultivars.

Symptoms

Alternaria brassicae, along with *A. alternata*, *A. brassicicola*, and *A. raphani*, cause symptoms at the seedling stage on cotyledons (embryonic leaves) and at the adult stage on true leaves, leaf petioles, stems, inflorescences, siliquae, and seeds. The disease first appears on the cotyledons and hypocotyls in the form of small (pinhead size) light brown lesions which soon turn black due to the appearance of spore masses (under humid conditions) which act as a source of infection for the other parts of the plant. A zone of yellow halo around the spots and the characteristic concentric rings are formed in these lesions (Figure 4.1 A). The center soon dries and may drop out. When numerous spots form on leaves, these spots coalesce to cause leaf blight and ultimately lead to defoliation. Leaf spots incited by *A. brassicicola* appear as small, dark-colored areas which expand rapidly to form circular lesions up to 1 cm in diameter. These spots appear sooty black, velvety and copiously covered with black conidiophores and conidia. The spots caused by *A. brassicae* are similar to those caused by *A. brassicicola* but tend to remain smaller and lighter in color. These spots are grey, dense, and sparsely covered with brown conidiophores and conidia. The lesions caused by *A. raphani* are small, spherical to elliptical with raised margins and are surrounded by a yellow translucent halo. Black sporulation of the fungus may be seen on the spots. Dark brown, circular to linear lesions also develop on the stem and siliquae. Lesions on the stem usually remain elongated and pointed at later stages. On stem lesions minute, black microsclerotia can also be seen. Infection on pods results in the formation of small, discolored, and shriveled seeds with white to brown or black mycelial mats of the fungi (Figure 4.1 B). Seedborne inoculum could cause pre- and post-emergence damping off.

Figure 4.1 Alternaria Blight of Mustard. (A) Brown spots with concentric rings on leaves, (B) Mustard seeds covered with white to brown mycelial mat of *Alternaria* spp.

Causal Organism and Etiology

Alternaria leaf spot of mustard is caused by *Alternaria species* (*Alternaria brassicae*, *A. brassicicola*, *A. raphani*, and *A. alternata*). However, *Alternaria brassicae* is a major pathogen of oilseed Brassicas. In the host, the hyphae of *A. brassicicola* are branched, septate, and hyaline at first, later becoming brown or olivaceous brown, inter- and intracellular. Conidiophores arise through stomata singly or in groups of 2–12 or more. The conidiophores of *A. brassicicola* are olivaceous, septate, and branched. Conidia are linear to obclate, borne in chains of 8–10 or more spores, septate, muriform, pale or dark brown. Each conidium has 1–11 but usually less than 6 transverse septa and few, usually upto 6 longitudinal or oblique septa. The beak is about one- eighth of the length of the conidium. The conidiophores of *A. brassicae* arise in fascicles and the conidia are dark, obclate, muriform, born singly or in short chains of upto 4 spores. The conidia usually have 16–19 transverse septa. The beak is about one-third the length of the conidium. The species produces microsclerotia and chlamydospores. *A. raphani* is similar to *A. brassicae* in its morphology but the conidia possess a shorter beak.

Disease Cycle and Epidemiology

The pathogen survives by developing mycelium, conidia, chlamydospores, and microsclerotia on diseased plant debris in soil or as mycelium and conidia on allied crucifers or weed hosts. The leaves initially infect a few plants to serve as sources of secondary infection. The pathogen also perpetuates through infected seeds. In rapeseed mustard, 22%–28% bold and symptomless seeds are reported to carry heavy aggregations of mycelium in the seed coat epidermis, palisade, endosperm, and embryo. *A. raphani* can survive as chlamydospores in soil for 5 years. Secondary infection occurs during the crop season through the conidia produced on infected host plants (under high humidity), which spread through rain splashes and wind to complete the disease cycle. Conidia after germination produce a germ tube and penetrate through stomata, cuticle, or wounds with or without the formation of appressoria. Direct penetration through cuticles is mostly by virulent strains. The severity of Alternaria blight disease on leaves is positively correlated to a maximum daily temperature of 18°C–27°C, minimum daily temperature of 8°C–12°C, daily mean temperature above 10°C, more than 92% morning relative humidity, above 40% afternoon relative humidity and mean relative humidity of above 70% in the preceding week.

Management

Cultural Methods

- Use certified and healthy seeds for sowing.
- Adopt timely sowing between 10–25 October.
- Follow proper field sanitation.
- Remove weeds, particularly collateral host plants.
- Apply a recommended dose of potash to reduce disease incidence.

Physical Control

- Hot water treatment of seed at 40°C for 30 minutes or 50°C for 20 minutes has been recommended against seedborne inoculum.

Host Resistance

- Grow resistant/tolerant varieties such as Coral-432, NRCHB 5-6, NPJ 112 (Pusa Mustard 25), NRCDR 601, RYSKS-2, DMH-I, Indian mustard-91, Toria-16, Yellow sarson-11, Gobhi sarson-11, Brown Sarson-3, Karan Rai-4, Taramira-5, Black mustard-1, and Saurabh.

Chemical Control

- Treat the seed with Thiram or Carbendazim + Mancozeb @ 3 g per kg of seeds.
- Foliar sprays of Carbendazim + Mancozeb (0.2%) or Chlorothalonil (0.2%), or Difenoconazole (0.1%) or Tebuconazole (0.1%) or Metalaxyl 8% + Mancozeb 64% WP @1000 g in 400 litres of water/acre can prevent the crop from *Alternaria* blight.
- Two to four sprays of Dithane M-45 (0.2%), Dithane Z-78 (0.2%), Difolatan (0.2%), and Blitox-50 (0.3%) at 15 days interval starting 45 days after sowing, give good control of the diseases.

4.2 WHITE RUST

White rust disease is also known as a white blister, stagheads, and white blister rust, on the basis of symptoms in the host plants. The disease causes yield losses from 1%–90% in oil-yielding Brassica and cruciferous crops. Formation of staghead accounts for the complete loss of seed formation and causes upto 90% yield loss. Combined infection of leaves and inflorescence may cause upto 60% yield loss in Indian mustard (*Brassica juncea*) in late crops the yield loss due to floral infection may be from 23%–54.5%. *A. candida* also causes white blisters of crucifers such as cabbage, turnip, swede, horseradish, etc. *Albugo candida* alone or in combination with co-infecting *Hyaloperonospora parasitica* can occasionally cause significant crop losses.

Symptoms

All plant parts except the root show symptoms. The disease is characterized by the expression of both local and systemic symptoms. In local infection, distinct pustules or sori develop on leaves and stems. These pustules or blisters are raised, shiny white or creamy yellow areas (Figure 4.2 A). They are very variable in shape and size and later coalesce to form larger patches. The host epidermis ruptures and exposes sori consisting of a white powdery mass of spores. Systemic infection results in abnormal growth, swellings, curved, twisted, and distorted inflorescence, and sterility of flowers and therefore referred to as 'stagheads', formed as a result of hypertrophy and hyperplasia (Figure 4.2 B). The petal may become sepal-like and stamens become leaf-like or occasionally like carpels; ovules and pollen grains are usually atrophied resulting in the sterile condition of the flowers. The pistil also becomes malformed and larger in size and sometimes the pods transform into leaf-like structures. Such malformed pods lack seed development. These swollen or hypertrophied parts are full of oospores of the pathogen and cells are starch-rich. The mixed infection of white rust and downy mildew is conspicuous at the floral stage of crop growth. On the malformed floral parts, white rust produces pustules, and downy mildew produces a coating of the fine white mass of sporangia. At maturity, under high humidity, malformed inflorescence may be covered with saprophytes turning into dark brown to black. Plants may remain stunted due to heavy infection.

Figure 4.2 White Rust of Mustard. (A) White blisters on leaf) and (B) Distortion of stem and inflorescence (stag head).

Causal Organism and Etiology

White rust of mustard is caused by an obligate parasite *Albugo candida* (formerly *Cystopus candidus*). It belongs to Chromista, Oomycota, class–Oomycetes, order–Peronosporeales, family–Albuginaceae. Mycelium is intracellular producing knob-shaped haustoria in the host cells. The pathogen is characterized by a unique asexual stage, which consists of subepidermal, erect, club-shaped sporangiophores, budding off sporangia. Spherical sporangia are formed in basipetal succession in chains (i.e. youngest sporangium at the base and the oldest at the top of the chain). The accumulating sporangia eventually rupture the raised epidermis and the white or creamy masses of conidia become visible, hence named "white rust" disease. The sporangia are hyaline, nearly spherical, and germinate by means of germ tubes or producing zoospores. In between the successive sporangia develop a mucilaginous pad or separation disc, called a disjunctor. The oogonia and antheridia are formed from the mycelium in the intracellular spaces of systemically infected host tissues. The oogonium is globose, terminal or intercalary and consists of about 100 nuclei. The antheridium is clavate, paragynous, and contains 6–12 nuclei.

Disease Cycle and Epidemiology

Albugo candida perpetuates as oospores in infected plant debris in soil and seed, functions as the source of primary inoculum of the pathogen. Perennial weed hosts may also serve as a source of primary inoculum. Oospores are spread by splashing rain and irrigation water, and possibly by blowing soil. On germination, biflagellate, remiform zoospores are formed and held in a vesicle on the oospore, before they are set free. These swim, encyst and germinate on the host leaf and cause infection by entering through stomata. The mycelium grows intercellularly, forming haustoria in the host cells. In *Albugo*, the mycelium in the host tissues is intercellular with only small spherical haustoria which contrast sharply with the coarsely lobed haustoria of *H. parasitica*. Infections may be localized or systemic. Diseased plants produce windblown sporangia that serve as secondary inoculum, infecting other plants and beginning the disease cycle again. The sporangia are visible externally as a white powdery mass dispersed by the wind. Sporangia germinate on reaching to a suitable host. Their germination depends on the availability of water and temperature. In presence of water and low temperature, produce about 4–8 biflagellate, and reniform zoospores per sporangium. After swimming for a time, a zoospore encysts and then forms a germ tube that penetrates the host epidermis. In the presence of high temperature and dry conditions, the sporangium behaves like a conidium. It germinates directly to form a germ tube which penetrates the host through a stroma or the epidermis. The asexual disease cycle may be completed within 10 days. The repeated production of sporangia spreads infection rapidly to more and more plants. When the crop season approaches its end, sexual reproduction ensues. The hyphae penetrate deeper into the tissue, become systemic, and form sex organs (oogonia and antheridia). The onset of sexual reproduction, typically oogamous, is usually indicated by hypertrophied plant parts. Oospores are formed in the hypertrophied tissues (leaves, stems, inflorescences, pods, and roots) of infected host plants. Oospores overwinter in infested crop debris and soil. The multinucleate oospores, after prolong resting period, germinate to form zoospores that eventually encyst and germinate by germ tube and cause infection, to complete the life cycle. Disease development is favored by moist, cool to moderate temperatures (10°C–20°C) and abundant rainfall. A film of water is essential for sporangial germination. Maximum sporangial germination occurs at 12°C–14°C. Oospore production can occur at 10°C–27°C, the maximum being at 23°C.

Management

Cultural Methods

- Use pathogen-free seed.
- Use certified seeds of the resistant or tolerant variety.

- Destroy weeds around the field.
- Follow crop rotation to avoid the soil-borne primary inoculum.
- Follow timely sowing of the crop.
- Adopt proper field sanitation.
- Follow proper crop rotation.
- Destroy crop debris, particularly stag heads from the previous crop.
- Avoid over-irrigation or water stagnation.
- Apply potash in the recommended dose

Biological Control

- Treat the seeds with freshly prepared garlic bulb extract @ 2% (w/v).

Host Resistance

- Grow resistant/tolerant variety such as Pusa Karishma (LES-39), Jawahar Mustard-1, PBR 91, Indian mustard-91, Toria-16, Yellow sarson-11, Gobhi sarson-11, Brown Sarson-3, Karan Rai-4, Taramira-5 and Black mustard-1.

Chemical Control

- Treat the seed with Metalaxyl 35% WS @ 0.75–1.0 kg/100 kg of seed.
- Spray Metalaxyl 8% + Mancozeb 64% WP @ 1000 g in 400 litres of water/acre.

4.3 DOWNY MILDEW

Downy mildew disease causes heavy yield loss to cruciferous crops and is distributed worldwide. In India, the disease is responsible for upto 66% yield loss in *B. juncea.* Mixed infection of downy mildew and white rust can cause a loss of 17%–32% in *B. juncea.*

Symptoms

The pathogen can infect all aerial plant parts, but symptoms are usually more conspicuous on leaves, stems, and inflorescences (Figure 4.3). At the seedling stage, small angular translucent light green lesions appear on cotyledons and the first few true leaves. Later, these lesions enlarge and develop into grayish-white, irregular necrotic patches on the upper surface of the leaf, while purplish-brown, downy mycelial growth appears on the undersurface. In a severe attack, diseased leaves dry up and shrivel. Symptoms of mixed infection of downy mildew and white rust diseases of *Brassicas* can be seen at both stages, i.e. leaf phase and stag head phase of plant growth. On leaves, white rust pustules are surrounded by white or creamy white sporulation of downy mildew pathogen.

Figure 4.3 Downy Mildew of Mustard.

In the systemically infected plants, the symptoms are very conspicuous and almost similar to those produced by *Albugo candida*. In downy mildew, the greatest deformities appear on the stem while in white rust the inflorescence shows maximum deformities. Stem swellings may be limited often with abrupt bending of the stalk, or swelling may be several inches long. The young ovary may be very much elongated into a twisted body with stamens being shrunken. This leads to the sterility of the flowers. On the malformed floral parts, *A. candida* produces pustules, and *H. parasitica* produces a coating of the fine white mass of sporangia or conidia. Conidia initially appear at the tip of the stag head, and later on covered the entire inflorescence. At crop maturity, and under high humidity, malformed inflorescence may be covered with saprophytes turning into dark brown to black color.

Causal Organism and Etiology

The disease is incited by a biotrophic fungal pathogen *Hyaloperonospora parasitica* (syn. *Peronospora parasitica*). The mycelium is hyaline and coenocytic. It grows intercellularly in the host tissues and produces haustoria to penetrate the host cells. The haustoria are large, lobed, elongated, or club-shaped. The conidiophores are hyaline and uniform with a flattened base and stout main axis. They are dichotomously branched, six to eight times, tips bifurcate, branching acute, and slightly thickened above each fork. The terminal branches are long, slender, and pointed and end in a single conidium. The sterigmata are slender and acutely pointed. The conidia are hyaline, broadly elliptic, or nearly globose, and are delimited from sterigmata by cross walls at maturity. A single conidium is borne at the tip of each branch and is deciduous. Oogonia are pale yellow, irregularly round, and swollen into crest-like folds. Antheridia are tendril-like and are produced on separate hyphae. The protoplasm of the oogonium becomes differentiated into a central vacuolated ooplasm and a peripheral multinucleate granular periplasm.

Disease Cycle and Epidemiology

H. parasitica survives as oospores in malformed inflorescence and senesced host tissues, as conidia on leaves and inflorescence, and as latent systemic mycelium in seeds or infected plant debris. Transmission by infected seed is possible but its importance is not significant. Primary infection occurs from soil-borne oospores. Further spread of the pathogen is by dissemination of conidia released from conidiophores formed on the cotyledons or hypocotyls. Detachment of conidia is possibly caused by hygroscopic twisting of the conidiophores which in turn is related to fluctuations in humidity. Conidia germinate in free water by a lateral germ tube, not by zoospores. Infection occurs both by direct penetration of the epidermis and through stomata. The conidia produce germ tubes that penetrate anticlinal cell walls often on the lower surface of the leaves. The true leaves are usually infected through wind-borne conidia, resulting in the spread of the disease through secondary infection. The pathogen dispersal over short distances in water droplets can also occur. The disease is favored by cool, moist conditions, temperatures between 10°C–15°C, and by high atmospheric humidity following rain or heavy dew. The conidia germinate most rapidly at 8°C–12°C, whereas penetration was optimal at 16°C and haustoria formation at 20°C–24°C.

Management

- Similar to the white rust disease.

4.4 SCLEROTINIA STEM ROT

Sclerotinia rot is known under a variety of names and the most common of these are cottony rot, white mold or watery soft rot, stem rot, white rot, white blight, stem blight, stalk break, stem break, and white canker. In India, *Sclerotinia* stem rot was reported from Pusa (Bihar) in 1915. The *Sclerotinia sclerotiorum* is a necrotrophic fungal pathogen causing stem rot disease in *B. juncea* worldwide. Severe outbreaks of stem rot disease under cool and moist weather conditions cause heavy yield losses ranging between 5%–100%. Sclerotinia stem rot is also now recognized as a major yield-limiting disease. Amongst the fungal diseases, earlier considered to be a minor problem in India has become a serious problem of rapeseed-mustard over the years in some parts of the country like Rajasthan, Haryana, Punjab, Utter Pradesh, and Uttarakhand. Since the pathogen is soilborne, the severity of *Sclerotinia* rot has been increasing year after year. The maximum *Sclerotinia* rot incidence recorded in the field of mustard growers of Rajasthan was 90%, whereas in Haryana, the maximum incidence was 30%. At the time of harvesting and threshing sclerotia of the fungus gets mixed with seed and this represents objectionable seed contaminants for export.

Symptoms

Generally, the symptom *Sclerotinia* rot appears on leaves, stem, and siliquae, during the flowering and pod development stage. Initially elongated, soft, water-soaked lesions develop on the stem especially at the base or at the internodes and later grayish-white fluffy mycelial growth (white mold) covers these lesions (Figure 4.4). Stems girdled with lesions are weaker and lodge sooner or later. The plant with girdled stem, ripe prematurely and becomes conspicuously

Figure 4.4 Sclerotinia Stem Rot of Mustard. White mycelial growth with black sclerotia on infected tissue of stems.

straw-colored in a crop which is otherwise green. As the disease progresses, large hard, black, irregular shaped, resting bodies (sclerotia) develop on these lesions. On ripening the dry portion of the stem, the pith can be seen full of sclerotia or they may also develop on the stem surface as a white mycelial web sticking to the host surface. If the sclerotia are absent, the pith of the infected stems turns darker in color as compared to the healthy stem. The airborne ascospores infect flower petals which fall and are deposited on leaves, leaf nodes, main stem, and side branches and infect them. Generally, lesions on leaves are grayish, irregular in shapes and size and often associated with adhering petals. Under severe infection, defoliation, shredding of stem, wilting, and drying of plants occur. Infected plants will ripe earlier and stand out among green plants. At harvest, the sclerotia are either threshed out with the seed or left in the field along with crop residues.

Causal Organism and Etiology

Sclerotinia stem rot of mustard is caused by a necrotrophic polyphagous fungus, *Sclerotinia sclerotiorum*. The systematic position of the pathogen is as follows: Kingdom–Fungi, Division–Eumycota, Subdivision–Deuteromycotina, Class–Gonomycetes, Order–Agonomycetales, Family–Mycelia sterilia, Genus–*Sclerotinia*, Species–*sclerotiorum*. However, the perfect stage of the pathogen belongs to Ascomycotina, order Helotiales, and family Sclerotiniaceae that is characterized by the formation of hard blackish sclerotia which germinate and produce cup shaped and brown-colored apothecia. The hyphae are closely septate inter as well as intracellular. The sclerotia are pinkish white when young and become dark brown to black at maturity. They vary in size and shape, being flattened, elongated, or roughly spherical. On germination, each sclerotium develops 2–5 stipes and each stipe produces apothecium at its tip. Apothecia are brownish or flawn colored, cup-shaped. Several asci develop in each apothecial cup. The asci are cylindrical, unicellular, and ovate. Ascospores are the chief means of spread of the disease and they are discharged violently.

Disease Cycle and Epidemiology

The life cycle of *Sclerotinia sclerotiorum* is relatively simple and various sexual and asexual forms help in the spread of the disease year after year. There are four stages in the life cycle of *S. sclerotiorum*: sclerotium, apothecium, ascospore, and mycelium. *S. sclerotiorum* survives up to 5 years as sclerotia, present on the soil surface or in crop debris or admixture with the seeds, and serves as primary inoculum. During crop harvesting and threshing, sclerotia fall in the ground, mix up in the soil and provide primary inoculum for the next years. *Sclerotinia sclerotiorum* infection of rapeseed-mustard is primarily through airborne ascospores discharged from apothecia. Seeds also play an important role in the survival, perpetuation, and transmission of disease from one area to another. Under favorable environmental conditions, these sclerotia germinate either myceliogenically (forming mycelium) or carpogenically (forming apothecia). Sclerotia of *S. sclerotiorum* most commonly germinate carpogenically by producing apothecia (mushroom-like fruiting bodies) which requires the sclerotia to be wet in soil for 1–2 weeks prior to germination. Apothecia produce ascospores as a result of sexual reproduction. In myceliogenic germination, the sclerotium produces mycelium when the soil remains wet and cool and the optimal conditions for germination are near-saturated wetness with soil temperatures of 12°C–24°C. The fungal mycelium causes infection at or below the soil level.

Large quantities of ascospores are forcibly discharged from apothecia into the air and are carried by air currents for distances ranging from a few centimeters to several kilometers. Ascospores discharged at the base of the plants in soil and cause primary infection. On coming in contact with susceptible host tissue, the ascospores germinate and produce germ tubes and an appressorium. The ascospores give rise to infection hypha, and the infection within the host takes place; directly by mechanical pressure through the cuticle, or the infection hypha may penetrate already wounded or injured tissue also. After entering the host plant, the fungus grows through the host tissues causing the cells to die in advance of the invading hyphae. Secondary infection results from green tissue coming in contact with an infected area but no secondary infection propagules are produced. The mycelium produces sclerotia externally on affected plant parts and/or internally in the stem pith. During harvesting and threshing, these sclerotia remain in fields with crop debris. The sclerotia survive in the soil and in the plant debris to complete the disease cycle. *Sclerotinia sclerotiorum* primarily survives from one crop period to another in the soil through sclerotia. Such sclerotial bodies get mixed with the soil through affected plant debris after the crop is harvested, or when seeds contaminated with the sclerotial bodies are sown in the soil. Infection of above-ground plant parts results from ascosporic inoculum whereas soilborne infection may result either from ascospores or sclerotia. In the presence of a thin layer of water, ascospores germinate in less than 24 hours at 5°C–30°C, with optimum range being 5°C–10°C. Cool temperatures, prolonged periods of high soil moisture and comparatively high relative humidity are required for apothecial formation. No apothecial initials are produced at either 5°C or 30°C. Continuous leaf wetness is required for infection by ascospores for approximately 48–72 hours. For successful infection, the process of ascospore release and petal fall should occur at the same time. Petals of rapeseed are known to stimulate ascospores germination. Flower petals serve as a source of nutrients for the germination of ascospores and invasion. After abscission infected petals are deposited on leaves, leaf nodes or axils. The epidemiology of *Sclerotinia sclerotiorum* is dependent on several factors viz., soil inoculum, soil type, soil moisture, rainfall, soil and environmental temperature, host susceptibility plant density, and cultural practices adopted. Development of the disease is preferred by average cool temperature (18°C–25°C) and high humidity (90%–95%) along with winter rains and wind currents particularly at the flowering to pod initiation stage.

Management

Cultural Methods

- Use healthy, certified, and clean seeds free from the sclerotial bodies for sowing.
- Deep ploughing of the field during summer.
- Follow crop rotation with nonhost crops like wheat, barley, rice, maize, etc.
- Timely sowing of the mustard crops in between October 16 to 31.
- Remove broad-leaf weeds like *Chenopodium* spp.
- Maintain optimum plant population in the field.
- Avoid overcrowding of plants in a row to minimize plant-to-plant contact through root and stem to reduce disease spread.
- Flooding of soil minimizes the sclerotial load in the soil.

- Need-based irrigation of mustard field.
- Use only recommended dose of fertilizers @ N: P: K: S (60:40:40:40) kg/ha.

Mechanical Control

- Destroy infected crop debris of the previous crops by either burying or burning to eradicate the sclerotia.
- Rogue out Sclerotinia rot affected plants, before the production of sclerotia in order to reduce soil inoculum.
- If the sclerotia are mixed with seed during harvesting and threshing, remove these sclerotia by handpicking or using a winnower or with the help of different size sieves, as these are objectionable seed contaminants.

Host Resistance

- Grow resistant/tolerant varieties such as NPC-9, Kiran, Pusa Karisma, RLM-619, Hyola 401, PBN 9501, PWR 9541, Kiran, RH 9401, RH 492, RW 8410, PCR-10, RW-8410, RW-8006, RGH-8006, and PAB 9511.

Biological Control

- Treat the seed with a mixture of *T. harzianum* and *T. viride* @ 10g/kg of seeds.
- In soil, incorporate a mixture of *T. harzianum* and *T. viride* based product @ 2.5 kg/ha pre-incubated in 50 kg of FYM/ha at the time of preparation of the field.
- Give need-based foliar spray of a mixture of *T. harzianum* and *T. viride* product @ 0.2% at 20 days intervals, soon after *Sclerotinia* stem rot appearance, which helps to control the disease spread.
- Integration of the seed treatment followed by foliar sprays of Carbendazim @ 0.1%, *Trichoderma viride*, and *Allium sativum* aqueous bulb extracts at 45–50 and 65–70 days after sowing (DAS) provide better reduction of the disease.

Chemical Control

- Treat the seed with Carbendazim at the rate of 2 g/kg seed.
- Drench Carbendazim @ 2 g/L of water.
- Spray Carbendazim @ 2 g/L of water, if needed, in standing crop.

5

Chickpea (*Cicer arietinum* L.)

5.1 WILT

Chickpea wilt is one of the most devastating, widespread, and important diseases throughout the world, causing heavy losses (20%–100%) depending upon the stage of infection, wilting and climatic conditions. It is also known as "Mar disease" in Marathi. In India, chickpea wilt was first reported by E.J. Butler in 1918 later on its etiology was determined by Padwick in 1940. In India, wilt is one of the major diseases of chickpea.

Symptoms

The wilt observed within 25 days after sowing is designated as "early wilt" and during early flowering stage to podding stage (7–8 weeks after sowing) as "late wilt". Sometimes only a few branches of the chickpea plant are affected, resulting in partial wilt. Wilting at the early stage of growth causes more loss than at the later stage.

Early wilt (seedling stage): Seedlings retain their dull green color. Leaves of plants develop greyish-green chlorosis, typically affecting lower leaves first. Seedlings may collapse and lie on the ground. The roots do not exhibit external discoloration and appear healthy. When roots split vertically shows a brown discoloration of the internal tissues (xylem). This is a diagnostic feature of Fusarium wilt.

Late wilt (adult stage): Late wilted plants exhibit drooping of the petioles, rachis, and leaflets, followed by yellowing, necrosis of foliage, drying, collapsing and death of whole plants. Nevertheless, seeds from late-wilted plants are lighter, rougher, and duller than those from healthy plants. Initially, symptoms appear on upper leaves, flowers, and twigs which show discoloration, desiccation, drooping of leaflets, rachis, and petioles. Internally, the xylem tissues show dark-brown to almost black. Affected plants do not show external root discoloration.

However, co-infection with other soil-borne pathogens may cause external root discoloration. Seeds harvested from the diseased plants are light, rough, and dull.

Causal Organism and Etiology

Chickpea wilt is caused by *Fusarium oxysporum* f. sp. *ciceris*. It belongs to Kingdom: Fungi Phylum: Ascomycota, Subphylum: Pezizomycotina, Class: Sordariomycetes, Order: Hypocreales, Family: Nectriaceae. Hyphae are septate and profusely branched. The pathogen produces macroconidia, microconidia, and chlamydospores. The conidia are produced on pinkish sporodochia. The microconidia are oval or cylindrical, straight or curved forms abundantly on short cylindrical, unbranched phialides. Macroconidia are fusiform, curved, with a tapering, pointed, sometimes hooked, apical cell and distinctly pedicellate basal cell. Chlamydospores are globose, terminal or intercalary, single, or in small groups or chains.

Disease Cycle and Epidemiology

In chickpea, *F. oxysporum* f. sp. *ciceris*, can be internally seedborne and the pathogen is found as chlamydospores-like structures in the hilum region of the seed. *Fusarium oxysporum* f. sp. *ciceris* can survive as mycelium, and chlamydospores in seed, soil, and infected crop residues, roots and stem tissue buried in the soil for more than 6 years. In fallow soils, infection of symptomless dicotyledonous weeds can enhance the survival of the pathogen. Thus, infested soil is a main source of primary inoculum for Fusarium wilt development. After the germination of chlamydospore, hyphae, conidia, and new chlamydospores are produced. Invasion of the roots is followed by the penetration of the epidermal cells of the host or the non-host and the development of systemic vascular disease in host plants. Penetration occurs either through wounds or directly. The most common sites of direct penetration are located at or near the root tip of both taproots and lateral roots. Pathogen penetrates the host plant through root apices or wounds and multiplies to plug the vascular bundles. Penetration may be influenced by a combination of different factors such as plant surface structures, activators or inhibitors of fungal spore germination, fungal compounds, and germ tube formation. During colonization, the mycelium advances intracellularly through the root cortex until it reaches the xylem vessels and enters them through the pits. The fungus then remains exclusively within the xylem vessels, using them to colonize the host. Wilting is most likely caused by a combination of pathogen activities such as accumulation of fungal mycelium in the xylem and/or toxin production, host defense responses, including production of gels, gums, tyloses, and vessel crushing by the proliferation of adjacent parenchyma cells. Desi type of chickpea is less susceptible than the Kabuli type. Root exudates influence wilt incidence. Exudates of some resistant cultivars contain antifungal compounds that suppressed spore germination of the pathogen. The fungal spores are dispersed by water flow, rain splash, and by the movement of infected soil or seed externally contaminated during threshing. Wilt development was greater at 25°C than at 20°C or 30°C. Root knot nematode may also have a role in the Fusarium wilt of chickpea.

Management

Cultural Methods

- Always use disease-free seeds.
- In fields having an incidence of wilt, the cultivation of chickpea should be avoided.

- Follow deep summer ploughing.
- Follow crop rotation for 4 years continuously with unrelated crops, e.g. wheat, barley, or oats.
- Avoid sowing when the temperature is high.
- Avoid late sowing. Sowing should be done during 3rd week of October.
- Deep planting (8–10 cm) in light soil.
- Apply castor cake 500 kg/ha at the time of sowing.

Host Resistance

- Grow resistant/tolerant varieties like C-214, Avrodhi, Uday, BG-244; Pusa-362, JG-315, Phule G-5, JGK2, JG322, KAK2, Vijay, Vaibhav, JG63, Birsa channa-3, etc.

  Desi: JG315, JG 322, JG 11, JG12, JG14, JG 16, JSC37, JSC55, JAK19218.

  Kabul: JGK1, JGK2, JGK3 (Gulabi)-JGG1.

Biological Control

- Soil solarization along with soil amendments with Neem cake or Mustard cake help in the reduction of inoculum.
- Soil application of bioagent like *Trichoderma viride* @ 2.5 kg/ha mixed with well rotten farmyard manure (FYM) or neem seed kernel cake (NSKC) of about 100 kg. After mixing the bioagent with manure, incubate it for 21 days, in shed before application.
- Treat seeds with *Pseudomonas fluorescens*, *Trichoderma harzianum*, *T. viride*, *Bacillus subtilis*, etc. @ 4 g/kg seed. Do not use bioagent and fungicide together as the chemical will kill the bioagent.

Chemical Control

- Seed treatment with Carbendazim @ 2.5 g/kg of seed or Benomyl @ 3 g/kg.
- Seed treatment with Thiram @ 2 gm/kg + Carbendazim @ 2.5 g/kg of seed.

5.2 BOTRYTIS GREY MOLD

Grey Mold or Botrytis grey mold (BGM) is the second most potentially important disease of chickpea after Ascochyta blight, especially in areas where cool, cloudy, and humid weather persists. The first occurrence of Botrytis grey mold on chickpea was reported from India by Shaw and Ajrekar (1915) and later by Butler and Bisby (1931). The first epidemic of BGM was reported by Carranza (1965) in Argentina, which resulted in a crop loss of 95%.

Symptoms

All the aerial parts of chickpea are susceptible to the disease with growing tips and flowers being the most vulnerable. Symptoms usually become more apparent in the dense crop canopies. The initial symptoms on stem, flowers, leaves, and pods develop as water-soaked lesions which later turn grey or dark brown covered with erect hairy sporophores and masses of single-celled hyaline spores (Figure 5.1). In the early morning, the grey fungal growth is evident on the leaf, stem, pods, flowers, and petioles. Grey fungal growth and profuse sporulation occur if the conditions within the canopy are moist or humid and disease spread rapidly through the canopy resulting in patches of dead plants. Stem lesions (10–30 mm long) completely girdle the stem and branches break off at the rotting point and the affected leaves and flowers turn into a rotting mass. Drooping of the affected tender terminal branches is a common field symptom. In cloudy weather, flower-drop and rotting of plant parts are conspicuous. All affected foliages are discolored. The pod lesions are water-soaked, irregular in shape, and may have black sclerotial bodies scattered throughout the infected areas. When the disease affects pods, no seeds or only small, shriveled, discolored seeds are formed, and sometime the seeds may be covered with a greyish-white fungal mat. Sometimes, tiny black sclerotial masses appear on the dead tissue. These dark brown sclerotia should not be confused with the larger dark or dark brown sclerotia of *Sclerotinia sclerotiorum* that are embedded in a white mycelial mat.

Figure 5.1 Botrytis Grey Mold of Chickpea.

Causal Organism and Etiology

Grey mold of chickpea is caused by *Botrytis cinerea*. The asexual stage of the necrotrophic fungus *B. cinerea* (Moniliaceae, Hyphales) is dominant in chickpea crops. The young hyphae are thin, hyaline, and 8–16 µm wide, and they become brown and septate with age. The conidiophores are light brown, septate, and erect, ramified pseudodichotomically with slightly enlarged tips bearing small pointed sterigmata bearing 1–2 celled, hyaline, oval, or globose conidia forming clusters. Sporodochia are produced on the host tissues which produce small, unicellular, and round microconidia. The microconidia do not germinate. The sporodochia may soon turn into a hard sclerotial mass. Sclerotia, which germinate asexually by producing conidiophores bearing conidia, can form on crop stubble. The teleomorphic stage of *B. cinerea*, *Botryotinia fuckeliana* (Family: Pezizales, Ascomycotina) is formed following fertilization of sclerotia with uninucleate microconidia followed by their exposure to cold temperatures. The sexual structure "apothecia" develops from the germination of fertilized sclerotia that release sexual spores, ascospores. The sexual state of *B. cinerea* rarely occurs in nature on chickpea stubbles.

Host Range: *Botrytis cinerea* is a non-specialized pathogen well known for its global distribution and extensive host range of more than 100 plant species from different genera including ornamental plants, vegetables, fruit, field and glasshouse crops, several weeds, and post-harvest produce. The host range includes species such as black gram, strawberry, grapevine, apple, cabbage, carrot, cucumber, eggplant, lettuce, lentil, mungbean, mustard, paddy, pea, pepper, pigeonpea, squash, tomato, chrysanthemum, dahlia, lily, rose, gladiolus, and tulip.

Disease Cycle and Epidemiology

Botrytis cinerea can survive through infected seeds, crop debris on the soil surface, and other host plants, as mycelium and sclerotia, and are considered the major source of primary inoculum. *Botrytis* can grow on crop debris either as parasitically or saprophytically. It can survive on seeds externally as well as internally upto 5 years at a temperature 18°C. Small infected plant debris mixed in the seed also plays an important role in the survival of the pathogen. *B. cinerea* is a facultative parasite having a wide host range. Due to the wide host range, the role of alternative hosts is likely to play an important role in the survival of the *B. cinerea*. Sclerotia-bearing debris is often carried with seed. Sclerotia develop on the previous season's chickpea stubble after exposure to cold (above 10°C) winter temperatures. The sclerotia germinate mycelliogenically (asexually) by producing conidiophores bearing conidia, on crop stubble. Asexual sporulation of the fungus occurs on stubble under warm (above 20°C), moist conditions associated with prolonged periods of high relative humidity. These conidia are disseminated to several hundred metres from their source, indicating that plant debris could also be a major source of primary inoculum. Conidia can also be dispersed by the fruitfly (*Drosophila melanogaster*), grape berry moth (*Lobesia botrana*), New Zealand flower thrips (*Thrips obscuratus*), Mediterranean fruit fly (*Ceratitis capitata*), and other insect vectors. Rarely, apothecia (carpogenic germination) emerge from the fertilized sclerotia and wind-dispersed sexual ascospores are released. When airborne conidia land on a wet plant surface they adhere and germinate quickly by producing a germ tube in presence of water (dew). When the germ tube emerges, it secretes a polysaccharide-based matrix that acts as much stronger glue. Germ tube enter the host through the wounds or directly. *Botrytis cinerea* can kill its host's tissue rapidly and then carries on growing on the dead remains. It is thus a classical necrotrophic pathogen. If the necrotrophic *B. cinerea* attempts to infect a host plant, the hypersensitive response (cell death) also takes place, but it fails to control the infection because *B. cinerea* can exploit the dead cells for nutrition and initial growth. If the conidia fall on dead plant parts such as flowers, leaves, etc. There also they grow and produce enormous numbers of conidia which are disseminated by the wind to healthy leaves and blossom. Chlamydospores also serve as structures of survival and infection. The chlamydospores germinate to produce mycelium, which either directly or after the production of macroconidia, serves as secondary inoculum. The pathogen can complete the entire disease cycle in 7 days under favorable conditions. *B. cinerea* can attack the chickpea plant at any growth stage, but the disease usually appears at flowering, when the canopy is fully developed and the weather is warm (20°C–30°C), cloudy, and humid (70%–100% RH). A closed chickpea canopy with reduced light penetration and air movement creates ideal conditions for the fast spread of the disease. Under such conditions, there is abundant sporulation of the fungus on dead plant parts, particularly, on flowers and pods.

Management

Cultural Methods

- Late sowing.
- Adopt wider spacing.
- Intercropping with linseed or wheat.
- Avoid excessive vegetative growth.

- Avoid excessive irrigation. Use compact varieties of chickpea.
- Deep summer ploughing.
- Reduce plant density to increase air passage between the plants.
- Plant the crop late i.e. first fortnight of November.

Host Resistance

- Grow resistant varieties such as BG 276, GL 90159, GL 91040, GL 91071, and GL 92162.

Chemical Control

- Treat the seed with Carbendazim + Thiram (1:1) @ 3 g/kg of seed.
- Spray the crop with Captan @ 5–6 kg/ha at 15 days intervals or spray Carbendazim @ 1.5 g/L of water or spray Mancozeb at the rate of 3 g/L of water.

5.3 ASCOCHYTA BLIGHT

Ascochyta blight of chickpea is one of the most serious diseases of chickpeas and the major obstacle for chickpea production in many parts of the world where cool and humid weather prevails during the flowering to the podding stage. In India, it was first described by E.J. Butler in 1911 and it occurred in epidemic form in Punjab, Bihar, and Jammu. The disease can lead to heavy losses ranging from 10%–100% by reducing the seed yields and quality. In India, Ascochyta blight is very common in Punjab, Haryana, Himachal Pradesh, Northwest U.P., Bihar, Rajasthan and M.P.

Symptoms

The symptoms of Ascochyta blight can develop on all the above-ground plant parts including leaves, petioles, stems, pods, and seeds and plants can be attacked at any growth stage (Figure 5.2). Lesions that develop on leaves and pods appear circular with dark brown margins and a grey center that contains pycnidia, whereas lesions developing on the petioles, stems, and branches are elongated (3–4 cm long). The symptoms first appear as small water-soaked pale or gray areas on the leaves, stems or pods that quickly turn into brown lesions with dark borders. As the disease progresses, small, circular, brown-black dots (pycnidia) develop in the center of the lesions and are frequently arranged in concentric circles and resemble a bull's-eye. Concentric rings of pycnidia are the most important diagnostic characteristic of this disease. The lesions that develop on apical twigs, branches, and stems differ in size and in later stages girdle the affected plant parts. The regions above the girdled portion are killed and may break off. Round and sunken lesions, with pale centers and dark margins, form on the pods. The fungus can penetrate the pod wall and infect the seed. Severe pod infection can result in a reduced seed set and discolored seed. Infected seeds may be discolored, small, shrunken or shriveled and, when severe, may also exhibit irregular cankers and lesions with dark pycnidia on the seed. As the disease progresses, the patches of drooping and wilting plants become prominent in the field and expand gradually, covering the entire area.

Figure 5.2 Ascochyta Blight of Chickpea. Circular to oval lesions with pycnidia arranged in concentric rings on leaves (A), and pods (B), and elongated lesion on chickpea stem with pycnidia arranged in concentric rings (C). Courtesy: Dr. Michael Wunsch, Plant Pathologist, North Dakota State University Carrington Research Extension Center, USA.

Causal Organism and Etiology

Ascochyta blight of chickpea is caused by an ascomycete fungus *Ascochyta rabiei* (syn. *Phoma rabiei*), teleomorph *Didymella rabiei* (syn. *Mycosphaerella rabiei*). Mycelium appears hyaline to brownish and septate. The anamorph, *A. rabiei*, is characterized by the formation of spherical or pear-shaped black fruiting bodies called pycnidia on stems, leaves as well as in pods. Pycnidium may be flat, round or globose dark brown with prominent ostiole. A pycnidium contains numerous hyaline one-celled and occasionally two-celled conidia, develop on conidiophores embedded in a mucilaginous mass. Conidia are oval to oblong, straight, or slightly bent at one or both ends. The teleomorph, *Didymella rabiei* (Syn. *Mycosphaerella rabiei*), is a bipolar heterothallic ascomycete and is characterized by pseudothecia developing on chickpea crop residues that have over-wintered in the field. Pseudothecia are dark brown to black, subglobose, erupting from the host tissue and without a conspicuous ostiole. Binucleate asci are cylindrical to subclavate surrounded by paraphyses and contain eight ascospores. Ascospores are hyaline, ellipsoid to biconic with a constriction at the septum. These are unequally two-celled with one cell larger than the other.

Disease Cycle and Epidemiology

The life cycle of *D. rabiei* consists of a single sexual generation per season, which develops on infected overwintering field stubble, followed by several asexual generations during the parasitic phase of the disease cycle. Primary inoculum sources for *A. rabiei* are infected seeds and infested chickpea debris from the previous growing season from which conidia (asexual spores) produced by pycnidia and/or ascospores (sexual spores) produced by pseudothecia are splash- and air-dispersed to plants, respectively. *Ascochyta rabiei* can survive in crop debris for several years and seed. Infected seeds serve as a major source of primary inoculum, long-distance dispersal, and initiation of new infections. Seed infections can be either internal or external on the seed surface. *A. rabiei* can retain its viability in seed for at least 5 years and seed transmission is a more effective source of infection than infected plant debris. Ascospores plays role in the initiation of disease epiphytotics. In India, seed is considered to be the main source of primary infection. Primary inoculum multiplies on hypocotyls, epicotyl, and near the stem base. During the growing season, the secondary spread of the disease is driven by splash-borne conidia produced in pycnidia that develop on Ascochyta lesions. Conidia spread through raindrop splashes, insects, contact with healthy and diseased leaves, and by the movement of humans and animals and cause secondary infection. Disease is favored by humid, cool, and cloudy conditions. The disease is favored at 10°C night and 20°C day temperatures respectively. Dense canopy also favors blight development. Wet weather with strong winds having 22°C–26°C temperature is most favorable for the spread of the disease. High rainfall during crop growth also favors the development of disease in epiphytotic form. Ascochyta blight is problematic at cooler temperatures with disease development being most rapid at temperatures of 20°C with 17 hours of leaf wetness. Little infection occurs at temperatures outside the range of 5°C–30°C or without leaf wetness, even when humidity is more than 95%. Under conditions of high humidity or moisture, conidia are easily seen oozing from pycnidia in slimy, wet masses known as "cirrus".

Management

Cultural Methods

- Sow disease-free seed to reduce inoculum pressure.
- Follow crop rotation preferably with crops other than Fabaceae.
- Intercrop with wheat, barley, and mustard.
- Select healthy seeds and adopt a low seed rate.
- Remove and destroy debris of dead plants
- Bury infected crop residue by tillage to reduce the primary source of inoculum. The breakdown of infested chickpea residues and the loss of pathogen viability are greater at the soil depths above 5 cm than in the soil surface. Moreover, conidia and ascospores produced on buried residues are not available for either wind or splash dispersal.
- Always disinfect machinery, vehicles, and boots.

- **Sowing rate**: Maintain a plant population of 35–50 plants per square metre (depending on region, sowing time, crop type (kabuli/desi), and variety), adjusting the seeding rate according to seed size and germination.
- **Sowing date**: Return to original sowing dates with Ascochyta blight-resistant varieties, but do not sow too early.
- **Sowing depth**: If using an Ascochyta blight susceptible variety, sow deeper than normal.

Physical Control

- **Hot water treatment**: Treat the seed with hot water at 52°C for 10 minutes to destroy seedborne inoculum.

Host Resistance

- Grow resistant varieties such as G-543, Pusa-256, GNG-146, PBG-1, GL 260584, GL 26069, GL 23094, IPC 2005-66, Him Chana-1, Gaurav, Vardan, Samrat, PBG1, etc.

Chemical Control

- Treat the seed with Thiram or Carbendazim at the rate of 3 g/kg of seed.
- Spray Zineb or Chlorothalonil or Maneb or Mancozeb or Captan at the rate of 2.5 g/L of water or spray wettable Sulphur at the rate of 2.3 g/L of water if noticed during the growth period.
- Post-infection application of Tebuconazole or Difenconazole to moderately resistant varieties gives good control in mild epidemics.

6 Lentil (*Lens culinaris* L.)

6.1 RUST

Rust disease is regarded as the most important foliar disease of lentil. Complete crop failures can occur due to this disease. Rust disease is a potential threat to lentil cultivation and causes substantial yield losses ranging from 60%–69%. A severe outbreak of lentil rust was recorded in the Narmada Valley of Madhya Pradesh in 1978 and Uttarakhand during 2008–09, resulting in yield losses upto 100%. In Tarai region and nearby areas of Uttarakhand, rust is a major constraint in lentil production and has appeared in epiphytotic form.

Symptoms

The disease symptoms start as minute, slightly raised whitish to yellowish pustules (pycnidia and aecial cups), singly or in small circular groups on the leaflets, leaf blade, stem, petiole, and pods (Figure 6.1). Later light brown uredial pustules appear on both the surfaces of the leaves and other aerial parts of the plant. The pustules finally become dark brown (telial pustules). Pustules are oval to circular and up to 1 mm in diameter. They may coalesce to form larger pustules. The plants give dark brown or blackish appearance visible as patches in the field. The rust causes partial defoliation and under heavy infection leads to premature plant death. The affected plant may also die without forming any seeds in pods or with small shriveled seeds.

Causal Organism and Etiology

Lentil rust is caused by the fungus *Uromyces viciae-fabae*. It is an autoecious fungus, completing its life cycle (five spore stages) on one host (lentil). Spermogonia are mostly abaxial, amphigenous in small groups associated with aecia. Aecia are mostly abaxial, predominantly along veins, surrounding spermogonia or scattered, peridium cupulate, and white. Aeciospores are broadly ellipsoid, hyaline, and finely verruculose. Uredinia are amphigenous, yellowish brown.

Figure 6.1 Rust of Lentil.

Urediniospores are broadly ellipsoid, uniformly echinulate, with 3–5 germ pores equatorial or scattered, and a light golden brown wall. Telia adaxial or amphigenous, exposed, blackish brown, compact, 1–2 μm diameter. Teliospores are ellipsoidal, obovoidal or cylindrical, with a rounded or sub-acute apex; wall chestnut-brown, smooth, 1–3 μm thick at the sides and 5–12 μm at the apex; pedicels brownish.

Host Range: There are 70 recorded hosts of *U. viciae-fabae* including lentil, chickpea, field pea, *Lathyrus* spp., and *Vicia* spp.

Disease Cycle and Epidemiology

Uromyces viciae-fabae is an autoecious fungus, completing its life cycle on lentil. The disease generally starts from low-lying patches in the paddock and radiate towards the border. The resting stage (telia or teliospores) perpetuates in crop residues and seeds. The seed may be contaminated with pieces of rust-infected leaf, stem, and pericarps, which can act as primary inoculum for the recurrence of the disease in most years. Rust may also survive on weed hosts from where it may infect lentil crops by windborne teliospores. Teliospores produced on crop residues are dispersed by the wind and infect seedlings and volunteer plants. The infection leads to the production of aecia (specialized fruiting bodies). Aecia produce aeciospores, which infect leaves and spread the rust within the crop and to other crops. Aeciospores germinate at 17°C–22°C and infect other plants forming either secondary aecia or uredia at 25°C. In turn, this leads to the production of urediniospores, which can be carried long distances to produce new infections. Most rust fungi usually enter their host through natural openings such as stomata except the urediniospores of the tree rust fungus *Puccinia psidii*, *Ravenelia humphreyana*, and soybean rust fungus *Phakopsora pachyrhizi* which enter their hosts through the cuticle. Urediniospores give rise to a series of infection structures such as germ tubes, appressoria, substomatal vesicles, infection hyphae, and haustoria. Severe epidemics are caused by the production of several generations of urediniospores. Uridinia develops late in the season and are rapidly followed by the production of telia and teliospores on stems and leaves. High humidity, cloudy or drizzly weather with temperatures 20°C–22°C favors disease development. The disease generally occurs during the flowering or early podding stage.

Management

Cultural Methods

- After harvest, the affected plant trash should be burnt.
- Normal and early sowing reduces the intensity of rust disease.
- Remove volunteer plants from the field.
- Remove and destroy the stubbles of the old host crop.

Host Resistance

- The use of host plant resistance is the best means to manage rust diseases.
- Grow resistant/tolerant varieties like Pant L-639, Pant L-406, Pant L-6, pant L-7, Pant L-8, DPL15, Narendra Lentil-1, IPL 406, Haryana Masur 1, LL-931, IPL 316, etc.

Chemical Control

- Seed treatment with Agrosan and Triadimefon is also effective.
- Spray fungicides such as Mancozeb @ 0.2% (2 g/L), Bayleton (0.05% a.i), and Calixin (0.2% a.i.) are found effective against the pathogen. Foliar sprays of Benomyl, Carboxin, Metalaxyl, Oxycarboxin, Thiram, and Triademafon either alone or in the combination with Dithane M-45 are also effective. One or two sprays at 50 days after sowing is good for managing rust.

6.2 WILT

In India, *Fusarium* wilt or vascular wilt disease in lentil was first observed in experimental plots by Padwick (1941) in Delhi and Karnal. Vasudeva and Srinivasan (1952) first authenticated and published a record of vascular wilt of lentil and noted about 67% disease incidence. In India Khare *et al.* (1979) reported that *Fusarium oxysporum* is associated with seed. Yield loss due to wilt ranges between 25%–70%. However, yield losses vary with the time of infection. Infection at the early stage results in 100% loss. Wilt disease causes more damage at the flowering and pod formation stage.

Symptoms

The disease appears as patches in the field. The symptoms (Figure 6.2) can appear at three stages i.e., pre-emergence, post-emergence (seedling stage), and at maturity (adults) stages.

Pre-emergence: Seedling does not emerge above the soil surface as they die during or immediately after seed germination (seed rot).

Post-emergence (seedling stage): Fusarium infection is characterized by sudden drooping of the plant (more like wilting and damping off) followed by drying of leaves, collapse, and the eventual death of the seedlings. The root system appears healthy, but with a reduced proliferation and nodulation rate. In most cases, there is no discoloration of the vascular system.

Figure 6.2 Wilt of Lentil.

Adult stage: In the adult stage, plants are seen in isolated patches. Symptoms usually appear from flowering to the late pod-filling stage. The symptoms are characterized as sudden drooping of the top leaves, epinasty, leaflet closure without premature shedding, the dull green color of leaves and eventually wilting and drying or yellowing of the whole plant or individual stems then death. Internal root or stem symptoms show dark discoloration of the vascular tissues. The fungal mycelium can also be observed in the stems of the infected plants. Root nodules poorly develop. Seeds from plants affected mid to late pod filling stage are small and often shriveled.

Causal Organism and Etiology

Wilt of lentil is caused by *Fusarium oxysporum* f. sp. *lentis*. The mycelium of the pathogen is hyaline, septate, and much branched. *F. oxysporum* f. sp. *lentis* produces three kinds of spores: microconidia; multi-septate macroconidia, which have a distinct foot cell and a pointed apical cell; and chlamydospores. Sporodochia are rarely produced. Microconidia are abundant, produced on short, unbranched monophialides (microconidiophores) in small, dry, false heads. Microconidia are hyaline, ovoid, or kidney-shaped, and usually single-celled. Macroconidia are fewer than microconidia and are produced on branched macroconidiophores. They are long with pointed apical cells and notched basal cell and 3–5 septate. Chlamydospores are globose to sub-globose, one-celled, thick-walled, and smooth surfaced, produced singly in macroconidia or in chains, usually apically or intercalary in the hyphae.

Disease Cycle and Epidemiology

The disease cycle of the Fusarium wilt of chickpea and lentil is essentially similar, varying only in the hosts they infect. Like many other formae speciales of *F. oxysporum*, the pathogen is host-specific i.e. it has a very limited host range as it infects only lentil in nature. The pathogen survives in soil and infected crop debris for up to 3–4 years and spores are spread by mechanical means and seed. Following the infection of host roots, the fungus crosses the cortex and enters the xylem tissues. It then spreads rapidly up through the vascular system, becoming systemic in the host tissues, and may directly infect the seed. Seed infestations and infections are common.

The movement of infected seeds plays an important role in the long-distance dispersal and transmission of fusarium wilt diseases into new areas. The root tips of healthy plants growing in contaminated soil are penetrated by the germ tube of spores or fungal mycelium. Entry is either direct, through wounds, or opportunistic at the point of formation of lateral roots. The mycelium takes an intercellular path through the cortex, and enters xylem vessels through the pits. The pathogen is primarily confined to the xylem vessels in which the mycelium branches and produces microconidia. The microconidia detach and are carried upward in the vascular system until movement is stopped, at which point they germinate and the mycelium penetrates the wall of the adjacent vessel. Lateral movement between vessels is through the pits. The water economy of infected plants is eventually severely compromised by blockage of vessels, resulting in stomatal closure, wilting, and death of leaves, often followed by death of the whole plant. The fungus then invades all tissues of the plant, to reach the surface where it sporulates profusely. Spores may then be dispersed by wind and water or the movement of soil or plant debris. *F. oxysporum* can survive as mycelium and chlamydospores in seed and soil, and also on infected crop residues, roots and stem tissue buried in the soil for more than 6 years. Chlamydospores can survive in soil either in dormant form or saprophytically without a suitable host. The presence of the root-knot nematode (*Meloidogyne javanica*) significantly increases wilt incidence and causes a significant reduction in shoot length, root length, and nodulation in both susceptible and resistant cultivars. The disease causes complete failure of the crop, especially in warm and dry soil conditions with an optimal temperature of 20°C–25°C. Lentil wilt is also favored by soil pH 7.5–8.0, soil moisture of 25%–30% water holding capacity, and increasing plant maturity and no disease occurs at 75% or more soil moisture.

Management

Cultural Methods

- Use of clean and certified seed.
- Rotate lentil crops with cereals (maize, wheat, sorghum) for 4–5 years. This will reduce the disease inoculum in the soil.
- Keep the field clean.
- Weeds control.
- Delayed sowing can reduce disease incidence.
- Avoid planting a new crop on previously affected fields with wilt or nearby it and ensure the soil is well drained.
- Summer deep ploughing.
- Remove infected crop debris before planting to reduce inoculum levels of Fusarium wilt.
- Soil amendment with organic matter enhances antagonism with other soil microflora.
- Avoid injuring plant stems and roots when weeding.
- Minimize movement of soil and plant material between affected and unaffected areas.
- Clean and sterilize farm tools by heating them with fire.
- Control run-off water from infected areas to clean areas to avoid spreading the pathogen.

- Rouge out the diseased plants and healthy plants 5–10 metres in the area surrounding the patch and burn them in the patch.
- Ensure that seed production is not taking place in affected farms as the disease is seedborne.

Physical Control

- Solarize the soil by covering the soil with a transparent polythene sheet (75 gauge) for 6-8 weeks during the summer for effective management.

Host Resistance

- The best method of managing lentil wilt is the use of resistant/tolerant varieties such as PL 406, PL 639, PL 234, Pant Lentil-4, Pant Lentil 5, Pant Lentil-6, Pant L-8 and Noori, IPL-316, RVL-31, Shekhar Masoor 2, Shekhar Masoor 3, etc.

Biological Control

- Using antagonistic microflora like *Bacillus subtilis, Trichoderma harzianum, T. viride* @ 4 g/kg seed, etc.

Chemical Control

- Treat the seed with Benomyl (0.3%) or Thiram + Benomyl (1:1) @ 0.3%.

7 Cotton (*Gossypium hirsutum* L.)

7.1 ANTHRACNOSE

Anthracnose is known to farmers as boll spot or boll rot. In India, it was first reported from Bihar in 1918 by E.J. Butler. The average yield loss due to disease is about 1%–3%.

Symptoms

Symptoms occur on all aboveground plant parts. The symptoms on seedlings appear as small reddish circular spots on the cotyledons and primary leaves. On young hypocotyls and stems, oblong, dark brown lesions develop. Severe disease results in stem girdling, wilting, and death of seedlings. In mature plants, the disease causes stem splitting and shredding of bark. The most common symptom is boll spotting which appears initially as small, water-soaked, circular, slightly sunken reddish-brown lesions on the bolls. Later these lesions enlarge and become black to dark brown with a reddish margin. Infection through the dead pistil results in internal rot of the boll. The lint is stained yellow or brown, and becomes a solid brittle mass of fibre. The infected bolls cease to grow and burst and dry up prematurely. Under humid or wet weather, lesions of all infected parts turn pink due to fungal sporulation (slimy masses of conidia). Small black dot-like acervuli develops in the necrotic lesions. Perithecia may also develop in dead tissue, but they are difficult to observe, as they appear only as tiny, dark bumps. These are the "beaks" of the perithecia that protrude above the plant epidermis. Most of the perithecial structures are submerged in the plant tissue. As the disease progress, lesions enlarge and coalesce, causing necrosis of the whole leaf. Severe infection often results in defoliation.

Causal Organism and Etiology

Anthracnose of cotton is caused by *Colletotrichum gossypii* (teleomorph: *Glomerella gossypii*) and *Colletotrichum capsici*. In most parts of India *C. capsici* is the causal organism whereas in some parts of Maharashtra *C. gossypii* also cause the disease. The fungus is both inter-cellular and intra-cellular, producing numerous acervuli on the affected leaves, stems, and bolls. The conidiophores are hyaline, single-celled and club-shaped, and the conidia are falcate, thin walled and single celled. Both species produce thick-walled, many septate seate which are dark colored at the base and lighter towards the tips.

Disease Cycle and Epidemiology

Glomerella gossypii is seedborne as mycelium in or on seed and as conidia contaminating the outside of seeds. *Glomerella gossypii* also survives in infected crop debris as acervuli and perithecia. *C. capsici* can also survive on several weeds that serve as sources of primary inoculum. These weed species include cocklebur (*Xanthium strumarium*), Johnson grass (*Sorghum halepense*), smooth pigweed (*Amaranthus hybridus*), pitted morning glory (*Ipomoea lacunosa*), prickly sida (*Sida spinosa*), sicklepod (*Cassia obtusifolia*), small flower morning glory (*Jacquemontia tamnifolia*) and spotted spurge (*Euphorbia maculate*). Primary inoculum for *G. gossypii* and *C. gossypii* may originate from three different sources. The first source is from infected seed that gives rise to diseased cotyledons on which conidia are produced and are disseminated by wind to host plants. The second primary inoculum originates from infested residue on which perithecia produce ascospores that are disseminated by the wind to host plants. A third source is acervuli and mycelia which produce conidia, which are disseminated by splashing rain and by wind. Disease is favored by prolonged rainfall at the time of boll formation, closer planting, and dense canopy with warm (29°C–33°C) humid conditions. Seedling blight is most severe at temperatures of 20°C–26°C.

Management

Cultural Methods

- Use pathogen-free seeds.
- Remove and burn the infected plant debris and bolls in the soil.
- Rogue out the weed hosts to reduce the chance of pathogen survival in the field.
- Rotate cotton with other crops.
- Plough under infested residue.
- Skip row planting provides an open stand and prevents the buildup of high humidity around lower bolls.
- Fertilize properly.
- Practice timely defoliation.
- Control insect pests.

Host Resistance

- Grow resistant/tolerant varieties.

Chemical Control

- Treat the seed with Carbendazim or Carboxin or Thiram or Captan @ 2 g/kg, helps to reduce disease spread at seedling stage.

- Foliar spraying of Mancozeb @ 0.2% or Copper oxychloride @ 0.3% or Carbendazim @ 0.1% at the boll formation stage suppresses the aggressiveness of the disease.

7.2 VASCULAR WILTS

Vascular wilt in cotton is caused by two soil-borne pathogens, *Fusarium* and *Verticillium* which frequently cause severe economic losses not only in India but also worldwide.

7.2.1 Fusarium Wilt

Fusarium wilt is a major disease of cotton capable of causing significant economic loss. The disease is prevalent in black, light sandy and acidic soils. In India, the disease was first reported from Nagpur by Evans in 1908. The seedborne nature of Fusarium wilt of cotton was first described by Elliott in 1923. Fusarium wilt is sporadic in nature.

Symptoms

The disease affects the crop at all stages. The earliest symptoms appear in the cotyledons of the young seedlings which turn yellow and then brown. Base of the petiole shows a brown ring, followed by wilting and drying of the seedlings. In older plants, symptoms include stunting, wilting, chlorosis and necrosis of leaves, dieback often beginning at the top of the plant, and plant death. The stem shows dark brown discoloration of the vascular tissues, which is the most diagnostic symptom of Fusarium wilt. In the transverse section, a discolored ring is seen in the woody tissues of the stem. The vascular bundles of the basal stem are commonly filled up with a gummy substance and fungal hyphae. Losses result from poor stands, stunting, small bolls, and poor quality of lint. Symptoms at the seedling stage can be confused with the symptoms of damping-off caused by *Pythium* spp., *Rhizoctonia solani*, and *Fusarium* spp., but the brown vascular system of the hypocotyls distinguishes Fusarium wilt from seedling disease.

Causal Organism and Etiology

Fusarium wilt of cotton is caused by *Fusarium oxysporum* f. sp. *vasinfectum*. Macroconidia are 1 to 5 septate, hyaline, thin-walled, and falcate with tapering ends. The microconidia are hyaline, thin walled, spherical or elliptical, single- or two-celled. Chlamydospores are thick-walled, mostly globular, dark-colored, and terminal or intercalary.

Disease Cycle and Epidemiology

The fungus survives in the soil as chlamydospores and in association with the roots of susceptible, resistant, and non-cotton hosts as well as in seed also. The fungus can survive in soil as a saprophyte for upto 25 years. It is a soil-inhabitant and has a good "competitive saprophytic ability". The pathogen is both externally and internally seedborne. The primary infection is mainly caused by dormant hyphae and chlamydospores in the soil. The conidia and chlamydospores are disseminated by wind and irrigation water. The fungus enters the young host roots through incidental wounds and openings caused by insects (e.g. grubs of Ash weevil, *Myllocerus pustulatus*) and plant parasitic nematodes (e.g. root knot nematode, *Meloidogyne incognita*). The other nematodes also have been reported to increase the severity of Fusarium wilt of cotton these includes: reniform (*Rotylenchulus reniformis*), sting (*Belonolaimus gracilis* and *B. longicaudatus*), lance (*Hoplolaimus seinhorsti*), and lesion nematode (*Pratylenchus brachyurus*). The wilting and senescence occur as a result of water stress induced by the combined effect of mycelial growth in the xylem, fungal toxins (fusaric acid), and vascular occlusion by the host in an attempt to prevent the systemic spread of the fungus. There is no secondary infection and the disease is monocyclic. The disease is favored by hot and dry periods followed by rains, soil temperature of 20°C–30°C, heavy black soils with an alkaline reaction, increased doses of nitrogenous (ammonia form) and phosphatic fertilizers and potassium deficiency in the soil. Spore production and germination are the maximum at 100% relative humidity. No germination has been observed below 80% relative humidity. Mycelial growth is maximum in the soil at 40% moisture holding capacity and pH 5.6–7.2.

Management

Cultural Methods

- Use pathogen-free cotton seed.
- Fields having a long history of the disease should be avoided.
- Follow mixed cropping with non-host plants.
- Follow crop rotation with *G. hirsutum* or non-host crops.
- Remove and burn the infected plant debris.
- Fields should be deeply ploughed during June-July and left for solarization.
- Apply increased doses of potash with a balanced dose of nitrogenous and phosphatic fertilizers. Use of nitrogenous fertilizers, particularly ammonium nitrate should be discouraged and calcium ammonium nitrate (CAN) should be used in place of urea or ammonium sulphate.
- Apply heavy doses of farm yard manure (FYM) or other organic manures.
- Cultivate cotton in alkaline soil.
- Eliminate the nematodes which predispose the plants to Fusarium wilt.

Host Resistance

- Grow disease resistant/tolerant varieties:

Cotton Species	Resistant/Tolerant Varieties
G. hirsutum and *G. barbadense*	Varalakshmi, Vijay Pratap, Jayadhar and Verum
G. herbaceum	Jayadhar, Vijalpa, Digvijay, and Sujaya with hybrids; G Cot DH7 and G Cot DH9)
G. arboreum	Girnar, Daulat, G-22, G-46, Y-1, and Sanjay

Chemical Control

- Treat the acid-delinted seeds with Carboxin or Carbendazim @ 2 g/kg.
- Spot drench with Carbendazim @ 1 g/L.

7.2.2 Verticillium Wilt

Verticillium wilt is also regarded as "the cancer of cotton." It is a serious soil-borne vascular disease that causes major losses to upland cotton worldwide every year. This disease was first observed on Upland cotton in Virginia in 1918 by Carpenter. In India, the disease occurs mainly in Tamil Nadu. Verticillium wilt causes significant losses in cotton yield and fiber quality worldwide. The losses may reach up to 80% of the lint cotton yield.

Symptoms

Cotton infected with Verticillium wilt shows several symptoms including marginal flagging, irregular yellow patches, browning of veins, cupping, leaf wilting, drying, defoliation, and vascular discoloration or necrosis. The characteristic diagnostic feature of Verticillium wilt is the drying of the leaf margins and areas between veins, which gives a "Tiger stripes" or "Tiger claws" appearance. In severely infected plants, younger bolls may abscise or become malformed. Other typical symptoms include stunting, lack of lateral growth, and decreases in yield, fiber quality, and seed quality. Symptoms of Verticillium wilt can be confused with Fusarium wilt and may require laboratory diagnosis. See the difference in Fusarium wilt and Verticillium wilt of cotton in Table 7.1.

Table 7.1 The difference in Fusarium wilt and Verticillium wilt of cotton.

Features	Fusarium wilt	Verticillium wilt
Causal organism	*Fusarium oxysporum* f. sp. *vasinfectum*	*Verticillium dahliae*
Vascular discoloration in cross-section of stem	Vascular discoloration of the internal tissue of the stem is similar to that of Verticillium wilt but usually appears as continuous dark-brown rather than flecking.	The discoloration is similar to that of Fusarium wilt but usually appears as less intense, irregular, flecking, rather than continuous browning.

Causal Organism and Etiology

Verticillium wilt of cotton is caused by *Verticillium dahliae*. Conidiophores are hyaline, verticillately branched, and have 2 to 4 phialides per node. Microsclerotia are dark brown to black, spherical, and compact.

Disease Cycle and Epidemiology

The pathogen, *Verticillium dahliae*, has a very broad host range of more than 400 plant species and can survive extremely long periods of time (up to 14 years) in host debris or in soil as microsclerotia. It is a soil invader and unlike Fusarium, cannot grow in soil. The fungus is not seed transmitted. The initial infection occurs early in the growing season when microsclerotia germinate under the influence of the root exudates of the host. The fungus penetrates and infects through the roots; invades the vascular system, resulting in systemic infection. Prolific growth in the xylem vessels disrupts the plants ability to transport water and nutrients. Microsclerotia germination increases with periodical wetting and drying of the soil. *Verticillium dahliae* first colonizes the root and then moves upward through the vascular system of the plant. Verticillium wilt is most severe during cool, extended wet weather and/or waterlogging and in late-maturing crops. Cotton disease resistance is temperature sensitive. Varieties that are resistant at 25°C to 27°C are susceptible at 20°C–22°C. Several factors, including variety selection, plant density, pathogen aggressiveness, inoculum density (microsclerotia per gram of soil), and excessive use of nitrogen in late-season growth, potassium deficiency, and environmental conditions influence Verticillium wilt development. Moisture is also an important component for conidia dispersal, along with root infection and colonization. Drought-stressed cotton typically suffers less infection by *V. dahliae* compared with irrigated cotton.

Management

Cultural Methods

- Avoid fields with a long history of disease occurrence.
- Fields should have a good drainage system.
- Follow crop rotation with non-hosts such as cereals or sorghum or paddy or lucerne or chrysanthemum for 2–3 years.
- The crop should not be over-irrigated.
- Deep summer ploughing of fields.
- Remove and destroy the infected plant debris.
- Good field sanitation reduces disease incidence.
- Apply heavy doses of FYM or compost @ 100 t/ha.
- Avoid late-season irrigations.
- Remove all alternative weed hosts.

Host Resistance

- Use resistant or tolerant cultivars like Sujatha, Suvin, CBS 156, and MCU 5 WT.

Chemical Control

- The seed should be acid-delinted.
- Treat the delinted seeds with Carboxin or Carbendazim @ 2 g/kg.
- Spray Potassium chloride (KCl).
- Spot drench with Benomyl (0.05%) or Carbendazim (0.1%).

7.3 BLACK ARM

The disease is also known as bacterial blight and angular leaf spot of cotton. It is the most serious disease of cotton. It was first reported from Alabama, USA in 1891 by Atkinson, who named it black rust. In India, the disease was first observed in Tamil Nadu in 1918. In India, the disease occurs in all the cotton growing areas and the losses range from 5%–25%.

Symptoms

The disease has four distinct types or phases of symptoms depending upon the plant part affected: seedling blight (seedling infection), angular leaf spot (leaf infection), and black arm (stem infection).

Seedling Blight: The seedling blight mostly results from seedborne infection. Small, water-soaked, circular, or irregular lesions develop on the undersurface of the cotyledons. Later these lesions enlarge, turn brown to black, and ultimately the seedlings collapse and die. The bacterial ooze is often present in the form of crust or scale on the lower surface of cotyledonary leaves.

Angular leaf spot: The necrotic lesions on leaves initially, appear as angular, dark green, water-soaked, restricted within veins and veinlets of the leaf which later turn brown to black (Figure 7.1 A). These spots are visible on both the surface of the leaves. The spots enlarge forming irregular large patches. The infection extends along the veins turning the veins black and infected veins give a finger-like appearance, hence called vein blight or black vein. From veins, the infection reaches the petioles through the midrib and infects the stem. Petiole infection leads to severe defoliation. Large quantities of bacterial yellow slime exuded which form a dry film or crusts on leaf lesions.

Black Arm: Elongated, sunken, dark brown to black lesions appear on the stem petioles and twigs of the plants and eventually girdle them. Severely affected stems and twigs show premature drooping off of the leaves, deep cracking, and gummosis, resulting in breaking off the stem by wind which hang typically as dry black twig to give a characteristic "black arm" symptom.

Boll rot: Water-soaked, circular spots appear on developing bolls, which later on, turn brown and become sunken (Figure 7.1 B). These bolls can rot or may fail to open properly. Invasion of secondary organisms leads to extensive boll rotting and destruction of lint. Severely infected bolls drop. The infected boll produces low-quality cotton lint. From the boll, infection progresses inside the boll and lint gets stained yellow because of bacterial ooze. The pathogen also infects the seed and causes discoloration, a reduction in seed size, and viability.

Causal Organism and Etiology

Black arm of cotton is caused by *Xanthomonas axonopodis* pv. *malvacearum*. The bacterium is non-endospore forming, encapsulated, gram-negative, motile by a single polar flagellum and rod-shaped measuring 0.3–0.6 × 1.3–1.7 μm. It is a facultative aerobe.

Disease Cycle and Epidemiology

The bacterium, *Xanthomonas axonopodis* pv. *malvacearum*, survives in seed, soil, and diseased crop debris. The bacterium remains viable in dry leaves for 17 years and in dry or moist soil for 8 days at 21°C–33°C. The infected seeds serve as the main source of primary inoculum. The bacterium may be present as the slimy mass on the seed coat (externally seedborne) or in the embryo (internally seedborne). On germination, infected seeds the bacteria move to cotyledons and then multiply and maintain a resistant population. Under favorable weather

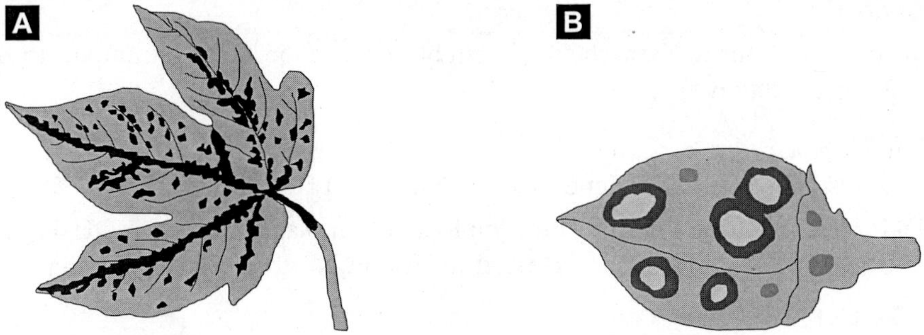

Figure 7.1 Black Arm or Bacterial Blight of Cotton. (A) Angular leaf spots with veinal necrosis, and (B) Ball Rot.

conditions, the inoculum from this source is disseminated to new leaves by wind-splashed rain and dew and further spread continues. Surface irrigation and overhead-irrigation also spread the bacterium. In some areas, wind-blown detached leaves and wind-blown dust may serve as means of dissemination. The bacterium is known to be transmitted by two cotton pests, namely, *Dysdercus koenigii* and *Earias* spp., but their role in the spread and development of disease is still not clear. The bacterium enters the host mainly through stomata. Insect damage also serves as entry point for boll infection. Once inside the leaves, the bacterium spread from sub-stomatal cavity to the vascular system and spread to leaves, bolls, and seed. The bacterium grows on several collateral hosts such as *Eriodendron unfructuosum*, *Thurbaria thespisioides*, *Lochnera pusilla*, and *Jatropha curcas*, but their role in the disease cycle is not confirmed. The disease development and spread are favored by frequent rainfall, high humidity, and warm temperature. Primary infection occurs at a relatively low temperature (30°C) and secondary infection occurs at relatively higher temperature (35°C). The presence of moisture is essential for the secondary spread of the pathogen during the first 48 hours of infection. For abundant infection high relative humidity (85% or above) is essential. Dry and hot weather retard the normal development of the disease.

Management

Cultural Methods

- Use disease-free seeds.
- Sow seeds on raised beds with maximum row spacing.
- Remove and destroy the infected plant debris from the previous crop before sowing.
- Avoid stacking infected plants.
- Avoid dense cropping.
- Rogue out the volunteer cotton plants, collateral, and weed hosts.
- Follow crop rotation with non-host crops.
- Early thinning and early earthing up with potash.
- Flood the field with water before sowing.

Physical Control

- **Hot water treatment:** Treat the seed with hot water at 56°C for 10 minutes to eradicate seedborne infection.

Host Resistance

- Grow resistant/tolerant varieties such as Sujatha, 1412, and CRH 71.
- Field tolerant cultivars derived from diploid cotton plants, such as LDH-11, LDH-91, BTR and Pusa 8–6 have been released for use in Northern states of India.

Chemical Control

- Delint the cotton seeds with concentrated Sulphuric acid or Hydrochloric acid @ 100–200 ml/kg of seed followed by seed soaking in Carboxin or Oxycarboxin @ 2g/kg or Bavistin (@ 2 gm/kg of seed) or 100 ppm of Streptomycin sulphate solution or Streptocycline solution (@ 100 ppm) overnight to eradicate internal infection.
- Spray Streptomycin sulphate @ 500 ppm or Copper oxychloride 50% WP @ 4 g/L of water, at intervals of three weeks, to check the secondary spread of the disease.
- Spray Paushamycin/Plantomycin 40 gm + Copper oxychloride 50% WP 500 gm/acre, three to four times at 15 days interval or Spray Streptomycin sulphate + Tetracycline mixture 100 g + Copper oxychloride 1.25 kg per hectare.
- Spray neem-based formulations (Plantolyte or Agricare @ 5 µg/ml) to prevent secondary infection.
- Soak seeds in 40–50 ppm Streptomycin solution before sowing.
- Treat the seed with Carboxin 37.5% + Thiram 37.5% DS @ 3.5 g/kg of seeds is also effective for bacterial blight management.
- On the first appearance of field symptoms, the crop should be sprayed with a mixture of Streptocycline (Streptomycin sulphate 90% + Tetracycline hydrochloride 10% SP). Repeat the spray at 10–15 days intervals to check the secondary spread.

8

Pea (*Pisum sativum* L.)

8.1 DOWNY MILDEW

Downy mildew of pea was first discovered by Berkeley in England in 1846. In India, the disease is common in pea growing areas of the Indo-Gangetic plain and causes considerable yield loss. The downy mildew of peas is widely distributed all over the world and can cause yield losses ranging from 30%–45%.

Symptoms

Scattered pale yellow-green blotches appear on the upper surface of leaves with profuse, fluffy, and white to mouse-grey cottony moldy growth (spore masses) in equivalent positions on the lower surface of the leaflets (Figure 8.1). The blotches turn greenish-brown, shrivel, and finally the leaf dies and dry up. As the disease progress, these lesions enlarge and cover the whole surface of the lamina. The pathogen produces abundant aerial sporangiophores and sporangia on infected plant surfaces which causes secondary infections. Secondary infection is localized in upper leaves, stems, tendrils, and pods and results in the appearance of isolated greenish-yellow to brown blotches on the upper leaf surface. Systemically infected plants are pale yellowish-green, stunted and distorted, and usually die before flowering. Infected pods deform, develop yellow blotchy patches on the surface, and a profuse cottony growth develops from the inner pod wall. The seeds beneath the lesions fail to develop, abort, undersized, and pods fill unevenly.

Causal Organism and Etiology

Downy mildew of field pea is caused by the obligate parasite *Peronospora viciae* (syn. *P. pisi*). Downy mildew of pea is caused by *Peronospora viciae*. The pathogen belongs to the Kingdom–Straminipila, Division–Oomycota, Class–Peronosporomycetes, Order–Peronosporales, and

Figure 8.1 Downy Mildew of Pea. (A) Pale yellow-green blotches on the upper surface, (B) Fluffy mouse-grey moldy growth (spore masses) on lower surface of the leaves, and (C) Severely affected pea leaf.

Family–Peronosporaceae. It can infect pea, sweet pea (*Lathyrus odoratus* L.), dry bean (*Phaseolus vulgaris* L.) and faba bean (*Vicia faba* L.). The downy mildew pathogen usually infects plants within the genus *Pisum*, thus delineating its forma specialis as *P. viciae* f. sp. *pisi*. Mycelium of the pathogen is hyaline, aseptate, intercellular, and branched with finger-shaped haustoria. Sporangiophores first appear as simple elongating hyphae from stomata on the abaxial leaflet surface then branching from a single axis to produce multiple terminal sporangia. Gametangia developed extensively on inner surface of pods from smooth, bulbous hyphae, adhering to the host epidermis. Species of *Peronospora* produce conidia that germinate by germ tubes. *P. viciae* also produces oospores, which have a typical reticulate pattern of the exosporium. Both heterothallic and homothallic isolates of these two species have been found. Sexual reproduction is probably important for the adaptation of the fungus to various host genotypes by recombination of virulence genes.

Disease Cycle and Epidemiology

The pathogen overwinters in infected seed, plant debris, and as oospores in the soil. In the seed, the pathogen may be present as a contaminant, or the seed infection may extend beneath the seed coat. Oospores in the soil survive for 10–15 years and serve as primary inoculum. Infection from these sources can lead to systemic and local leaf infections in volunteer pea seedlings. Soilborne oospores germinate following stimulation from germinating pea seeds. Germinating oospores produce a germ tube that infects germinating seeds and infection progresses to the emerging seedlings. Abundant sporangia develop on infected plant parts which are disseminated by the rain splash or wind and cause the secondary infection. The disease thus spreads in the entire field. Close cropping of peas is the main cause of severe infection. Oospores are produced within infected tissue and once returned to the soil they remain viable for many years. The disease is most common soon after the emergence but may affect plants at any growth stage during periods of moist, cool weather. For sporangial germination and infection, leaf wetness is required for upto 4 hours. High humidity (above 90% RH) and low temperatures (5°C–15°C) for 4–5 days are ideal for infection and the rapid development of disease. High humidity also helps in the production and dissemination of sporangia. Rain is the major means of spore dispersal and infection. Dry, warm weather is unfavorable for the disease. Systemic infection of plants can occur late in the season if conditions are favorable.

Management

Cultural Methods

- Field sanitation.
- Remove and burn diseased plants soon after detecting them in the field.
- Destroy plant debris left in the field from the previous crops as the pathogen survives for a longer duration in the form of oospores.
- Follow crop rotation with non-host crops for 3–4 years.
- Proper drainage in the field.
- Wider spacing creates an unfavorable micro-climate for disease development.
- Watering plants early in the morning so that leaf surfaces dry out rapidly. Avoid watering in the evening which can lead to high humidity or leaf wetness and persists throughout the night.

Physical Control

- **Hot water treatment**: Treat the seed with hot water at 50°C for 25 minutes to eradicate the seedborne inoculum.

Host Resistance

- Grow resistant/tolerant varieties.

Chemical Control

- Eradicate seedborne inoculum by seed-dressing with systemic Acylalanine fungicides like Metalaxyl, Cymoxanil, and Fludioxanil which are very effective against systemic seedling infections.
- Spraying and dusting pea plants with fungicides (e.g. Metalaxyl @ 0.1%) are effective in limiting the spread of the disease.
- Check the secondary spread of the disease by spraying fungicides like Carbendazim + Chlorothalonil and Carbendazim + Mancozeb. To avoid the buildup of fungicide-resistant strains, it is always better to rotate the fungicide and use mixtures of systemic and non-systemic fungicides.

8.2 POWDERY MILDEW

Powdery mildew is one of the most important worldwide distributed diseases of garden pea. The disease was first reported in India from Dehradun by Butler in 1918. It is damaging particularly in late sowings or in late maturing varieties. The disease can cause 25%–50% yield losses in pea production. Severe infection can result in yield loss of up to 80% in susceptible cultivars. The disease also affects green pea quality.

Symptoms

The symptoms of powdery mildew are characterized as small irregular areas of white powdery (flour-like) patches that occur on both surfaces of the leaves, tendrils, stems, and pods (Figure 8.2). In advanced stages of the disease, large areas of the host surface are covered with white floury patches. Severe infection can cause chlorosis, drying, shriveling, and drop off of the leaves and pods fail to fill, are small in size, and malformed. Brown to coffee-colored spots also appears on severely affected pods.

Figure 8.2 Powdery Mildew of Pea.

Causal Organism and Etiology

Powdery mildew of pea is caused by an obligate pathogen, *Erysiphe pisi* var. *pisi*. Other species such as *Erysiphe trifolii* and *Erysiphe baeumleri* have also been reported causing this disease in pea. The pathogen belongs to Eumycota, Fungi, Ascomycota, Ascomycetes, subclass–Erysiphomycetidae, order–Erysiphales, and family–Erysiphaceae. The mycelium of *Erysiphe pisi* var. *pisi* is generally fine, persistent, and rarely thick. It is ectophytic, the entire thallus, except the haustoria in epidermal cells, is present on the host surface. Conidia are hyaline, ellipsoid to ovate with vacuolated cytoplasm, and form singly or in short chains. Cleistothecia (sexual fruiting body) develop as short, black, minute bodies scattered in the mycelial web. Cleistothecia usually contain 2–3 ovate, nearly sessile asci, and myceloid-type appendages. The asci contain 3 to 8 ascospores which are hyaline, elliptical, and unicellular.

Disease Cycle and Epidemiology

Powdery mildew of pea is an airborne disease. The fungi overwinter in crop debris and alternative host plants such as vetches and other wild legumes. Cleistothecia develop on dead plant debris and serve as a source of primary inoculum. The primary infection takes place through ascospores released from the disintegration of the wall of the fruiting bodies and blown by the wind. Under favorable conditions, the ascospores, when come in contact with the lower most leaves of pea, germinate by producing germ tube, and cause "primary infection" and produce powdery mass of spores. The primary infection may also caused by the conidia produced on other hosts as the pathogen has a wide host range. The secondary infection of the host is caused by the conidia which are produced in large numbers and are disseminated by wind, animals, and even humans. During the growing season, more than one secondary infection may occur which results in the severe epidemics. Cleistothecia (perennating structures) are produced on fallen infected debris in the soil. These contain asci and ascospores which serve as the primary inoculum of the disease. Rainfall suppresses the disease by washing off the spores and causing them to burst instead of germinating. Free moisture on plants also restricts the germination of the spores and does not promote the epidemic. Powdery mildew develops quickly in warm (temperature 15°C–25°C), dry, humid (over 70% RH) conditions for 4–5 days late, particularly at the flowering and podding stage. The disease develops best when there is prolonged drought. The fungus does not need moisture to reproduce. Powdery mildew transmits more frequently in very dense planting.

Management

Cultural Methods

- Burn infected pea stubble soon after harvest.
- Avoid sowing field pea crops adjacent to last season's stubble.
- Control volunteer field peas which can harbor disease.
- Adopt a bower system.
- Field sanitation.
- Rotate the crop with non-host crops such as potato, maize, wheat or other grains.
- Crop sown in rainy seasons is less affected by powdery mildew.
- Avoid close spacing (less than 50 cm between rows).

Host Resistance

- Grow resistant/tolerant varieties such as Rachna, Arpana, Pant Matar 25, etc.

Chemical Control

- Spray Benomyl 50% WP @ 80 g in 240 litres of water per acre or Carbendazim 50% WP @ 100 g in 240 litres of water per acre or Dinocap 48% EC @ 120 g in 300 litres of water per acre or Fenarimol 12% EC @ 0.04% (40 ml/100 litres of water).
- Second spray after 15 days of interval or Sulphur 40% WP 2.26–3.00 kg in 300–400 litres of water per acre or Sulphur 52% SC 800 ml in 120–160 litres of water per acre or Sulphur 80% WG @ 750 gm to 1.0 kg in 300–400 litres water per acre or Sulphur 85% DP @ 6–8 kg per acre or Triadimefon 25% WP @ 0.1% in 300 litres water per acre. Second spray after 25 days of intervals.
- Spray with lime sulfur at 500 ml per 20 litres knapsack before and after flowering to prevent flowers from falling.
- Apply preventative fungicides such as Dithane-45 or Copper oxychloride @ 3 g/L of water.

- Apply curative fungicides such as Triazoles @ 2 g/L of water or Triazoles + Strobirulins @ 1 g/L of water.
- Spray inorganic Sulphur @ 0.25% or Tridemorph @ 0.1%.

8.3 RUST

Rust is a serious and very common disease of garden pea, particularly in wet areas of pea cultivation throughout the world. In India, pea rust pathogen *U. viciae fabae* on *Vicia faba* was reported from Pusa, Bihar in 1906 by Sydow and Butler. The first report of its occurrence on peas in India was reported by Butler (1918). The disease may cause epidemic proportions under favorable weather conditions resulting in considerable yield losses.

Symptoms

The first symptoms appear with the development of aecia which initially appear as minute, whitish, slightly raised spots on the undersurface of the leaves, petioles, tendrils, stems, and pods (Figure 8.3). These pustules enlarge and rupture the epidermis. The formation of the aecial stage is preceded by a slight yellowing, which gradually turns brown. Light brown pustules contain abundant, powdery urediniospores, but eventually, they turn dark brown to black when overwintering teliospores are produced. Most commonly teleutopustules develop on stem and petioles. All four stages (pycnia, aecia, uredinia, and telia) develop on the green part of the host, including the pods.

Figure 8.3 Pea Rust. (A) Yellow pustule with uredospores on leaves, (B) Dark brown pustules containing teliospores on stem, and (C) Yellow pustule with uredospores on pod.

Causal Organism and Etiology

Two species of *Uromyces viz. U. viciae-fabae* and *U. pisi* cause pea rust. *U. viciae-fabae* (=*U. fabae*) is autoecious and the most common rust of pea and lentil. *Uromyces pisi* is a heteroecious species having its aecial stage in *Euphorbia cyparissias* and rarely occurs in India. The mycelium in intercellular, branched, septate having yellowish or orange-red oil drops in the cytoplasm knob-like haustoria are formed in the host cell. The pycnia of *U. viciae-fabae* are inconspicuous. Aecia are cupulate, short and whitish, amphigenous or hypophyllous, usually in groups surrounding the pycnia. The aeciospores are yellow, round to angular, or elliptical with fine warts. Uredospores are round to ovate, light brown, spiny. Teliospores (sexual spores) are smooth, brown, subglobose, ovate, or elliptical, with rounded or flattened apexes. The teliospores on germination give rise to a promycelium. Each nucleated cell of the promycelium then produces short sterigmata at the tip of which swells to form a globular basidiospore.

Disease Cycle and Epidemiology

Uromyces viciae-fabae is a macrocyclic and an autoecious rust fungus, completing its lifecycle on a single host. While *U. pisi* is a heteroecious rust pathogen, completing its life cycle on two host plant species. The sexual stages of *U. pisi* are produced on *Euphorbia cyparissias* (cypress spurge), while the asexual stages on leguminous crop hosts such as *Lathyrus*, *Orobus*, *Pisum*, and *Vicia* spp. *Uromyces viciae-fabae* can be carried with seed as contaminations. It may also perpetuate on weed hosts from where it may infect lentil or field pea crops by windborne spores. Teliospores produced on infected plant residues are blown by the wind and infect volunteer plants and seedlings. Urediospores are short-lived while teliospores can survive in plant debris from one season to another and cause primary infection after germination. Germination of teliospores takes place between 17°C–22°C temperatures and at the start of the next season producing basidiospores which initiated a new infection cycle. Infection leads to the production of aecia. Aecia produce aeciospores, which infect leaves and spread the rust within the crop and to other crops. Aeciospores germinate at 17°C–22°C and infect other plants forming either secondary aecia or uredia at 25°C. In turn, this leads to the production of urediniospores, which can be carried long distances to produce new infections. Severe epidemics are caused by the production of several generations of urediniospores. Uredinia develop late in the season and are rapidly followed by telia. Teliospores develop in late summer on stems and leaves. After harvest, aecia and uredia present on the plant die out, but teliospores are more persistent. Teliospores geminate and cause infections through the sporidia. Urediniospores and aeciospores both cause secondary infections and spread the disease. The pycnia and aecia appear as the earliest symptoms. The disease is favored by high humidity, cloudy and rainy weather conditions. In the field conditions, disease development is favored between 20°C–22°C.

Management

Cultural Methods

- Destroy all diseased plant debris after harvest.
- Follow suitable crop rotation with non-leguminous crops.
- Mixed cropping.
- Apply vermicompost to the soil.

Host Resistance

- Grow resistant/tolerant varieties.

Biological Control

- Neemazal, a natural product of neem, induces resistance to pea powdery mildew.
- Several fungi such as *Ampelomyces*, *Cladosporium*, *Tilletiopsis*, *Verticillium*, and insects (*Thrips tabaci*) can be used as a natural biocontrol agent against powdery mildew.

Chemical Control

- Spray Sulphur 80% WP @ 1.252 kg in 300–400 litres of water per acre or Sulphur 85% DP @ 6–8 kg per acre or Triadimefon 25% WP @ 0.1% in 300 litres of water per acre. Second spray after 25 days of intervals.
- Spray the crop with Dithane M-45 @ 2 kg/ha in 1000 liters of water. Two to three sprays at 10 days interval are sufficient.

9

Mango (*Mangifera indica* L.)

9.1 ANTHRACNOSE OR BLOSSOM BLIGHT

Anthracnose is widespread and most important disease of mango wherever it is grown. Anthracnose on mango was first reported from Puerto Rico in 1903 and subsequently from various countries of the world. Losses due to anthracnose have been estimated to be 2%–39%.

Symptoms

Almost all plant parts viz., the young leaves, branches, inflorescence, and fruits are affected causing leaf spot/leaf blight, wither tip, blossom blight, and fruit rots.

Leaves: Leaf lesions of anthracnose disease start as small, dark brown spots with irregular margins and are surrounded by chlorotic haloes which are not limited by leaf veins. These spots are present on both sides of the mango leaf. Later, leaf lesions enlarge and coalesce to form irregular necrotic patches, frequently along the leaf margin. Centers of old lesions drop out from the leaves and give a shot-hole appearance. Young leaves are most susceptible to infection.

Twigs: Wither tip or die-back symptom appears on very young twigs. In severe cases, elongated, blackened lesions form on twigs that dry from the tip to downwards, accompanied by defoliation of branches.

Flowers: Panicle anthracnose or blossom blight can affect both the inflorescence stalk and the individual flowers. In the stalk, elongated dark grey to black lesions appear. The blighted flowers are dry, and their color ranges from brown to black.

Fruit: Fruits smaller than pea-size can be infected and aborted; whereas, larger fruits that are aborted because of normal self-thinning or other physiological causes are usually mummified. The mummified fruits are invaded saprophytically by the pathogen, and it sporulates abundantly on them. On ripening fruits, anthracnose is characterized by the development of irregular, dark brown to black lesions with slightly sunken (depressed) and the cracks on fruit surface (Figure 9.1 A). These spots coalesce to form large spots and ultimately cover the large surface area of fruits. Fruits can show "tear stain" symptoms, resulting in an "alligator skin" effect (Figure 9.1 B). Such infection causes fruits to develop wide and deep cracks on the skin that extend into the pulp. Lesions on fruit initially are superficial, and but in severe cases penetrate deeper than 5 mm into the mesocarp only. In the advanced stages of the disease, the fungus produces acervuli, and abundant orange to salmon pink masses of conidia appear on the lesions.

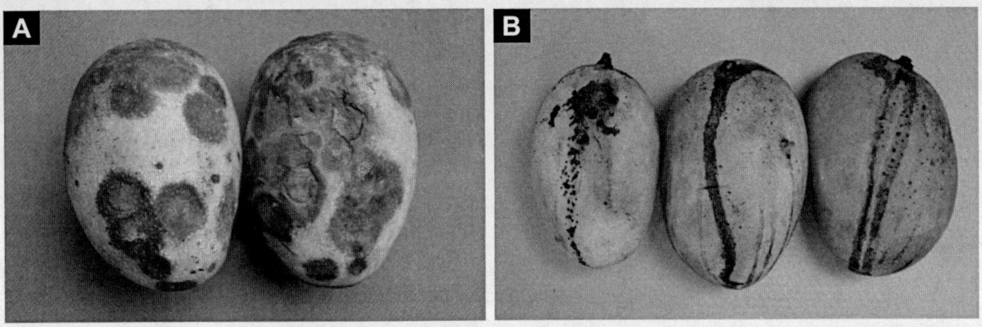

Figure 9.1 Mango Anthracnose. (A) Typical black sunken (depressed) lesions on fruits with orange to salmon pink masses of conidia (sporulation) and fruiting bodies of the fungus and (B) The "tear stain" symptoms with linear necrotic regions that finally lends an alligator-skin effect.

Causal Organism and Etiology

The anthracnose disease of mango is caused by *Glomerella cingulata* (*Colletotrichum gloeosporioides*). Mycelium consists of septate hyphae. Acervuli are formed abundantly on the affected host surface. One or more conidia are formed from the apex of each conidiophore. The conidia are hyaline but in mass appear pinkish. They are broadly oval to oblong, with rounded ends, non-sepate, and sometimes contain 1–2 globules. In the perithecial stage (*Glomeremella cingulata*) the fungus produces perithecia on stromatic cushions. Asci are subclavate, often slighty pedicellate, fugacious and 55–70 µm long. Ascospores are hyaline, one-celled, allantoids.

Disease Cycle and Epidemiology

Colletotrichum gloeosporioides survive on diseased dead twigs, branches, leaves, panicles, and mummified fruit and flowers and flower brackets, present on plants or fallen on the ground and serve as primary inoculum. The fungus has long saprophytic survival ability on dead twigs. The role of ascospores (sexual stage) in the disease cycle is unclear. Since the conidia are formed abundantly in the mango canopy which are considered as primary sources of inoculum. Spores (conidia) of the pathogen are dispersed passively by splashing rain or irrigation water. Spores landing on the infection sites penetrate into the tissues. The fungus can enter the pores of green fruits. The latent infection of mature fruits may take place through lenticels. On germination conidial germ tubes produce dark appressoria. Melanization of appressorium is essential for pathogenesis by conidial germ tube. Conidia can be rain-splashed to other leaves or flowers and cause secondary infections; thus the disease is polycyclic in these organs. In the case of postharvest anthracnose, developing fruit are infected in the field, but infections remain quiescent until the onset of ripening, which occurs after harvest. The disease is most serious after periods of wet weather. During the dry season, rain and humidity favor disease development. For conidial germination and appressorium formation, *C. gloeosporioides* requires free water or relative humidity above 95%. In general, infection is favored at temperatures ranging from 20°C–30°C and relative humidity of 95%–97%. The disease spreads rapidly in the rainy season. Cloudy and misty weather during flowering favors damage to the infected floral parts.

Management

Cultural Methods

- Diseased leaves, twigs, gall midge-infected leaves and fruits, should be collected, and burnt.
- Cover the fruits on the tree 15 days prior to harvest with new or brown paper bags.
- Wider plant spacing and yearly pruning of trees of diseased leaves, twigs and fruits.
- Optimize nutrition–high fruit nitrogen levels can exacerbate anthracnose. Timing of nitrogen fertilizer application can also be important.
- Do not harvest during wet weather.

Regulatory Measure

- Restriction on the movement of infected planting material/fruits.

Physical Control (Post-Harvest)

- Hot water treatment at 52°C ± 1°C for 20 minutes. The duration can be reduced to 10 minutes by supplementing Carbendazim or Prochloraz both (@ 0.05%) in hot water.
- Before storage, treat the mango with hot water, (50°C–55°C) for 15 minutes or dip it in Benomyl solution (500 ppm) or Thiabendazole (1000 ppm) for 5 minutes.
- Dip fruit after harvest in hot water (usually 52°C for 5 minutes) containing a recommended fungicide.
- Cool the fruit promptly after harvesting at 10°C–12°C. Store ripe fruit at this temperature for a maximum of 3 days. Lower temperatures may cause chilling injuries.
- Keep ripened fruit at 18°C–20°C with ethylene; reduce the amount of anthracnose development.

Biological Control

- Use biocontrol agent viz. *Streptosporangium pseudovulgare*.
- Spray (five to seven) *P. fluorescens* @ 5 g/L on flowers and branches at 3 weeks intervals commencing from October.

Chemical Control

- Spray with the recommended fungicides to reduce infection of new growth flushes, flowers, and fruit.

- Spray Azoxystrobin 23% SC @ 0.1% or 100 ml in 100 litres of water depending on the size of the tree canopy or Copper oxychloride 50% WG @ 0.24% or 240 g in 100 litres of water or Copper oxychloride (0.3%) or Bordeaux mixture (1%) or Carbendazim (0.1%) or Methyl thiophenate (0.1%) or Hexaconazole (0.05%) at 15 days interval can effectively manage blossom infection.
- The foliar infection can be managed by spraying Copper oxychloride (0.3%). Pre-harvest sprays of Hexaconazole (0.01%) or Carbendazim (0.1%) at 15 days intervals should be done in such a way that the last spray falls 15 days prior to harvest.

9.2 MALFORMATION

The term 'malformation' refers to the "abnormal growth of inflorescences" that occurs on affected trees. Mango malformation is a major constraint in mango production worldwide causing heavy economic losses depending on cultivar type and susceptibility. Mango malformation disease (MMD) occurs in Asia, Africa, and the Americas. Mango malformation was first recognized by Maries in 1891 from Darbhanga, Bihar. The malady has variously been ascribed to be acarological, viral, fungal, and physiological in nature. Malformation is widely prevalent in northern India, particularly in the states of Punjab, Delhi, and western U.P. It causes 50%–60% economic loss every year and in severe cases, it may extend up to 100%.

Symptoms

Mango malformation disease affects vegetative and floral meristematic tissues. Malformations in mango are of two types: (1) vegetative malformation and (2) floral malformation.

Vegetative malformation: It is most serious on seedlings, young trees in nurseries and mature trees. The major symptoms of vegetative malformation include hypertrophied, tightly bunched young shoots, with swollen apical and lateral buds. These buds produce misshapen terminals with shortened internodes and dwarfed leaves that curve from the tip back toward the adaxial portion of the petiole. The multi-branching of shoot apex with scaly leaves is known as "Bunchy Top", also referred to as "Witches' Broom" (Figure 9.2 A). The early malformed seedlings remain stunted and die while those infected later resume normal growth above the malformed areas.

Floral malformation: It is the malformation of panicles. It is characterized by the compact and clustery appearance of flowers (Figure 9.2 B). The flower buds transform into vegetative forms and leave the flower bud seldom opens and remains dull green in color. Some malformed panicles are not compact but both types of malformed panicles do not bear fruit. Primary or secondary axes on affected panicles are shortened, thickened, and highly branched. Malformed panicles may also produce dwarfed and distorted leaves (phyllody).

Causal Organism and Etiology

Mango malformation is a serious disease in tropical and subtropical areas of the world and has been attributed to various species of *Fusarium*, including *F. mangiferae*, *F. proliferatum*, *F. sacchari*, *Fusarium mexicanum*, *F. sterilihyphosum*, and *F. subglutinans* that belong to the *Gibberella fujikuroi* species complex. The vegetative form of mango malformation disease was first reproduced

Figure 9.2 Mango Malformation. (A) Vegetative malformation: "bunchy top" symptoms on a branch of mango tree and (B) Floral malformation: a malformed panicle with numerous small heads due to clustering of hypertrophied flowers.

in 1966 with *Fusarium moniliforme* and the floral form with isolates of *F. moniliforme* var. *subglutinans* from both vegetative shoots and floral tissue. Mango is the only known host of the pathogen *Fusarium mangiferae* var. *subglutinans*. It is anamorph of *Gibberella fujikuroi* var. *subglutinans*. It is a result of hormonal imbalance in the trees, induced by the *Fusarium* infections and associated with bud mite (*Aceria mangiferae*) infestations. Discrete or confluent sporodochia are cream to orange. Conidiophores are sympodially branched and bear mono- and polyphialides. Microconidia are usually oval, and single-celled. Macroconidia are long and slender, usually 4–6 celled. Macroconidia have a slightly beaked apical cell, a foot-like basal cell, and are usually four- to six-celled. Sterile, coiled hyphae forms, but chlamydospores are not in culture media.

Disease Cycle and Epidemiology

The fungus, *Fusarium mangiferae* produces both macrospores and microspores in malformed inflorescences which are the infective propagules and serve as the primary source of inoculum. The main spread of mango malformation disease to new areas is by infected pruning equipment or grafted planting material. The conidia are disseminated passively in the air as conidia blown or fall from dry malformed inflorescences as dry debris. Long-distance spread of the disease relies on the movement of infected nursery stock, but the means by which within tree and tree-to-tree spread is accomplished is not known. The conidia, the infecting unit, of the pathogen, are produced profusely over dead necrotic malformed shoots and panicles which are carried to the infection site, developing buds, by the mite. Emerging buds are tender and a better site for invasion. The host produces abnormal metabolites and the pathogen various toxins. The resultant of such interaction is that the buds are ultimately transformed either into malformed shoots or panicles. Later mangiferin increases to a cytotoxic level. The malformed tissues undergo necrosis due to the effect of high content of mangiferin and phytotoxins. At this stage, mangiferin is oxidized into polymeric quinone which does not have any perceptible anti-*Fusarium* activity. The surviving propagules grow over the dead necrotic cells and produce new crops of conidia to initiate fresh infection. The mango bud mite, *Aceria mangiferae*, is an important factor in the natural development of malformation. It also spread the disease within a tree and not between trees. It feeds on epidermal cells of floral and vegetative buds of mango, and is often observed in high numbers on malformed trees. The severity varies considerably from year to year. A tree, once affected, cannot escape the disease in subsequent years. The disease is serious in the north-west region where temperatures lie from 10°C–15°C during December-January (winter) before flowering. The disease is mild in the areas where temperatures lie from 15°C–20°C, sporadic from 20°C–25°C, and nil beyond 25°C. The occurrence of malformation differed according to the age of the plants. They observed more diseases in young plants than in old ones. About 91% incidence in 4–8 years old plants and 9.6% in older plants was reported.

Management

Cultural Methods

- Use disease-free planting material for planting.
- Keep orchards in good hygienic condition.
- Regular pruning of malformed panicles and parts along with affected shoots.
- Considerable incidence reduction has been reported by spraying 100–200 ppm NAA during October.
- Deblossoming of early emerged/infested panicles.
- Prune and burn the malformed floral panicles or vegetative shoots to reduce the disease incidence.

Biological Control

- Simple, effective, and eco-friendly management can be done using leaf extracts of neem and common weeds, e.g. *Datura stramonium*, *Calotropis gigantean*, etc.

Chemical Control

- In areas with less than 5%–10% infection, pruning of diseased plants should be compulsory; pruning of diseased parts along the basal 15–20 cm apparently healthy portions. This is followed by the spraying of Bavistin (0.1%) or Captaf (0.2%).
- Application of NAA (200 ppm) or planofix (90 ml/20 L) in the first week of October (before bud differentiation stages) followed by deblossoming in the late December or January or bud burst stage reduces the incidence of malformation.
- Spray chelated Zn^{++} (100 ppm) and Cu^{++} (40 ppm) during flower bud differentiation and flowering stage combined with a spray of Carbendazim (0.1%) during November to December.

9.3 POWDERY MILDEW

Powdery mildew is one of the most serious and widespread diseases of mango in India and world. It is a sporadic but very severe disease of mango. Blossom infection is the most serious and is responsible for yield loss up to 90%. It affects fruit set and development and causes fruit drop, resulting in heavy yield loss. In India, it was first recorded by Me Rae (1924).

Symptoms

The first symptoms of the disease are scattered patches of superficial cottony, white growth that develops on young shoots, leaves, stalk of panicles, flowers, and small fruit. These white patches later turn grey in color. Young leaves are most susceptible and covered with the white powdery coating of spores forming on either or both the underside or topside of leaves. On infected panicles, the powdery growth can cover all tissues, resulting in brown, shriveled necrosis and premature fruit drop which can reduce the crop load considerably or might even prevent the fruit set. Fruitlets are covered entirely by mildew. As it grows, the epidermis of the infected fruits cracks and corky tissues are formed. Purplish-brown blotchy areas appear on the skin of the fruits. Such fruits may remain on the tree until they reach up to pea size and then drop prematurely. Severe infection can cause the death of the flowers leading to poor fruit set and premature fruit drop of young fruits at the peanut stage.

Causal Organism and Etiology

Powdery mildew of mango is caused by the fungus *Acrosporium mangiferae* (syn. *Odium erysiphoides* f.sp. *mangiferae*). Mycelium is septate, hyaline, branched, and ectophytic. Conidiophores are short, hyaline. Conidia are unicellular, hyaline, elliptical to barrel-shaped. Conidia are produced in a basipetal fashion and are sometimes singly or in pairs of two. The lengths of germ tubes vary depending upon relative humidity, and they terminate in appressoria. Globular haustoria form in host epidermal cells. Conidiophores are of the pseudo-oidium type, with two to four septa and a straight basal cell.

Disease Cycle and Epidemiology

Mango is the only known host of the pathogen, *Acrosporium mangiferae*. It survives as dormant mycelium or haustoria in affected leaves, buds, or in malformed green panicles under dense foliage. The fungal spores are wind-borne and spread from other affected trees or within the same tree. Spores blown by the wind from infected areas readily adhere to hairy, unopened flowers near tip of the inflorescence and germinate in 5–7 hours, cause infection, and conidia are produced within 5 days of infection. Thus, the life cycle from conidia to conidia is completed in about five days. Mildew also appears naturally on the vegetative shoot of the mango but the life cycle is somewhat longer (about nine days). Secondary spread occurs through airborne conidia. Role of the perfect stage of fungus, alternate host, and the collateral hosts in disease cycle is unclear. Rain or mist in the daytime accompanied by cooler nights, cloudy weather, heavy morning mists, and high wind velocity for 3–4 days during flowering is conducive for quick spread. Disease development occurs within a very broad range of temperatures (10°C–31°C) and relative humidity (23%–90%). The mildew is favored by high humidity, cloudy weather, and low night temperature between panicle development and fruit set.

Management

Cultural Methods

- Monitor regularly to check for infected plant parts. If diseased plant parts are detected, carry out phytosanitary pruning.

- Selective pruning of over-crowed trees and branches after harvest increases air circulation and reduces humidity and therefore decreases the ability of the fungus to survive in non-favorable conditions.
- Avoid excess application of nitrogen fertilizer to limit the production of succulent tissue.
- Remove and destroy infected plant parts as this reduces inoculum and the spread of the disease.
- Field sanitation: Collect and burn or bury fallen leaves and fruit to reduce inoculum and the spread of the disease.
- Prune plants regularly so as to improve air circulation in the canopy.
- Remove diseased leaves, severely infected panicles, and destroy them to reduce the primary inoculum.
- Plantations must be established far away from unattended mango trees.
- Clear weeds and prune trees to improve aeration to reduce the disease spread.
- Intercrop with onion or forestry trees to reduce the ease with which the disease affects the mango.
- Remove severely infected panicles and destroy them by burying or burning.

Chemical Control

- Dust the plants with fine sulphur (250–300 mesh) @ 0.5 kg/tree.
- The first application may be soon after flowering, the second 15 days later (or) spray Wettable sulphur @ 0.2% or Carbendazim @ 0.1% or Tridemorph @ 0.1%, or Karathane @ 0.1%.
- Use Azoxystrobin + Difenoconazol-based products (such as Amister Top 325 SC and others). Apply at the early sign of infection @ 40–70 ml/15 L of water.
- Addition of stickers like Teepol @ 1 ml/L to the fungicidal solution is essential to increase the effectiveness of the fungicide.
- Spraying with wettable Sulphur @ 0.2% or Carbendazim @ 0.1% or Tridemorph @ 0.1 % or Karathane @ 0.07%.
- After the occurrence of high humidity and low temperature for 4–5 days in disease-prone areas, two sprays with wettable Sulphur WP @ 0.2% or Tridemorph EC @ 0.1% or Penconazole @ 0.05% or Hexaconazole @ 0.05% at 15 days intervals could combat the disease very effectively.
- Apply Potassium dihydrogen phosphate @ 25 g/100 ml of water to prevent effectively the spread of powdery mildew.

9.4 BACTERIAL BLIGHT AND CANKER

Bacterial blight of mango is also known as bacterial black spot or bacterial canker. In India, it was first reported by Patel et al. (1948) from Pune (Maharastra) as bacterial leaf spot. It is a very serious disease of mango that may reduce 50%–80% of fruit production in areas with high winds and heavy rains. The disease is a limiting factor in mango production because it is very difficult to control it, particularly in susceptible varieties. It is prevalent in Uttar Pradesh, Bihar, Maharashtra, Tamil Nadu, Delhi, Haryana, and Punjab.

Symptoms

All aerial parts of the mango trees are affected by the disease including fruits. Leaf and fruit symptoms are the most common (Figure 9.3).

Figure 9.3 Bacterial Blight and Canker of Mango. (A) Black spots symptoms on a leaf, (B) Black spots surrounded by chlorotic halo, and (C) Stem canker with bacterial oozing.

Leaves: The initial symptoms on leaves appear as minute, water-soaked, irregular lesions with greasy margins, which enlarge, turn brown or dark brown, become angular due to delineation by veins, and finally cankerous and slightly raised. Most of the lesions are surrounded by a chlorotic halo, which are more pronounced on young leaves, and becomes grey and crack with age (Figure 9.3 A–B). Several adjacent lesions may coalesce to form larger necrotic (dead) cankerous patches and bacteria may ooze from the lesions during high-humidity conditions. Severe leaf infection results in yellowing of the leaves and defoliation. Lesions also occur on petioles and twigs. Compared to anthracnose lesions, bacterial black spots are darker and more angular with distinct, usually raised, margins.

Twig and stem lesions: In very severe infections symptoms on twigs and branches are also found. Twig and stem lesions are black and cracked and can be an important means of survival for the black spot bacterium. Stem canker can lead to shoot-tip dieback. Bacteria ooze out, from the cankerous lesions (Figure 9.3 C).

Fruits: Initial symptoms appear around the lenticels as small, irregular, water-soaked specks from which a bead of bacterial ooze may develop. Later these lesions become black, star-shaped, raised, cankerous, and often burst open exuding a highly contagious gummy substance containing bacterial cells. Fruit lesions are more raised and darker than those on leaves. Cracks develop frequently on the fruit skin from which bacterial ooze releases. Bacteria from fruit, twig or leaf lesions may infect the fruit and often, a "tear stain" infection pattern is observed on the fruit. Severely affected fruit drop prematurely. Lesions can also occur on flower and fruit peduncles. Fruits with cracked skin are often invaded by secondary pathogens and insect larvae which cause post-harvest rot of the fruits. Fruit symptoms often lead to downgrading or market rejection of the affected fruits and cause economic losses. Flowers are also infected. The symptoms of bacterial blight and canker may be confused with those caused by the common anthracnose disease. However, there are some key differences (Table 9.1).

Table 9.1 The difference in bacterial blight and anthracnose of mango.

Bacterial Blight and Canker (*Xanthomonas axonopodis* pv. *mangiferaeindicae*)	Anthracnose (*Colletotrichum gloeosporioides*)
Lesions of bacterial blight and canker are generally confined to the peel and upper pulp, are darker and more angular with raised margins. These lesions when coalesce, form cracks from which sap oozes out that is full of bacteria.	Anthracnose lesions are dark brown to black, larger than initial lesions of bacterial blight and canker, often semicircular in shapes, and flat or sunken (not raised); eventually anthracnose lesions penetrate into the fruit pulp.

Causal Organism and Etiology

Bacterial blight of mango is caused by the bacterium, *Xanthomonas axonopodis* pv. *mangiferaeindicae* (Syn. *X. campestris* pv. *mangiferaeindicae, X. citri* pv. *mangiferaeindicae*). The bacterium has all the characteristics of the genus *Xanthomonas* except the production of yellow pigment. It produces non-pigmented (creamy-white) colonies on agar media. The bacterium is gram-negative, aerobic, rod-shaped (0.4–0.5 × 1.0–1.5 μm), and motile by means of a single polar flagellum. The bacterial cells are single or in the short chains and do not form capsule and spores.

Disease Cycle and Epidemiology

The bacterium survives in leaf lesions and in cankers formed on the twigs and branches, stone of fruits and in weeds but the survival is more efficient in leaf lesions. The primary inoculum is brought by contaminated nursery-stock, wind-driven rain, water splash, and insects. The bacterial cells are rain-splashed from the source of survival (diseased tissue) to healthy parts. Long distance dispersal or introduction to the pathogen-free areas occurs through contaminated or infected propagating plant material. Many insects such as *Orthaga evadrusalis* (leaf webber), *Cantheconidia furcellata* (bug), and *Myllocerus discolor* var. *variegata* (ashy weevil), transmit the bacterium from the diseased to healthy plants. Bacteria enter through natural openings on leaves (stomata) and fruit (lenticels) or abrasions or injuries. Leaf susceptibility decreases with time while fruit susceptibility increases over-time and it is highest during the harvesting. The disease spread rapidly during rain. Disease development is favored by high relative humidity and a temperature range of 25°C–30°C.

Management

Cultural Methods

- Maintain clean orchards and use only certified seedlings.
- Remove diseased twigs and destroy them.
- Sterilize harvesting and pruning tools regularly after use.
- Do not spray during fruit set.
- Do not enter the orchard until one day after spraying. The pre-harvest interval is at least 7 days.
- Maintain proper aeration by training and pruning for reducing the leaf wetness period.

Chemical Control

- Spray Copper oxychloride–based products when the disease is first observed.
- Spray Streptocycline (100 ppm) or Agrimycin-100 (100 ppm) thrice at 10 days intervals.

10

Citrus (*Citrus* spp.)

10.1 CANKER

Citrus canker is also known as Asiatic canker or true canker or A-form canker. Citrus canker is one of the most devastating diseases of citrus trees. It is one of the most serious quarantine diseases worldwide. It is an extremely persistent disease. The presence of the disease on citrus specimens collected from India (on *C. medica* from Dehradun, Uttarakhand, India in 1827–1831). In India, Luthra and Sattar (1942) first reported citrus canker from Punjab. Although millions of dollars are spent annually on prevention, quarantines, eradication programs, and disease control, citrus canker remains a serious challenge. A fruit infected with canker is safe to eat but has reduced marketability as fresh fruit.

Symptoms

Citrus canker causes necrotic brown lesions or spots with a rough surface on citrus leaves, twigs, stems, thorns and fruit (Figure 10.1). On leaves, the disease starts as tiny, oily, water-soaked, slightly raised, brown, circular, blister-like lesions, surrounded by a yellow halo on young leaves, and twigs. Both epidermal surfaces of leaves may rupture by tissue hyperplasia. They are often clustered at the leaf margin, leaf tip, or areas where water tends to accumulate. On leaves, stems, thorns, and fruit, circular lesions become, raised and blister-like, growing into white or grayish or yellow spongy pustules. Later these pustules become dark, thick and a light tan to brown corky or spongy rough canker and the centers become sunken or crater-like. Symptoms on fruits are similar to those leaves except the yellow halo. Cankers on twigs cause them to break. Fruit lesions do not enter the flesh, but the cratered appearance makes them unmarketable. Severe infection leads to defoliation, dieback, deformation, and premature fruit drop. The most reliable diagnostic symptom of citrus canker is the water-soaked margin that develops around the necrotic tissue and citrus tissue hyperplasia (excessive mitotic cell divisions), resulting in cankers.

Figure 10.1 Citrus Canker. Canker lesions on leaves (A), stem (B), fruits (C), and leaf lesion with the damage of Asian leaf miner (*Phyllocnistis citrella*) (D).

Causal Organism and Etiology

Citrus canker is caused by the bacterial pathogen *Xanthomonas citri* subsp. *citri* (syn: *Xanthomonas axonopodis* pv. *citri*), a gram-negative, slender, rod-shaped bacterium. The bacterium *Xanthomonas axonopodis* pv. *citri* is a rod-shaped, gram-negative, and has a single polar flagellum. Colonies on laboratory media are usually yellow due to 'xanthomonadin' pigment production. The bacterium is obligated aerobic and growing well at a temperature range between 28°C–30°C.

Disease Cycle and Epidemiology

The bacterium survives and multiples in lesions of infected leaves, twigs and fruits. It may also survive in crevices in the bark tissues of the citrus trees. Bacterial cells ooze from existing lesions during wet weather to provide inoculum for further disease development. Wind, rain-splashes, flooding, insects, birds and humans movement spread the disease over short distances. However, long-distance dispersal of the bacterium can be through infected plants, seedlings, propagative material, budded trees, fruits, tropical storms, hurricanes, and tornadoes. Nursery workers can carry the bacterium to new locations on their hands, clothes, tools, and equipment. The bacterium enters the plant through the stomata, lenticels, wounds (caused by weather damage or insects), along with water and invades the intercellular space (apoplast). The infection also occurs through wounds caused by citrus leaf minor (*Phyllocnistis citrella* and *Thosconyrsa citriq*). Young leaves are the most susceptible. Citrus canker is favored by mild temperatures and wet weathers. Prevailing temperature and duration of leaf wetness influence the extent of infection of leaves by the bacteria. For 100% infection temperature of 25°C–35°C and leaf wetness duration of 4–24 hours is required.

Management

Cultural Methods

- Use of disease-free nursery stock for planting in new orchards.
- Destroy all infected and suspected trees by burning.
- Collect the fallen canker-affected twigs and leaves, and burn them.
- Disinfect worker's tools and clothes.
- Raise wind-breaks of tall trees around the orchard in windy areas.

Regulatory Measure

- Prevent the entry of infective materials in canker-free citrus-producing areas through strict quarantine measures.

Host Resistance

- Most cultivars of citrus are susceptible to canker disease.
- Use resistant cultivars like Citron (*Citrus medica*) or Mandarins (*C. reticulata*) and highly resistant varieties like Calmondin (*Citrofortunella* sp.) and Kumquat (*Fortunella* spp.).

Biological Control

- A strain of *Pseudomoans syringae* is reported to show antagonism to citrus canker bacterium and also prevent canker development. The antagonist probably stimulates phytoalexin (citrusnins) synthesis in tissues.
- *Pseudomonas fluorescens* also is a strong antagonist of *X. axonopodis* pv. *citri*.

Biological Control of Vector

- Spray Neem oil @ 5%.
- Many parasitoids are known as natural enemies of citrus leaf minor (*Phyllocnistis citrella*). The parasitoid *Semielacher petiolatus* (Hymenoptera: Eulophdae) has been found very effective which prefers to attack second- and third-instar larvae of the citrus leafminer, but it can also parasitize fourth instars (pre-pupae). In addition to killing citrus leaf miners by parasitism, adult females killed by the host feeding, in which the female sticks her ovipositor into the larva and then drinks, the hemolymph.
- *Bacillus thuringiensis* strains 04-1, 454, and HD-1 have all been found to cause 80%–97% mortality in *P. citrella*.

Chemical Control

- Prune the affected twigs in old orchards and spray fungicides like Bordeaux mixture (1.0%) or, Copper oxychloride (0.3%) at periodical intervals, particularly during the rainy season.
- Bordeaux mixture (1.0%) sprays on plants before they are planted in new orchards.
- Captan 75% WP @ 666.8 g in 6–8 litres of water/acre.

10.2 CITRUS GREENING OR HUANGLONGBING

Citrus greening, first observed more than a hundred years ago in Asia, is one of the deadest citrus diseases in the world. Huanglongbing (HLB) is the most serious disease threat to the citrus growing industry worldwide due to its complexity, destructiveness, and incalcitrant to manage. The disease is also known as "huanglongbing" or "yellow dragon". Huanglongbing is a Chinese word meaning "yellow shoot disease". Citrus greening disease has different names according to the expression of symptoms in different parts of the world. It is called "citrus dieback and decline" in India, "mottle yellows" in the Philippines and citrus vein phloem degeneration (CPVD) in Indonesia, "decline" (likubin) in Taiwan, "yellow branch", "blotchy mottle" or "greening" in South Africa. The name "citrus greening" is based on a fruit symptom only that refers to the green color of fruits which remains even after maturity. Huanglongbing (HLB) is the officially accepted name for the disease since it was first described by O.A. Reinking from China in 1919. On average, the disease can cause 30%–100% yield losses depending upon the severity of the disease.

Symptoms

The symptoms vary with the time of infection, age of host, host species/cultivar, and biotic and abiotic factors that affect the growth of trees. The symptoms appear very late, usually two years.

Leaf symptoms: Symptoms consist of smaller leaves, asymmetrical blotchy mottling of the leaves, swelling and corky veins, and visible psyllids or waxy psyllid droppings on leaves (Figure 10.2). The most characteristic foliage symptoms are the blotchy mottling and yellowing of leaves. Chlorotic blotchy mottlings develop on the leaf blades, which result from the disruption of chloroplast caused by the accumulated starch in leaves. The youngest leaves may show symptoms similar to zinc deficiency with yellow venation, while the older leaves have the characteristic green mottling. Both HLB and nutrient symptoms can be seen on the upper and lower side of the leaf.

Fruit symptoms: The most characteristic symptoms include poorly colored, under-developed, asymmetric, and lopsided fruit, sour bitter taste, hard fruit with small, dark aborted seeds, fails to ripe, premature fruit drop, fruits remaining green on stylar end even after maturity, hence the name greening. Grayish-white waxy marks appear on the rind surface when pressure is exerted with a finger.

Whole tree symptoms: Symptoms of HLB-infected trees include stunted plant growth, yellow shoots, severe die-back of twigs or limbs, poor flowering, reduced foliage, root decline, and eventually the death of citrus trees. Yellowing of leaves or its parts or usually, the entire canopy of the trees occurs. The appearance of a yellow shoot or shoots has given the name "yellow shoot disease".

The Difference between Huanglongbing and Other Diseases

Some symptoms help in differentiating HLB from a chronic citrus decline caused by the *Citrus tristeza virus*, blight, and citrus variegated chlorosis. These are the following:

- Midrib and veins yellowing and corky and slight swelling of leaf veins.
- Dark-stained aborted seeds in small and under-developed fruits.

Figure 10.2 Huanglongbing of Citrus. (A) and (B) Blotchy mottle pattern of HLB on leaf (C) Green islands on the leaf surface (D) Vein corking in leaf, and (E) Phloem feeding Asian citrus psyllid (*Diaphorina citri*), a vector of HLB.

The Difference in Huanglongbing and Nutrient Deficiencies

Huanglongbing of citrus is often confused with the symptoms of nutrient deficiencies and vice-versa. HLB-infected leaves often display small leaves, chlorotic patterns and islands, shoot yellowing, and swelling and corking of veins. Symptoms like small leaves and chlorotic islands can be confused with zinc and vein corking with boron deficiency. Therefore, the common, most popular and easy way to distinguish between HLB symptoms and nutrient deficiency is a "pen test". Pen test is a visual diagnosis tool which determines the symmetrical (homogenous–nutrient deficiency) or asymmetrical (nonhomogenous–HLB infected) pattern of the symptoms across the mid-vein on both surfaces of leaves (Figure 10.3). According to this test, take the leaf in question and draw a circle with a pen at the same position on both sides of the leaf blade. If the same pattern of chlorosis is observed in the area encompassed in the circle, then most likely it is a nutrient deficiency and if the chlorosis is random, then it can be HLB. The most common HLB leaf symptom is the asymmetrical (scattered) pattern of blotchy mottle across the mid-vein. Nutrient deficiencies will have various patterns for each deficiency, but the patterns will always be symmetrical (exactly the same pattern of circles on both sides) across the mid-vein.

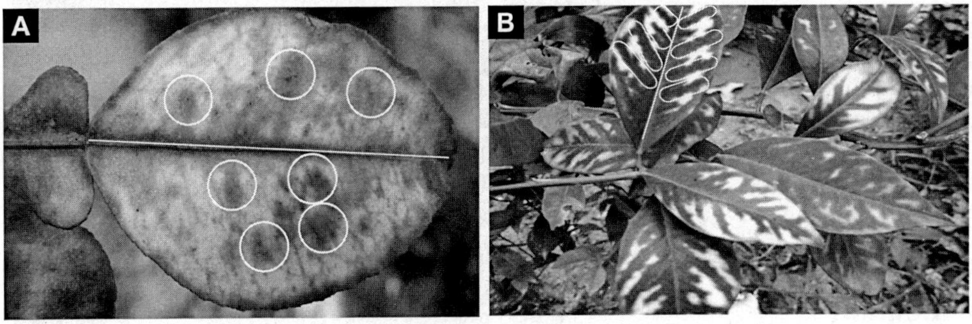

Figure 10.3 Pen Test for the Diagnosis of HLB at Field Level. (A) Symptoms of HLB showing asymmetrical blotchy mottle pattern on both sides of the leaves and (B) Symptom of zinc deficiency.

Causal Organism and Etiology

Huanglongbing disease of citrus is caused by the unculturable, phloem-restricted alpha-proteobacteria *Candidatus Liberibacter.* The name "*Candidatus*" refers to the inability to culture the bacteria and "*Liberibacter*" refers to the "liber" or "sieve elements" in which they were found. The genus has three species, viz. *Candidatus Liberibacter africanus* (Laf, found in Africa), *Candidatus Liberibacter americanus* (Lam, found in America), and *Candidatus Liberibacter asiaticus* (Las, found in Asia) are transmitted by psyllid vectors. The cells of the phloem-limited fastidious bacterium are walled, non-motile, gram-negative, non-pleomorphic, rigid rods measuring 1.0–2.0 × 0.2–0.5 μm. The bacterium is sensitive to different antibiotics including Penicillin and Tetracyclines.

Disease Cycle and Epidemiology

The pathogen (*Ca. Liberibacter* spp.) survive in the infected hosts as citrus is perennial in nature. The pathogen is spread by infected nursery plants, vegetative propagation, grafting diseased bud wood, infected dodder (*Cuscuta* spp.), and two highly moving psyllid vectors. Although huanglongbing of citrus is a bacterial disease, it is not disseminated through rain, wind or contact. Psyllid transmission is the only natural mode of dissemination of the bacteria, although spread by infected budwood during citrus propagation may occur locally. The bacterium, *Ca. L. asiaticus* and *Ca. L. americanus* transmitted by *Diaphorina citri* (Asian citrus psyllids), while *Ca. L. africanus* is transmitted by *Trioza erytreae* (African citrus psyllids). The Asian psyllid is found mostly on new shoots and its population increases during the active growth of citrus plants. In the absence of a vector, all *Ca. Liberibacter* species remain confined to their plant hosts. The disease is also transmitted through grafting but it is inconsistent. The bacterium moves in the phloem throughout the tree. *Ca. Liberibacter* spp. multiplies in the sieve tubes of infected plant hosts. The bacteria cause damage to the phloem (subsequently disrupting the internal movement of sugars). A single female can lay 800–1000 eggs during her life. The eggs hatch to release nymphs. Five nymphal stages are present in the life cycle of the insects. Nymphs of both species acquire the bacteria, which are transmitted by the adults in a persistent manner. Acquisition feeding period is 30 minutes or longer and the pathogen remains latent for 3–20 days. *D. citri* can acquire the pathogen in 15–30 min of feeding with a latent period of 8–12 days while one hour or more is required for 100% transmission. *T. erytreae* acquires the pathogen in 24 hours of feeding and transmits it in 7 days later with an exposure time of one hour. There is also evidence for transovarial transmission by *T. erytreae*. The temperature tolerance and sensitivities of these two vector species match those of HLB type they are naturally associated with i.e., *T. erytreae* is sensitive to high temperature whereas *D. citri* is heat tolerant. Within the temperature range of 15°C–30°C, the longevity of females increases as the temperature decreases. The temperature above 30°C may shorten the life span of adults to less than 30 days, otherwise at 15°C–20°C they can be expected to live 50°C–80°C days.

Management

The citrus HLB pathosystem represents one of the most daunting challenges for plant disease management. Once a tree is infected, there is no cure for HLB. Currently, no effective HLB management approach exists. However, antibiotics have been tried with partial success.

Cultural and Mechanical Methods

- Establish the orchard with plant, disease, and insect free.
- Remove and destroy infected trees at earliest.
- Regularly monitor the mother stock for citrus greening and keep under isolation to prevent the plants from insect.
- Eradicate and replace young trees (less than four-year old) and those not bearing fruits and showing symptoms, whereas trees with fruit should be pruned.
- The continuous foliar application of micronutrients such as ($ZnSO4$ + $MnSO4$ (3 : 1)) for 7 weeks is also effective to induce growth and production in infected trees.

Regulatory or Quarantine Measures

- Strict quarantine measures should be applied to restrict the imports of propagative citrus material originating only from certified regions or from certified plants free of psyllids and citrus greening bacteria.
- Restrictions on the movements of plants should be for the hosts of the HLB only.

Physical Control

- **Heat Therapy or Thermotherapy**: Another approach found to be very effective in the control of citrus greening. Graftwood is heated to 48°C–50°C for several minutes. Saturated hot air treatment to graftwood at 48°C–58°C eliminates the yellow shoot disease. Shoot tip grafting along with pre-heat treatment also reduced the disease with 100% efficiency. Treatment of budwood at 47°C for 2 hours reduces the disease incidence and longer treatments eliminate the pathogen. Treatment of infected young plants and seedlings budded with infected tissue at 38°C–40°C for 6 weeks also kills the pathogen.

Biological Control

- The major parasites of psyllids (*T. erytreae* and *D. citri*) are parasitic wasp (*Tamarixia/ Tetrastichus radiata* and *T. dryi*) have been found effective against the disease.
- Many predators cause the natural mortality of nymphs. Coccinellid beetles are the main predators.

Chemical Control

- **Management of bacterium**: For direct control of bacterium, trunk injection of Tetracycline hydrochloride (500 ppm solution), preferably continuously for 7 days gives good control but this antibiotic is phytotoxic for citrus budwood.
- **Management of vector**: Spray systemic insecticides like Diazinon, Endrin, Parathion, Dimethoate or Monocrotophos or Methomyl or Malathion or Carbofuran @ 0.02% on citrus trees at 10–12 days which gives good control of the psylla.
- Spray petroleum oil (025%–1.0%) to manage vector population. First and second instars are the most susceptible but eggs are highly tolerant.

10.3 GUMMOSIS

Citrus gummosis is one of the most economically important diseases of citrus worldwide. Losses occur from damping-off in seedbeds, from root or crown rot in nurseries foot rot, and from diseases known as gummosis, brown rot gummosis, brown rot, trunk rot, foot rot, collar rot, root rot, fibrous root rot, leaf fall and brown rot of fruit in orchards. Foot rot (gummosis) is one of the most serious diseases caused by *Phytophthora* in citrus. Gummosis is one of the longest-known diseases of citrus in the world. The first Phytophthora epidemic was observed during 1832–1836 in citrus orchards of Azore Ireland. Later in 1841, it was reported in France and during the 1845s it spread in Portugal and Mediterranean countries.

Symptoms

Root rot: Colorless or tan to dark brown decay of feeder roots, which may extend into the larger secondary and primary roots causing root-lesions and extensive discoloration and death of the root system. Yellowing of the leaves, defoliation, and twig dieback is a common consequence of Phytophthora root rot, particularly the loss of feeder roots.

Collar rot: The main symptom is dark, water-soaked areas of bark overlaying pale brown discolored cambium and sapwood tissue, often with gum exudation on the trunk near ground level.

Brown rot: Leaf symptoms include blackening or brown discoloration and death of young tip growth. Dark, water-soaked areas occur on leaves, particularly at the tips and along the margins. Affected leaves often fall while still green. Fruit symptoms are first seen as a light brown discoloration of the rind on one side of the fruit. The affected area remains more or less circular as it enlarges, and may eventually cover the entire fruit. The rot penetrates deep into the fruit, but generally remains firm and becomes darker in color and leathery in appearance. Infected fruit has a characteristic pungent odor and fall readily. In wet weather, a white, fungal growth appears on the surface of the rotted area. As symptoms develop one week after infection, brown rot can be a serious postharvest problem.

Gummosis: Gummosis is a common, although not a diagnostic symptom of Phytophthora infection in trees, and is characterized by the exudation of large amounts of gum on the outside of the bark (Figure 10.4). Infected bark remains firm with small longitudinal cracks through which abundant amber-colored gum exudation occurs. Under favorable conditions, gummosis can occur on the trunk, or even on branches in the canopy. During the rainy season the gum is washed off while during summer gum deposits, dry and adhere to the bark making the gummosis symptom evident.

Figure 10.4 Citrus Gummosis.

Causal Organism and Etiology

Citrus gummosis is caused by three species of *Phytophthora*, viz. *P. palmivora*, *P. parasitica* (*P. nicotianae* var. *parasitica*), and *P. citrophthora*.

- *P. palmivora*: Mycelium is intercellular with haustoria. Hyphae are large and often swollen at regular intervals. Sporangiophores are simple or branched with inverted pear shaped, rarely round and always terminal sporangia. Zoospores are large when encysted. Oospores are spherical, thick-walled, and produce secondary sporangium on germination.
- *P. citrophthora*: Hyphae are fairly coarse. Sporangiophores are delicate, short, scarcely widening at base of the sporangium, irregularly branched and with a swelling at the point of branching. Chlamydospores may be abundant, few or absent.
- *P. nicotianae* **var.** *parasitica*: Hyphae are tough, irregularly wide but without typical hyphal swellings. Sporangiophores are more slender than vegetative hyphae, irregularly or sympodially branched. Sporangia are papillate and broadly ovoid, ellipsoid, obyriform to spherical, and occasionally lateral or intercalary. Antheridia are amphigynous, spherical or oval. Ooginia become rough, thick-walled, and yellowish brown with age. Chlamydospores are produced abundantly.

Disease Cycle and Epidemiology

Phytophthora spp. survives as chlamydospores or oospores in soil or decaying roots. Chlamydospores germinate readily when moisture is present and quickly form sporangia. In presence of abundant free moisture, zoospores are released from sporangia. The motile zoospores swim to short distances or are carried by soil water to long distances. Dispersal of sporangia and zoospores is by wind, raindrop splashes, irrigation water, and even insects. Zoospores are attracted to roots by root exudates, especially when roots are damaged but living. Zoospores are attracted to wounds or to the region of elongation of root tips, where they encyst, germinate and cause infection by penetrating the cortex. A wound is necessary for infection of suberized bark. All species of *Phytophthora* that affect citrus can cause brown rot. Most commonly, this disease is caused by *P. citrophthora* where rainfall occurs during the winter. In humid, subtropical and tropical areas, it is often caused by *P. palmivora*. Both species produce abundant sporangia on the fruit surface that can be spread through the tree by wind-blown rain. In contrast, *P. nicotianae* is mostly soilborne and propagules must be splashed from the soil; consequently, most infections are within one meter of the ground. The primary means by which *Phytophthora* spp. are spread through citrus orchards is by use of infested nursery stock. The pathogen is also carried in soil on equipment when vehicles move from infested to non-infested groves or nurseries. Flooding methods for irrigation water may also move the pathogen from area to area. The disease is favored by rains, heavy to poorly drained soils, excessive and flood method of irrigation leading to prolonged contact of trunk and crown with water or moist soil, high water-table, deep planting, low budding, and injuries to roots or base of the stem. For brown rot of fruits caused by *P. parasitica* a minimum of 3 hours of fruit wetness at 27°C–30°C is essential for infection.

Management

Cultural Methods

- Prefer double-ring basin method of irrigation over the flooding method of irrigation.
- Disease spread can be controlled by removing diseased bark and a buffer strip of healthy, light brown to greenish bark around the margins of the infection.
- Select the proper site with adequate drainage.

Host Resistance

- Use of resistant rootstocks and avoiding contact of water with the tree trunk by adopting the ring methods of irrigation are effective.

Chemical Control

- Spray Aureofungin 46.15% w/v SP @ 300 (1%) gm per ml per acre.
- **Trunk Paint**: Scrap the affected bark portions of the tree trunk, branches, and limbs along with some healthy green bark with the help of sharp knife or khurpa. Properly collect and destroy by burning the diseased bark and do not allow it to fall on the ground. Apply the fungicidal paint of Mefenoxam MZ (available as Ridomil Gold) with the help of a painting brush twice in a year whenever symptoms appear.

- **Soil Drenching**: Drenching the root zone area of the infected tree with Mefenoxam MZ (25 g/tree in 10 litres of water) twice in February-March and July-August. Trunk paint can be combined with soil drench for effective disease management.
- **Foliar Sprays**: Spray Fosetyl aluminium (Aliette) @ 2.5 g/L can be done during September–October for the effective disease control. The spray will check the lesions on the tree parts as well as help the regeneration of feeder roots. Aliette spray can be combined with the drenching of the root zone area with Mefenoxam MZ.

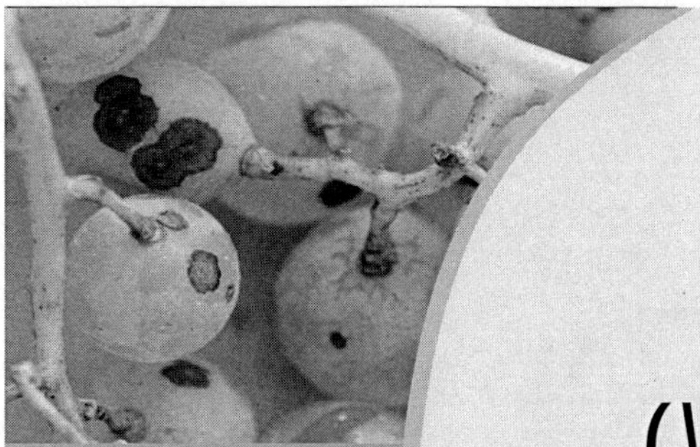

11 Grapevine (*Vitis vinifera* L.)

11.1 DOWNY MILDEW

Downy mildew is a highly destructive disease of grapevines in all grape-growing areas of the world. Downy mildew of grapes is one of the first diseases, that was controlled by a fungicide and the discovery was accidental. In 1885, P.M.A. Millardet first used the Bordeaux mixture (copper sulfate and lime) to control downy mildew in the vineyards of France, beginning the chemical era of disease control. The disease causes direct yield losses by rotting inflorescences, berries, clusters, and shoots. Indirect losses can result from premature defoliation of vines due to foliar infections. Under favorable environmental conditions, crop losses vary from 50%–100%.

Symptoms

Leaves: Circular to irregular yellowish-green or translucent "oily spots" located between the veins and surrounded with a brownish-yellow halo, appear on the upper surface of the leaves. As lesions expand, the affected areas turn brown, necrotic, or mottled. After humid and dew nights, sporulation of the pathogen occurs on the underside of foliar lesions that appear as downy growths (a delicate, dense, white to grayish, cottony growth) which is characteristic of this disease (Figure 11.1). Severely infected leaves eventually turn brown, wither, curl and drop.

Shoots and tendrils: Initial symptoms appear as water-soaked, shiny depressions on which a dense growth of whitish mildew appears. Infected young shoots, petioles, tendrils, or cluster stems often become stunted, distorted, thickened, or curled. The shoots and tendrils are also covered with whitish mycelial growth of the pathogen. Severely infected shoots and tendrils usually turn brown and die.

Inflorescences and fruits (berries): Infected young inflorescences distort into a curl or corkscrew. The flower clusters and young berries are highly susceptible. During humid weather,

Figure 11.1 Downy Mildew of Grapevine.

infected inflorescences and young berries appear yellow or grey, covered with white, cottony abundant sporulation. Infected flowers dry and drop and the fruits became grayish, hard, shrivel, mummified, and remain attached to the bunch.

Causal Organism and Etiology

The downy mildew of grapevine is caused by an oomycete *Plasmopara viticola*. The intracellular mycelium of the pathogen consists of coenocytic, thin-walled, hyaline, and hyphae. Haustoria are spherical or pear-shaped. Tree-like sporangiophores, bearing white, lemon-shaped sporangia, are produced from a mycelial mat within the host tissue and emerge through stomata. The sporangia are hyaline, thin-walled, ovoid or lemon-shaped, papillate. The branching of the sporangiophores is at right angles to the main axis and at regular intervals. From the apex of each branch, 2–3 sterigmata arise and bear lemon-shaped, papillate sporangia. The pathogen produces large number of asexual sporangia which are hyaline, ellipsoidal, and papillate. They are produced at the tips of acutely right-angled branched sporangiosphore that usually arise from mycelial cushions in the substomatal cavity. The pathogen, *Plasmopara viticola*, produces asexual, biflagellate zoospores and sexual oospores. Its mycelium is aseptate. It is an oomycete in the order Peronosporales. *Plasmopara viticola* is an obligate parasite, and it absorbs its nutrients from the living host tissue via globose haustoria.

Disease Cycle and Epidemiology

Plasmopara viticola is a biotroph. The pathogen survives as oospore in the infected leaves, shoots, and berries and also as dormant mycelium in infected twigs, buds, and leaves. Oospores serve as the primary inoculum. Oospores can survive in leaf debris for at least 2 years. The overwintering oospore germinates and produces a germ sporangium when the soil is wet and the temperature is above 10°C. These sporangia and or zoospores are spread by wind, splashing rain on vine surface and handling wet plants. When plant surface is covered with a film of moisture, the sporangia quickly germinate to produce large numbers of motile or swimming spores (zoospores). The zoospores splashed on the leaves and cause the primary infection. Usually, the primary infection starts on the leaves near the ground. They encyst near a stoma and form a germ tube that penetrates the leaf through the stomata or the epidermis. All green parts of the vine that have mature, functioning stomata, including fruit, leaves, and young shoots, can be infected. After infection, the pathogen grows intercellularly, producing numerous globose haustoria to obtain nutrients. Once inside the plant, the pathogen grows and spreads through tissues. The chlorophyll is degraded, cell contents disintegrate, cell walls collapse, and the entire cell turns brown. The symptoms usually appear after 7–15 days of infection. During humid weather, abundant sporangiophores and sporangia are produced that emerge through stomata of infected leaves and other grapevine tissues (e.g. bunches), which are wind-disseminated and zoospores cause secondary infections. These zoospores then infect the plant through the stomates and lenticels, and the disease spreads rapidly. Under favorable conditions, repeated secondary infections spread the disease. Presence of free water (rain) on the host surface is the principal factor in the development of the disease. High humidity (80%–100%) and temperature (18°C–25°C), favor the disease development. The pathogen grows out through the stomata of infected tissue and produces sporangiophores on the lower leaf surface at a temperature above 13°C.

Management

Cultural Methods

- Destroy the infected plant debris from the soil.
- Grow the crop in well-drained soils and in rows to provide good air movement to keep the leaves dry.
- Avoid overhead irrigation to prevent the wetting of leaves.
- Remove dead leaves and berries from vines and ground after the leaf drop.
- Improve air circulation.
- Remove weeds and tall grasses in the vineyard and surrounding areas.
- Prune and burn infected twigs.
- Any practice that speeds the drying time of leaves and fruit will reduce the potential for infection.
- Collect and burn fallen leaves and twigs.
- Sanitation of the orchard.
- Remove abnormal un-harvested bunches from vines and burn them.
- Remove all shoots emerging from the crown near ground.
- Shoots hanging from trellises towards the ground should be tied on trellises or can be removed if they are extra.
- Remove fallen leaves which are the source of overwintering inoculum.

Host Resistance

- Grow-resistant varieties like Amber Queen, Cardinal, Champa, Champion, Dodridge, and Red Sultana.
- Interspecific hybrid cultivars of *V. vinefera* and North American species have good resistance.

Chemical Control

- Remove dead wood present on arms and paste cut surfaces with Bordeaux paste.
- Spray the vineyard with 1% Bordeaux mixture.
- Spray Bordeaux mixture @ 0.5% or Copper hydroxide @ 2.0 g/L or 3.0 g/L at 15–20 days intervals to cover all rainy periods.
- The disease can be effectively managed by giving 3–5 prophylactic sprays with 1% Bordeaux mixture or Fosetyl-Al (Aliette) 0.2% or Metalaxyl + Mancozeb 0.3–0.4%.
- Other recommended fungicides are listed below with the rate of application:

Fungicides	Rate of application
Aureofungin 46.15% w/v SP	0.005%, 300 litres/acre
Captan 50% WP	1.0 kg in 300–400 litres of water/acre
Captan 75% WP	666.8 g in 400 litres of water/acre
Chlorothalonil 75% WP	0.2% or 200 g/100 litres of water
Copper oxychloride 50% WG	0.24% or 240 g/100 litres of water
Copper oxychloride 50% WP	1.0 kg in 300–400 litres of water/acre
Copper sulphate 2.62% SC	0.4 L in 200 litres of water/acre
Cymoxanil 50% WP	0.24% or 240 g/100 litres of water as required depending upon the crop stage and equipment used.
Cyazafamid 34.5% SC	80 ml in 200 litres of water/acre
Dimethomorph 50% WP	400 g in 300 litres of water/acre
Fosetyl-AL 80% WP	560–800 g in 300–400 litres of water/acre
Kresoxim-methyl 44.3% SC	240–280 ml in 200 litres of water/acre
Mancozeb 75% WP	0.6–0.8 kg in 300 litres of water/acre
Propineb 70% WP	0.30% or 300 g/100 litres water
Zineb 75% WP	0.6–0.8 kg in 300–400 litres of water/acre
Ziram 80% WP	0.6–0.8 kg in 300–400 litres of water/acre
Cymoxanil 8% + Mancozeb 64% WP	600–800 g in 200–400 litres of water/acre
Famoxadone 16.6 % + Cymoxanil 22.1% SC	200 ml in 200–300 litres of water/acre
Fenamidone 4.44% + Fosetyl-Al 66.7% WG	800–1000 g in 200–300 litres of water/acre
Fenamidone 10% + mancozeb 50% WDG	600 g in 200–300 litres of water/acre
Metalaxyl M 4% + Mancozeb 64% WP	200–400 litres/acre of 0.25%
Metalaxy l8% + Mancozeb 64% WP	200 litres/acre of 0.5%
Metiram 55% + Pyraclostrobin 5% WG	600–700 g in 300 litres of water/acre

11.2 POWDERY MILDEW

Powdery mildew is one of the most widespread and destructive diseases of grapes worldwide. It appears in epidemic form in all the vineyards in India. Losses in fruits yield due to the powdery mildew may be up to 40%–60%. The most significant effect of powdery mildew is on berry sugar levels, juice color, and acidity. Infected fruits become unsuitable for making the wines.

Symptoms

Powdery mildew can infect all green tissues. The characteristic symptom appears as a dusty ash-grey to white coating on leaves, stems, buds, tendrils, flowers, and young fruits of the grapevine. Severely infected leaves may exhibit discoloration, malformation, or deformity, including leaf curling and withering. Shoots develop small grey to oily black blotches that may enlarge until severely diseased shoots are covered by a layer of spores and become stunted. Infected fruit turn grayish-white at first and ultimately exhibit a brown, rusted appearance. Infected fruit may crack, shrivel, or drop from clusters. Oily, web-like pattern develops on infected berries and shoots where the fungal mycelium grows. On young berries and bunch stalks grey to white sporulation occurs. The diseased vines have a wilted appearance and remain stunted in growth. Late in the season, many black specks, and sexual fruiting bodies (cleistothecia), may develop on the surface of infected areas.

Causal Organism and Etiology

Powdery mildew of grapes is caused by the fungus *Uncinula necator* (syn. *Erysiphe necator*), an obligate fungus in the Vitaceae family. The mycelium is entirely superficial on the attacked plant parts to which it adheres by means of bilobate or multilobate appressoria. The hyphae are hyaline, slender, septate, and branched. The conidiophores are simple, multiseptate, and erect. They bear a chain of 3–4 hyaline, oval-shaped conidia. The oldest conidium is at the distal end of the chain. The conidia germinate by a short germ tube terminated by a bilobed or multilobed appressorium. Cleistothecia of the pathogen are found embedded in the superficial mycelium on the leaves, or shoots nodes or in buds among the scales and hairs. The peridium is covered with 8–25 septate appendages that are coiled (uncinate) at the distal end and brown at the base. Each cleistothecium contains 4–6, sometimes more, ovate to subglobose asci. Each ascus contains, 4–6, oval to ellipsoid, hyaline ascospores.

Disease Cycle and Epidemiology

The fungus *Uncinula necator*, survives as mycelium in dormant buds and as cleistothecia (fungal resting bodies) in cracks in the bark and/or leaf litter. Besides, the collateral hosts and volunteer plants provide conidia as the primary inoculum. The infected buds give rise to systemically infected shoots. In cool and moist conditions, the cleistothecia release ascospores which infect the leaves. Wet leaves are required for conidia germination, penetration, and infection. The conidia and ascospores germinate by producing a short germ tube that terminates into a bilobed or multilobed appressorium. The fungus penetrates only the epidermal cells, inserting haustoria (specialized feeding structures) inside the cells to absorb nutrients. After the infection from ascospores or the hibernating mycelium in the host buds, enormous conidia are produced to carry on the conidia-to-conidia life cycle. The presence of fungal mycelia with conidiophores and conidia on the surface of the host tissue gives infected vines a whitish-gray, dusty, or powdery appearance. Conidia cause secondary infection and spread the disease in the same and nearby fields. Cleistothecia are formed at the end of the season on the infected leaves, stems, and fruits. They are washed down and settle in the cracks of the bark. Powdery mildew is a dry weather disease and is favored by mild, cloudy conditions. Rapid germination of conidia, infection and development of the pathogen occurs at temperatures of 21°C–30°C although the fungus can grow at a temperature from 5°C–30°C. Ultraviolet light reduces the development of the disease that grows best in humid overcast or shaded conditions, which occur inside large canopies. Free moisture (leaf wetness) is not required except for cleistothecia germination. Free water causes poor and abnormal germination of conidia. Rainfall is detrimental because it removes conidia and disrupts the mycelium.

Management

Cultural Methods

- **Clean cultivation of vines**: Pruning after leaf shading, thinning out and cutting back of laterals, and removal and destruction of all diseased parts.
- Orient vineyards and use training systems to allow good movement of air and penetration of light in the canopy and reduce excessive shading.

Host Resistance

- Grow-resistant varieties like Chholth Red, Chholth white, Skibba Red, Skibba White, etc.

Chemical Control

- Dustings of vines with 300 mesh Sulphur @ 15 kg/ha (first dusting should be done when new shoots are two weeks old, second prior to blossoming, and third when the fruits are half-ripe). Sulphur dust is preferred in dry climates whereas in areas with high rainfall wettable powder or flowable formulations are recommended for their retention qualities. Sulphur causes phytotoxicity in some varieties of grapes (e.g. those of *V. labrusca* origin) and is thought to be phytotoxic to most grapes at high temperatures (more than 30°C).
- Spray wettable Sulphur @ 0.3%.
- In addition to Sulphur dust (15 kg/ha), Sulfex (0.25%), Topsin-M (0.1%), Thiovit (0.25 %), Karathane (0.05%), and Calixin (0.05%) have been found effective against powdery mildew of grapevines.
- Prophylactic spray with Bordeaux mixture (1%) or Lime Sulphur at the dormant stage delays the development of the disease by decreasing initial inoculum.
- Spray Morestan @ 0.03% at 4 days interval, starting from the last week of December to the 1st week of March.

11.3 ANTHRACNOSE

Anthracnose of grapes also called "grapevine black spot" and "bird's-eye rot" for the distinctive spots it causes on grape berries. The anthracnose of grapes is the most destructive disease in warm, wet seasons and results in the loss of fruit quality and quantity.

Symptoms

Anthracnose of grapes occurs on stems, leaves, petioles, rachises, pedicels, tendrils, young shoots, inflorescences and berries but lesions on shoots, and berries are most common and destructive.

Leaves: Initial symptoms appear as isolated small red-brown to black-brown spots on young leaves. In wet weather, the spots enlarge rapidly, becoming slightly sunken with grey-white centres and dark brown or purple edges, round or irregular shaped. As the leaf matures, the necrotic centers of older lesions usually fall off, forming a "shot-hole". On leaf veins, the lesions are depressed, grey or grayish brown with a dark brown edge, and cause the distortion of the leaf blade in young leaves. Severely infected leaves shrivel up and drop.

Inflorescences: Inflorescences are highly susceptible which can turn yellow, brown, and then dry out completely, during severe infections.

Berries: The disease appears first as dark red or brown spots on the berry. Later, these spots enlarge and become circular, sunken, with ashy-gray centre and dark-brown to purple-brown margins, giving it the "bird's-eye rot" appearance (Figure 11.2). The spots vary in size from 1/4 inch in diameter to about half the fruit. Several lesions may coalesce into large lesions and often cause berry cracking, favoring infection by opportunistic microorganisms that cause sour rot. Severely infected berries dry up and drop prematurely.

Figure 11.2 Anthracnose of Grapevine.

Petioles, tendrils, and stems: Symptoms on petioles, tendrils, and young stems are characterized by light-brown spots, which are initially circular but later elongate into elliptical, sunken, necrotic cankers. On stems, white-centered, bird's eyes spots develop, which can grow and girdle the stem resulting in dieback.

Causal Organism and Etiology

Grapevine anthracnose is caused by the ascomycete fungus *Elsinoe ampelina* whose asexual stage is *Sphaceloma ampelinum*. Acervuli containing short, cylindrical conidiophores bearing conidia are formed on necrotic lesions. Conidia are cylindrical to oblong with rounded ends, hyaline, and aseptate. When the sexual stage occurs, ascospores are produced in asci, which contain eight ascospores in each. Asci are globose or elliptical bitunicate and distributed irregularly in the upper part of the ascomata. Ascospores are hyaline, 1–3 septate. Sclerotia are formed on diseased grape tissue from late summer to winter.

Disease Cycle and Epidemiology

The fungus *Elsinoë ampelina*, overwinters as sclerotia and acervuli in infected leaves, berries, canes, woody tissue, and pruned canes on the ground. The acervuli (asexual fruiting bodies), produce conidia covered with a mucilaginous substance that enables them to stick to the site of infection. They are dispersed by free water or rain or overhead irrigation. Primary conidial sporulation can occur from overwintering sclerotia even at low temperatures. The conidia germinate and cause a primary infection when free water is present for 12 hours and the temperature is between (2°C–32°C). Ascospores are produced within a sexual fruiting body, also form on infected canes or berries left on the trellis or the vineyard floor. Besides conidia, the ascospores of *E. ampelina* also cause primary infection. Asexual fruiting bodies (acervuli) form on necrotic areas once the disease is established. These acervuli produce conidia in wet weather which are the secondary source of inoculum for the rest of the growing season. Anthracnose can be very damaging during heavy rainfall and hail. The longer the leaves and stems remain wet (more than 12 hours), the more severe the infection. The disease develops at temperatures of 10°C–35°C; the optimum temperature for disease development is between 20°C and 26°C. Symptoms appear 4–12 days after infection.

Management

Cultural Methods

- Remove infected fruit, twigs, or plant parts.
- Select the sites with suitable sun exposure and good air circulation.
- Make sure to keep a wide distance between vines.
- Use leaf mulch or fine bark to cover infected debris for homegrown grapes.
- Remove any wild grapes near the vineyard.
- Monitor the vines and remove those shows the symptoms of the disease.
- Prune the vines in early winter during dormancy.
- Remove plant residues from the vineyard.
- Plough the fields and bury the plant residues and berries.
- All affected twigs or canes showing cankers should be removed while pruning.
- The pruned twigs and leaves should be burnt or buried deep in the soil.

Host Resistance

- Use more tolerant varieties, if available.

Chemical Control

- Spray fungicides like Copper oxychloride @ 0.2% or Mancozeb @ 0.25%.

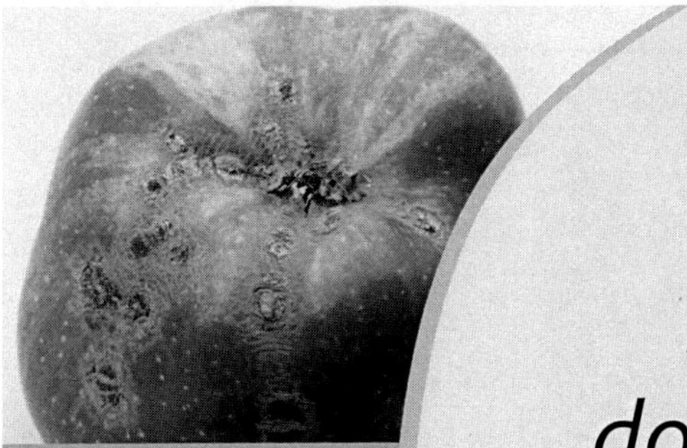

12 Apple (*Malus domestica* Borkh.)

12.1 SCAB

Apple scab also known as black spot is one of the most serious diseases of apple throughout the world where apple is grown and causes huge economic losses (up to 70% reduction in apple production). The first report on scab was published in 1819 by a Swedish scientist, Fries in Sweden, but the oldest clue to the existence of scab dates from 1600, in a painting by Michelangelo Caravaggio. Nusbaum and Keitt (1938) were the first to report the cytology of the interaction between *V. inaequalis* and apple. In India, the disease was first reported from Kashmir Valley in 1935 on a popular variety of apple indigenous apple cultivar, "Ambri" by P. Nath. Apple scab disease caused epidemic in Kashmir (1973) and it severely damaged the commercial cultivar "Red Delicious". The losses caused by the disease had been very heavy and are both types, qualitative and quantitative. The affected fruits have lower market value. Due to the polycyclic nature of the disease, more than 12 fungicide treatments may take place each season to manage the apple scab.

Symptoms

Symptoms of apple scab disease appear on leaves, petioles, blossoms, sepals, fruit, pedicels, and less frequently, on young shoots and bud scales. The characteristic symptoms include circular light or olive green velvety, necrotic or chlorotic lesions single or scattered on both surfaces of leaves. On sepals and pedicels, olivaceous spots are produced. On young fruits, dark colored sharply bordered brown and corky lesions (Figure 12.1) are produced while small black spots on the matured fruits which are termed as "pin-point scab." Infections of young fruit also cause fruit distortion. Splitting of fruit skin on the spot area is common with early infection. Severely infected fruit are malformed and often drop prematurely. The lesions can develop on fruits even in storage. The bark of twigs can split at the point of spots and give a silvery grey look.

Figure 12.1 Apple Scab.

Causal Organism and Etiology

Apple scab is caused by the fungus including two different states: *Venturia inaequalis* (perithecial stage or perfect stage or sexual) or *Spilocea pomi* (saprophytic state, imperfect stage or asexual or parasitic state). Apple scab fungus *Venturia* is hemibiotrophs, i.e., they must pass part of their lives on the host as parasites and part on dead tissues of the same host on the ground as saprophytes in order to complete their life cycle in nature. *V. inaequalis* is an ascomycete fungus. The mycelium of *V. inaequalis* is septate, and the nuclei are haploid. It produces sexual spores (ascospores) in a sac-like structure called an ascus (plural asci). Asci are elongated, sac-like structures, each of which contains 8 ascospores. The ascospores are brown, two-celled spores, and have a characteristic "footprint" shape. The shape of ascospores inspired the Latin name for apple scab, "inaequalis", which refers to the unequal size of the cells. *V. inaequalis* reproduces asexually by spores called conidia which are single-celled, uninucleate, and narrower at one end than the other. In mass, the conidia appear brown or olive in color.

Disease Cycle and Epidemiology

Venturia inaequalis is a hemibiotroph and overwinters as pseudothecia (sexual fruiting bodies) following a phase of saprobic growth in fallen leaf tissues. The scab pathogen completes its life cycle in two stages: (a) perfect or perithecial or saprophytic stage which is found on dead leaves and where the pathogen overwinters, (b) parasitic or stage is the second stage in which the pathogen completes its parasitic life cycle on living parts of the plants such as on leaves, flower, twigs, and fruits. The pathogen survives during cold winter in the diseased leaves that fall on the orchards. As the temperature rises above 10°C in late January and February, the fungus enters the sexual stage and produces pesudothecium which yields 800–1200 unequally bicelled ascospores (footprint or flame-shaped). The primary inoculum consists of ascospores that are released by rainfall from pseudothecia. Ascospores developed on dead leaves are the primary source of infection. They infect young leaves and buds during spring. This release is timed to coincide with host budburst and leaf unfurling. The ascospores (sexual spores) germinate by producing germ-tubes arising from ascospores that penetrate through the cuticle, either directly or via an appressorium, and develop into multilayered, pseudoparenchymatous structures, termed stromata, which are presumed to obtain nutrients from the subcuticular space. Cell wall-degrading enzymes are only produced late in the infection cycle. Conidia (secondary inoculum) arise from the upper surface of the stromata and are produced throughout the growing season, initiating multiple rounds of infection. Secondary spread of the disease is by conidia developed after the primary infection sets in conidia are the principle source (upto 1,00,000 conidia in each lesion) involved in the build-up of the disease. Conidia are disseminated by wind and rain from lesions and allow secondary infection to occur within the orchard throughout the fruit development period. Many conidial cycles occur and these also help in creating an epidemic throughout the fruit-growing season with the production of new crops of conidia. The conidial cycle (conidia-infection-conidia) of *Venturia inaequalis* completes within 8–10 days. The pseudothecia are formed within 30 days of leaf fall occurring from September to November. Weather conditions are very important for the development, release of ascospores, infection of the host, and repetition of the conidial cycle. The optimum temperature for ascospore maturation is about 20°C. Rains favor the ejection of ascospores. The ascospores germinate and cause infection at a temperature of 6°C–26°C and free water or 90% relative humidity. For infection to occur, the spores or leaf surface must be continuously wet for 28 hours at 6°C, for 14 hours at 10°C, for 9 hours at 18°C–24°C, or for 12 hours at 26°C.

Management

There are two approaches to manage the apple scab:

- Interrupting the life cycle of apple scab pathogen by reducing pseudothecial production.
- Protection against primary and secondary infection.

Cultural Methods

- **Clean cultivation**: Collect and destroy the infected plant debris, fallen leaves, and pruned materials in winter to prevent the sexual cycle.
- Trim and prune the infected twigs and branches, and burn them.
- Interrupt the life cycle of the scab pathogen: Spray of 5% urea in the autumn before leaf fall and 2% urea before bud burst or apply urea (@ 2 kg/acre at pre-leaf fall stage in spring and dolomitic lime (2.5 ton/acre) in autumn over fallen leaves to accelerate decomposition.

Host Resistance

- Use resistant/tolerant varieties such as Prima, Priscilla, Sir Prize, Jonafree, Florina, Macfree, Nova Easy Grow, Coop 12, Coop 13 (Redfree), Nova Mac, Liberty, Freedom, Firdous, Shireen, Emra, Red Free, Ambstraking, Ambroyal, Ambrich, and Ambred.

Biological Control

- Antagonists (*Athelia bombicina, Microspaeropsis ochracea*, *Cladosporium cladosporioides*, and *Chaetomium globosum* are promising bioagents against *Venturia inaeqalis*, and have been shown to reduce ascospore formation in leaf litter.

Chemical Control

In Himachal Pradesh, following schedule of fungicide sprays has been developed:

- First spray of Difolton @ 300 g in 100 litres of water or other Benzimidazole during March-April.
- Second spray of Dithane M-45 @ 200 g in 100 litres of water or Captan @ 200 g in 100 litres of water during April-May.
- One to three sprays of Dithane M-45 @ 250 g in 100 litres water at post blossom stage during summer at 10–15 days interval.

Post-Symptom Spray

- Even after the establishment of infection, certain measures can be taken. Fungicides like Myclobutanil, Fenarimil, Carbendazim, etc. have a high level of eradicant or antisporulent properties to inhibit spore production and reduce new infection.
- Post-harvest spray with 0.4% Benlate or other fungicides will reduce perithecia formation on fallen leaves.

For the latest spray schedule in apple (2023) given by the Department of Horticulture, Himachal Pradesh, Shimla-2 and Dr. Y. S. Parmar University of Horticulture and Forestry, Nauni, District Solan, (H.P.) see Table 12.1

Table 12.1 Spray schedule (2023) for the management of apple diseases.

Plant growth stage	Chemicals and Quantity Required for 200 litres of Water	Diseases managed
Green tip	Captan* (600 gm) or Dodine (200 gm)	Scab
	or Ziram* (600 [ml/gm]) or Fluxapyroxad 75 g/L + Difenoconazole 50 g/L SC (60 ml)	Scab/Powdery mildew
Pink bud	Mancozeb* (600 gm) or Propineb (600 gm) or Difenoconazole (30 ml)	Scab
	or Hexaconazole 4% + Zineb 68% WP (500 gm)	Scab/Powdery Mildew/Core Rot
Petal fall/ Pea Stage	Carbendazim* (100 gm) or Thiophanate Methyl* (100 gm)	Scab
	or Hexaconazole (100 ml) or Myclobutanil (80 gm) or Flusilazole 40% EC (50 ml) or Tebuconazole 50% + Trifloxystrobin 25% WG (80 gm) or Tebuconazole 8% + Captan 32% SC (500 ml)	Powdery mildew
	or Boscalid 25.2% + Pyraclostrobin 12.8%w/w WG (50 gm) or Metrafenone 500 g/l SC (20 ml) or Carbendazim 12% + Mancozeb 63% WP (500 gm)	Scab/Powdery mildew
	or Carbendazim 25% + Flusilazole 12.5% SC (160 ml)	Alternaria leaf spot
Fruit Development (Walnut size)	Mancozeb* (600 gm) or Propineb (600 gm)	Scab
	or Dodine (150 gm) or Metiram 55% + Pyraclostrobin 5%WG (200 gm)	Alternaria leaf spot/ Premature leaf fall Remarks: Do not spray Dodine if the temperature is above 30°C and slow drying condition prevails.

(Contd.)

Plant growth stage	Chemicals and Quantity Required for 200 litres of Water	Diseases managed
	or Tebuconazole 8% + Captan 32% SC (500 ml) or Fluxapyroxad 250 g/l + Pyraclostrobin 250 g/l 500 SC (20 ml) or Fluopyram 17.7% w/w + Tebuconazole 17.7% w/w SC (126 ml)	Alternaria leaf spot/ Premature leaf fall
	or Dodine 40% SC (150 ml) or Hexaconazole 4% + Zineb 68% WP (500 gm)	Alternaria leaf spot/ Premature leaf fall/Scab
Fruit Development (20 days after 4th spray)	Tebuconazole 50% + Trifloxystrobin 25% WG (80 gm)	Premature leaf fall, Scab
	or Propineb* (600 gm) or Zineb* (600 gm)	Scab, Black rot
	or Carbendazim 25% + Flusilazole 12.5% SC (160 ml)	Alternaria leaf spot/ Premature leaf fall
Pre-harvest (20–25 days Before harvest)	Captan* (600 gm)	Scab/Fly speck/Bitter rot,
	or Ziram* (600 ml/gm)	Scab
	or Metiram 55% + Pyraclostrobin 5% WG (200 gm)	Alternaria leaf spot/blight
	or Fluopyram 17.7% w/w +Tebuconazole 17.7% w/w SC (126 ml)	Alternaria leaf spot/Premature leaf fall/ Fruit Rot
	or Hexaconazole 4% + Zineb 68% WP (500 gm)	Alternaria leaf spot/ Premature leaf fall/ Scab

Note: *The pesticides are proposed to be banned viz. Captan, Ziram, Mancozeb, Carbendazim, Thiophanate methyl, Zineb, Malathion, and Chlorpyriphos by the Government vide Gazette Notification No. CG-DL-E-18052020-219423, Extraordinary Part-II- Section-3-Subsection (ii), dated 18th May 2020 regarding-Banning of Insecticides Order, 2020. (Source: Spray Schedule Year 2023 (Apple): Department of Horticulture, Himachal Pradesh, Shimla-2 and Dr. Y. S. Parmar University of Horticulture and Forestry, Nauni, District Solan, H.P.).

12.2 POWDERY MILDEW

Powdery mildew is one of the most important fungal diseases of apples which cause severe economic losses both in orchards as well as in nurseries on susceptible apple cultivars. In nurseries, the disease is more severe and after infection, seedlings are killed in one season. The disease is reported from almost all apple-growing areas. Biology of the pathogen is widely investigated all over the world in the past 100 years.

Symptoms

Powdery mildew of apple produces symptoms on blossoms, young terminal shoots, leaves, and fruit. In general, symptoms are most noticeable on the leaves and fruit.

Blossoms: Infected flower buds have a silver-gray appearance and open 5–8 days later than healthy ones, if at all. When they emerge, flower parts and leaves are usually covered with white mycelium. Petals are distorted and pale yellow or light green.

Shoots: The fungus colonizes the young, green tissue (terminal buds) as it emerges. The infected shoots have a silver-gray appearance and may exhibit defoliation, stunted growth, and die-back. As the disease progresses, the primary infections on the shoots produce inoculum, which causes secondary infections on leaves and fruit. Small, dark brown globular fruiting bodies known as cleistothecia (chasmothecia) are sometimes formed in the mycelial mat on stems and petioles in late summer.

Leaves: Leaves are colonized as they emerge from the buds. White fungal colonies develop on leaves. Lesions first appear as whitish, felt-like patches of fungal mycelium and spores on the undersides and along the margins. The lesions spread rapidly and may cover the entire leaf. Infected leaves are narrower than normal, are folded longitudinally, and become stiff and brittle with age.

Fruit: Fruits are rarely infected unless the disease is built up to high levels. Infected fruits are russeted and sometimes stunted. Infected receptacles may shrivel and fail to produce fruit or may mature to produce fruit that is discolored, russeted, dwarfed, and/or distorted.

Causal Organism and Etiology

Powdery mildew of apple is caused by *Podosphaera leucotricha.* It is an ascomycete fungus in the Erysiphaceae family. The ectophytic mycelium forms saccate haustoria in the epidermal cells. Aerial conidiophores arise from this mycelium on leaves and shoots. Each conidiophore bears a chain of oval, hyaline conidia which contains distinct fibrosin bodies. Fibrosin bodies are refractive inclusion bodies that exhibit varied shapes including rods and cones, and that are helpful in the recognition of this group of powdery mildews. Cleistothecia may form on one-year-old shoots, petioles, and midrib. The cleistothecia are globose, black, and partially embedded in the mycelial web. Two types of appendages (apical and basal appendages) are formed over the surface of the cleistothecia. Apical (top) appendages are long stiff and divergent while basal appendages are rudimentary, pale brown, short, tortuous, and simple or irregularly branched. Each cleistothecium contains a single ascus with eight elliptical ascospores.

Disease Cycle and Epidemiology

Probably the cleistothecia of the fungus *Podosphaera leucotricha*, produced within the mycelial mat on heavily infected leaves and shoots, do not play any major role in the disease cycle (survival and infection). The sexual spores (ascospores) contained in these cleistothecia are seldom viable and do not germinate readily. The disease cycle is mostly conidia-mycelium-conidia. At most places, the fungus survives in the form of dormant mycelium or encapsulated haustoria in the terminal and lateral shoot buds and blossom buds produced and infected in the previous growing season. Buds are most susceptible at the 3–6 unfolded leaf stage. The youngest leaves are the most susceptible but become increasingly resistant as they mature. Fruit infection occurs at the time of blossoming. When the infected buds break dormancy in the next season the surviving fungal structures resume growth and colonize the developing shoots and young leaf tissue. From these primary infections, abundant asexual conidia are produced on conidiophores which are wind-borne and serve as secondary inoculum. In addition to contributing toward a rapid inoculum buildup, secondary disease cycles are also responsible for infecting lateral and terminal buds that carry the fungus through the winter. As the mildew colony expands or as secondary infections lead to new colony formation, the infection process (hyphal outgrowth

> appressorium > infection peg > haustorium) is repeated until susceptible tissue is no longer available. Conidia germinate at high relative humidity (more than 70%), temperatures between 10–25°C; in contrast to most foliar fungal pathogens, leaf wetting is a deterrent to infection. Conidia are dispersed through wind and do not require free moisture to germinate. If they land on susceptible tissue, they initiate infection and produce colonies of mycelium. Conidia germinate to form hyphal outgrowths, which traverse the leaf surface, swell and then flatten to form appressoria. These structures release enzymes, which allow fungal infection pegs to penetrate the plant's epidermal cells and then enlarge to form haustoria.

Management

For the management of powdery mildew of apple reduction of primary inoculum and the protection of leaves, fruit, and buds from secondary infections are important.

Cultural Methods

- Sanitation of orchard.
- Remove over-wintering infected terminals by pruning (6–8 inches below the infection) and burn them to reduce the inoculum.
- To reduce the humidity practice trimming and pruning.

Host Resistance

- Grow resistant/tolerant varieties such as Maharaji Chunth, Golden Chinese (apple cultivars), Yantarka Altaskya, and Dolgoe (crab apple cultivars).

Biological Control

- *Ampelomyces* (pycnidial fungi) are the most common natural antagonists of powdery mildews worldwide. On apple trees, *Ampelomyces* (mycoparasites) overwintered as resting hyphae in the dried powdery mildew mycelia covering the shoots and in the parasitized ascomata of *P. leucotricha* on the bark and the scales of the buds.
- Prophylactic and therapeutic activity of 3 high linoleic acid oils (corn, soybean, and grape seed) and 3 low-linoleic acid oils (sunflower, olive, and canola) has been found effective against *P. leucotricha*, providing 99% control of the disease.

Chemical Control

- Spray Aureofungin 46.15% w/v SP @ 0.5 ml/L of water.
- Use lime sulphur 22% SC @ 800–2000 ml/acre (use 2% pre-blossom and 1% post-blossom in conventional sprayers) or sulphur 80% WP @ 1000–2000 g in 300–400 litres of water/acre.
- Give four sprays of Bavistin @ 0.05% or Morocide @ 0.1% or Triadimefon @ 0.05% starting at the bud swell stage and repeat at 15 days interval.

12.3 FIRE BLIGHT

Fire blight is a devastating disease of apples and pears, causing enormous economic losses around the world. Fire blight may kill flowers and twigs and it may girdle large branches and trunks, thereby killing the trees. The disease "fire blight" was first coined by William Coxe in 1817, when he described a disorder that "in a few hours, turn the leaves suddenly brown, as if they had passed through a hot flame". Fire blight is of historical importance because this was the first plant disease shown to be caused by a bacterium. In 1985, J.C. Arthur, proved Koch' postulates, but the concept that a bacterium can be a plant pathogen was given by T.J. Burill in 1878 from the University of Illinois, USA. *Erwinia amylovora* is also the first bacterium to be associated with an insect vector.

Symptoms

The infected blossoms, leaves, twigs, and branches give a burnt or scorched appearance, hence the name fire blight. The first sign of disease consists of droplets of creamy bacterial ooze or fine, hair-like strands on the surface of infected tissues during wet conditions. The ooze turns dark after exposure to air, spreads to form a thin glistening layer, or dries down to form hard beads or films. There are at least five distinct kinds of infections associated with fire blight which are blossom, canker, shoot, trauma blight, collar, and rootstock blight.

Blossom blight: It is the most devastating phase of the disease, resulting from the direct infections of open flowers with intact petals and usually appears within 1 to 2 weeks after bloom. Initially, the infected blossoms appear water-soaked and then they shrivel, wilt, and eventually turn brown to black. The infected blossoms may fall, but often remain hanging on the trees. Infection of a single flower in the cluster usually results in the death of the whole spur. Droplets of amber-colored exudates having masses of bacterial cells appear on the pedicels. Early infection of inflorescence can cause complete loss to the crop. But if the infection sets in after the fruit set, brown circular lesions develop on unripe fruits which may drop prematurely. Blossoms and young fruits are later aborted, followed by necrosis and wilting of leaves and shoots on infected branches.

Fruit Blight: Symptoms on fruit appear as water-soaked, turn brown, shrivel, and finally turn black. In warm and humid weather, bacterial ooze comes out from the surface of infected fruits. The infected fruits usually remain clinging to the trees for several months. Usually immature and unripened fruits are infected.

Shoot or Twig Blight: it is also called "blight strikes". The earliest symptom of shoot blight is wilting of the growing tip, which causes the tip to curve downward and form a characteristic "shepherd's crook" (hook) at the end. The bark of infected twigs turns brownish-black and is soft in the beginning, but shrinks and hardens later on. The leaves on the infected twigs turn black and remain clinging to the twig.

Leaf Blight: The disease also attacks leaves and spurs. The necrotic, brown-to-black lesions appear along the midrib or main veins, between the veins, and along the margins. As the blackening spreads, the leaves curl and shrivel, hang downwards, and generally remain attached to the hooked and blighted twigs. In the worst case, the entire tree may die.

Trauma Blight: The symptoms are similar to shoot blight. The symptoms result from the damage caused by traumatic events like hail, frost, or storm. The wounds caused by these traumatic events allow bacteria to penetrate the host, bypassing the normal defenses even in the resistant hosts.

Canker Blight: It is also called by other names like "body blight", "limb", "trunk blight", depending on which part the infection occurs. Cankers are formed as a result of bacteria traveling systemically into the wood tissue of the tree. From the infected twigs, and fruit spurs, the infection moves down to the branches, scaffold limbs, and trunks, forming the cankers. In the beginning, the bark where canker is formed appears water-soaked and later on, becomes discolored, sunken, and dry.

Collar and Root Blight: When the cankers develop at the base of the trunk, it is called the collar blight. The formation of cankers at collar or ground level results in the early death of plants. The infection of the root by the bacterium results in root blight or rootstock blight. The primary route of entry of the bacterium into the rootstock is internally, often through apparently healthy limbs and trunks of trees that had only a few infected blossoms or shoots of the scion. After reaching the susceptible rootstock, the bacteria initiate the formation of new cankers that completely girdle and kill the tree in one to a few months.

Causal Organism and Etiology

Fire blight, caused by the bacterium *Erwinia amylovora*, is a most devastating disease of rosaceous plants, including apple, and pear. It is a facultative anaerobe, gram-negative, rod-shaped, peritrichous flagella of the *Enterobacteriaceae* family.

Disease Cycle and Epidemiology

The fire blight bacterium, *Erwinia amylovora,* overwinters in the cankers of infected plants and serves as a primary inoculum for infection in the next season. During early spring, bacteria multiply at canker margins, in adjacent bark tissues, and ooze out and transmitted by insects, honeybees, flies, birds, wind, and rain to flowers, new leaves, small wounds, and natural openings on succulent new growth. The sugary exudates having a large number of bacterial cells attract aphids, ants, bees, beetles, flies, bees, and ants, pick up the bacteria on their bodies and carry the bacteria from place to place from infected to healthy trees. Stigmas are the principal site of epiphytic colonization and growth by *E. amylovora*. The primary infection of the flowers initiates the disease-causing blossom blight. More flowers are infected by secondary infections, caused by the transmission of bacterial ooze by rain, and insects (honey bees). Bacteria gain entry into flower parts through nectaries (nectarthodes), lenticels, and stomata. The shoot blight starts with the infection of young and tender stems by bacteria through stomata or wounds. Infection of fruit occurs through lenticels, wounds, and infected flower spurs (pedicels). The secondary infections multiply the disease and finally result in infections of the tree trunk and rootstock. Bacteria that move systemically in the plant also infect the rootstock. Open wounds, caused by pruning, insect feeding and ovipositing, hail, frost, or severe wind also serve as entry points for the pathogen. From the infected blossoms, shoots, or fruits, the disease spreads systemically through the spurs to larger twigs, and branches to cause cankers. During the warm and humid weather, bacteria come out on the surface of infected blossoms in the form of ooze and are disseminated by pollinating insects. This phase is critical for the outbreak of the fire blight epiphytotic because the bacteria can be readily disseminated throughout the orchard by pollinating insects. Bacteria can be spread by wind and wind-driven rain within and between trees as ooze, strands (polysaccharide threads), and aerosols. This disease is most severe during flowering, when warm spring weather (21°C–27°C), coupled with rainfall and hail, provides optimum conditions for fire blight development. An extended period of warm (25°C–28°C) and wet weather increases the multiplication rate of bacterium and infection. For the prediction of infections event accurately, fire blight forecasting models, namely, MARYBLYT™ and CougarBlight can be effectively used.

Management

Cultural Methods

- Pruning and destruction of cankered limbs during the dormant phase (in winter) reduce the amount of primary inoculum.
- Nitrogen status in the soil should not be high as it makes the trees disease-prone.
- Remove fire-blight infected branches during summer.
- Remove all blighted twigs and cankered branches. Prune twigs and branches 8 to 12 inches below the edge of visible infection.
- During the post-bloom period, manage the insect populations that are responsible for spreading the bacteria to avoid the disease spread.

Biological Control

- Two applications of Blight-Ban A506 (a commercial product of *Pseudomonas fluorescens* strain A 506) can reduce 40%–60% blossom blight.

Host Resistance

- Grow moderately resistant and resistant varieties, if available, in areas where the disease is destructive. Some resistant varieties are Red Delicious, Northwest Greening, Coxs Orange Pippins, Stayman, Winesap, Britemac, Carroll, Primegold, Pricilla, Quinte, Splendour, and Viking.
- Avoid blight susceptible apple rootstocks, especially when grafted to susceptible scions.
- Use of transgenic apple plants: Transgenic apple lines, overexpressing the apple gene *MpNPR 1*, have shown increased resistance to fire blight pathogen.

Chemical Control

- Pruning of affected trees during off season, and applying 0.1% Bordeaux mixture can control the disease.
- Spray Streptomycin sulphate @ 100 ppm in 1% glycerine during the pre-blossoming and blossoming stages.
- Aluminum tris is a bactericide used prior to and during bloom.
- **Plant growth regulators**: Growth-regulating Acylcyclohexanedione compounds, such as Prohexadione-calcium (trade names Apogee® and Regalis®) and Trinexapacethyl (trade names Moddus® and Primo®) offer a new, promising, and safe method for induction of resistance in pome fruit trees against fire blight and other pathogens and also for maintaining vegetative and reproductive equilibrium.

12.4 CROWN GALL

Malformations of the apple are variously termed "aerial crown galls," "stem tumors," and "burr knots".

Symptoms

Crown gall is identified by overgrowths appearing as galls and most commonly occur on roots and at the base or "crown" of apple plants (Figure 12.2). Initially, the galls appear light-colored and spongy. As the galls mature, they turn dark brown and woody. Galls, varying in size, form on the crown, roots, trunk, or limbs. The texture of a gall can range from soft and spongy to hard, depending on the amount of vascular tissue it contains. Careful diagnosis of smaller galls is important, as they may be confused with excessive callus growth around wound sites or with nematode or insect-induced galls.

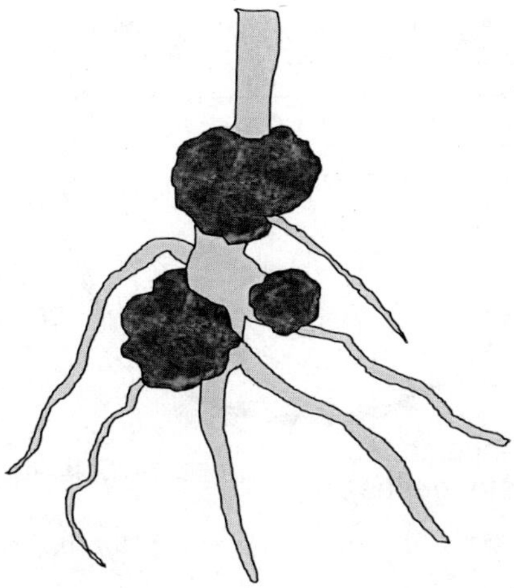

Figure 12.2 Crown Gall

Causal Organism and Etiology

Crown gall disease is caused by the bacterium *Rhizobium tumefaciens* (formerly *Agrobacterium tumefaciens*). It causes crown-gall disease in over 100 dicots, conifers, and some monocot plant species. It is a gram-negative, strictly aerobic, non-sporing rods measuring 0.6–1.0 × 1.5–3.0 μm, and motile by 1–6 peritrichous flagella. Only 1% of the isolates are pathogenic containing tumor-inducing plasmid, which is essential for pathogenesis and tumorization. *Agrobacterium* species are soilborne gram-negative bacteria exhibiting predominantly a saprophytic lifestyle. Only a few of these species are capable of parasitic growth on plants, causing either hairy root disease (*A. rhizogenes*) or crown gall disease (*A. tumefaciens*, *A. rubi*, and *A. vitis*). Both diseases are characterized by neoplastic growth at the site of infection, resulting from the oncogenic transformation of the plant cell. A large extrachromosomal piece of DNA, commonly referred to as a tumor-inducing (Ti) plasmid, is carried by *R. tumefaciens*. It is an excellent genetic engineering tool. *Agrobacterium* species that are pathogenic on plants (*A. tumefaciens*, *A. vitis*,

A. rubi, and *A. rhizogenes*) all carry megaplasmids. The DNA transformation capabilities of the bacterium have been extensively exploited in biotechnology as a means of inserting foreign genes into plants. Nester (2008) has rightly called it a "natural genetic engineer" and also a household name in plant genetic engineering and plant molecular genetics. *R. tumefaciens* cell contains a plasmid known as the Ti (tumor-inducing) plasmid (200–800 kbp). All Ti plasmids code for functions associated with (a) plasmid replication and maintenance, (b) conjugative transfer, (c) virulence, (d) opine utilization, and (e) sensory perception of exogenous signals released by the plant host and neighboring agrobacterial cells at the site of infection. Most of the Ti plasmids (Figure 12.3) have the following regions:

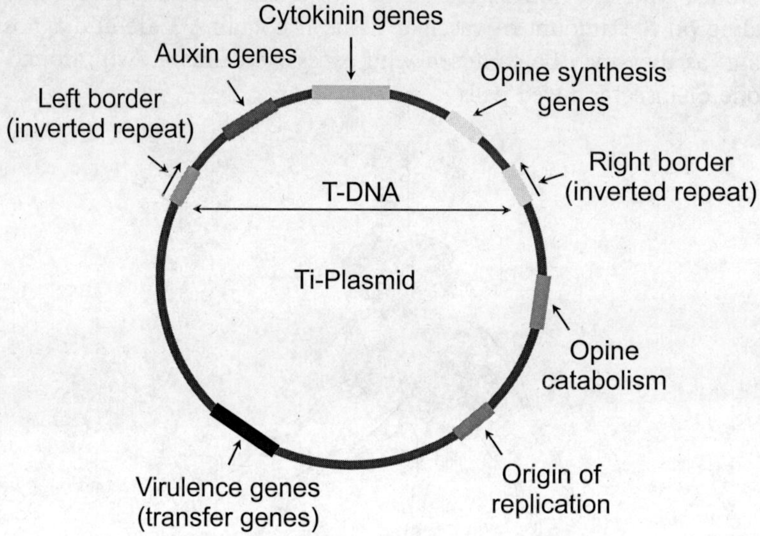

Figure 12.3 Detail Structure of Ti-Plasmid of *Agrobacterium tumefaciens.*

Regions in Ti plasmids	Functions
(1) T-DNA region	Responsible for tumor induction. T-regions generally represent less than 10% of the Ti plasmid. Some Ti plasmids contain one T-region, whereas others contain multiple T-regions. T-regions are defined by T-DNA border sequences.
(2) An opine synthesis (OS) region	Responsible for the synthesis of unusual amino acids or sugar derivatives, which are collectively called opines.
(3) T-DNA border sequences	On either side of the T-DNA, there is a short sequence of 25 bp called the border sequence. Both the left and right border sequences are essential for tumor induction and these are highly homologous in sequence. They flank the T-region in a directly repeated orientation.
(4) Origin of replication	Responsible for replication.
(5) *Tra* and *Trb* region	Responsible for conjugative plasmid transfer.
(6) *Inc* region	Responsible for incompatibility among plasmids in a bacterium.
(7) *Vir* region	Virulence region and plays a crucial role in the transfer of T-DNA into the plant nuclear genome.

Disease Cycle and Epidemiology

Agrobacterium tumefaciens is basically a soil inhabitant, growing as a saprobe in the rhizosphere, and remaining viable in soil for several years. It is a gram-negative soil bacterium with the ability to infect plants by delivering a specific segment of its genome to the nuclei of susceptible plant cells. It can also survive in infected plant debris, on and around root surfaces utilizing nutrients leaked, and in the outer portion of primary galls. The dissemination of the bacterium occurs through wind-splashed rain and chewing insects. Abundant bacterial cells present on the surface of galls are washed down by rainwater and carried away by the flowing water. Long-distance spread of the pathogen occurs through the infected nursery and planting material, bacterial cells present on the surface or inside the host tissue. The bacterium naturally infects plant wound sites and causes crown gall disease via delivery of transferred (T)-DNA from bacterial cells into host plant cells through a bacterial type IV secretion system (T4SS). The bacterium penetrates the host through wounds, fresh (less than 24 hours old) and deep wounds are more conducive for infection. The wounding of the plants results from pruning, cultural operations, growth cracks, grafting, and chewing insects. The bacteria stimulate the tree to produce plant hormones that cause a tumor or gall to form. It does not invade the host cells; rather it releases the T-DNA of its Ti-plasmid, which integrates with the host genome (Figure 12.4). The transferred DNA (T-DNA) is delimited by 23 base-pair repeats on the tumor-inducing (Ti) plasmid. The T-DNA is important for infection because it codes for genes that,

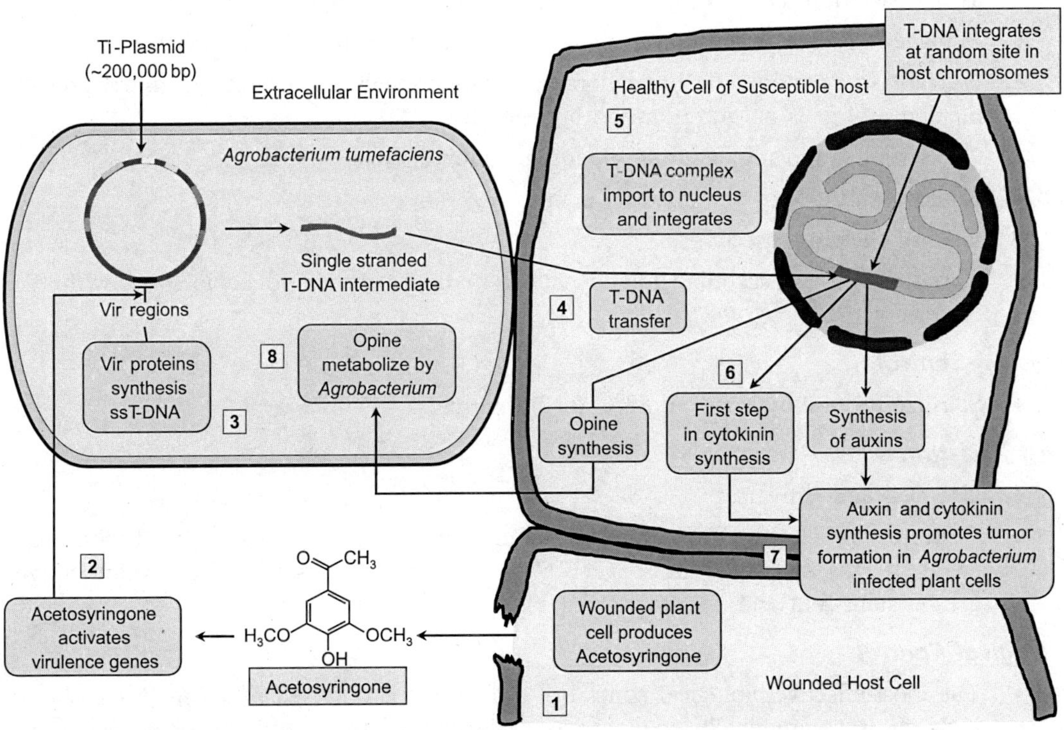

Figure 12.4 Mechanism of Agrobacterium-Mediated T-DNA Transfer into the Susceptible Host Cell. Adapted from Hammond-Kosack, K.E. and Jones, J.D.G. 2015. Responses to plant pathogens. pp 984-1050. Biochemistry & Molecular Biology, of plants, John Wiley & Sons, Ltd.

when expressed in the plant cell, disrupt plant cell growth and division events. The wounding of host cells results in the release of cell sap containing phenolic and sugar compounds. The bacterial cells are attracted by these chemicals such as Acetosyringone, swim through soil water with the help of flagella, and attach to the wounded plant cells. The subsequent binding of the bacteria to specific plant cell wall components initiates a seven-step process involving: (i) wounded plant cell produces acetosyringone, (ii) acetosyringone activates virulence genes, (iii) Vir proteins synthesize single-stranded T-DNA (iv) T-DNA transfer (v) T-DNA complex import to the nucleus and stably integrates into the plant genome and transform its target cell (vi) synthesis of opines, cytokinin and auxin, and (vii) auxin and cytokinin synthesis promotes tumor formation in *Agrobacterium* infected plant cells. Synthesized opines are metabolized by *Agrobacterium*. Once the T-DNA integrates into the plant chromosome, the altered plant cells start dividing rapidly and uncontrollably, and the infected tissue develops a tumor-like swelling. The tumor formation occurs between 10°C and 33°C. A single bacterial cell can cause infection under favorable conditions.

Management

Cultural Methods

- Use certified and disease-free planting materials.
- Avoid wounding the crowns and roots during transplanting, cultivating, and mowing as the bacterium enters only through relatively fresh wounds.
- Manage root-chewing insects in the nursery to reduce crown gall incidence.
- Nursery stock should be budded rather than grafted because of the much greater incidence of galls on graft than on bud unions.
- Use sterilized tools for grafting, budding, pruning, etc.
- Remove and destroy the diseased plants.
- Control chewing insects.
- Avoid planting susceptible nursery stock in fields known to be infested with the bacterium.

Physical Control

- Sterilize the soil by steam at 82°C for 30 minutes.

Host Resistance

- Use resistant cultivars.
- **Genetic engineering**: Biotechnology companies, such as DNA Plant Technology are applying sense strand messenger RNA or small-interfering RNAs to develop crown gall-resistant fruit and nut crops.

Biological Control

- The most effective biological control of crown gall is obtained by using *R. radiobactor* K-84. It is commercially available as Galltrol-A, Gallex, and the latest and most effective NoGall. Soaking germinated seeds or dipping nursery seedlings or rootstocks, spraying nursery seedlings, treating non-germinated seeds, or by drenching the soil with a suspension of the above formulations provides 100% control. *R. radiobactor* produces the bacteriocin agrocin 84 which is inhibitory to most virulent *R. tumefaciens* strains. Unfortunately, some strains of *A. tumefaciens* inherited from strain 84 its resistance to agrocin 84. Therefore, a new strain (K-1026) is now used because it lacks the ability to transfer its resistance gene to pathogenic *Agrobacterium* strains.

Chemical Control

- Creosote-based compounds, copper-based solutions, and strong oxidants such as Sodium hypochlorite are transiently effective in eradicating crown gall.
- Most of the bacterial plant pathogens cause necrotic or hypoplastic effect on the host tissue. However, crown gall bacterium (*Agrobacterium tumefaciens*) has an opposite effect, i.e., it causes hyperplasia and hypertrophy of the affected tissue.

13 Peach (*Prunus persica* L.)

13.1 LEAF CURL

Peach leaf curl is an important plant disease that causes significant losses in fruit production. It is found worldwide and is a serious disease of peach. Leaf curl is a springtime disease of peach, nectarine, almond, and related ornamental species. It is caused by fungus and aphid.

Symptoms

The peach leaf curl pathogen can infect leaves, flowers, fruit, stem, young green twigs, and shoots (Figure 13.1 A–E).

Leaves: Infected leaves become severely distorted, puckered, swollen, and curl downwards, initially with a distinct reddish or purple coloration. Red plant pigments accumulate in the distorted cells. As the disease progress, thickened areas turn yellowish and then grayish-white, as velvety fungal spores are produced on the surface. Later, affected leaves turn yellow or brown and can remain on the tree or may fall off. Defoliation can result in sunburn injury on exposed branches.

Stem: Abnormal reddish swelling (hypertrophy) develops on peach stems.

Twigs and Shoots: Infected twigs and shoots become swollen, stunted, distorted, exude gum, usually with deformed leaves at their tips, and often die. Leading branches may develop a 'witches broom' appearance due to lateral branching.

Fruit: Irregular, raised, green (initially), or reddish blotches, wrinkled to distorted, or hypertrophied, often with gummosis and surface cracking areas develops on fruit surfaces. Infected fruit becomes deformed and drop prematurely.

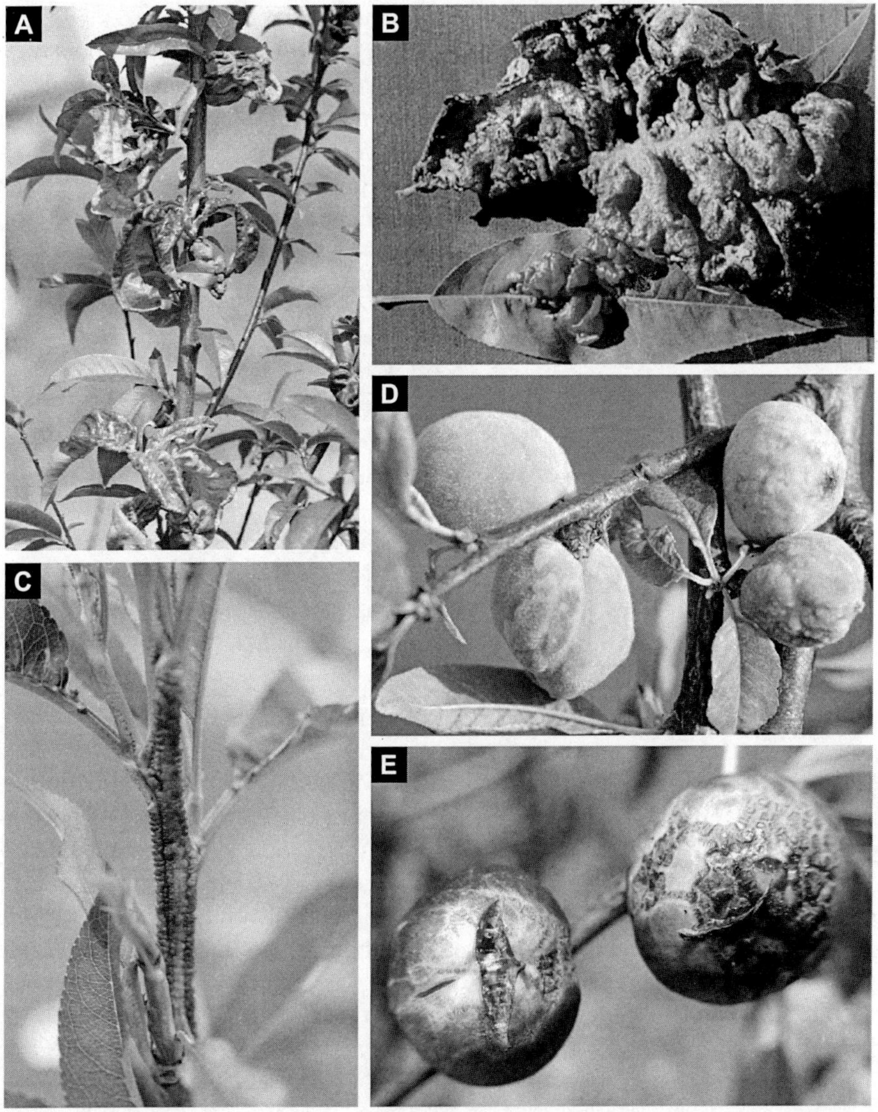

Figure 13.1 Leaf Curl of Peach. (A–B) Typical leaf symptoms: curling, puckering, deformation and change in color (yellow to reddish-brown), (C) Reddish swelling (hypertrophy) on peach stem, (D) Diseased fruit with shiny, raised, warty, discolored areas, and (E) Gummosis and cracking on fruit surface of nectarine.

Diagnosis Peach Leaf Curl Caused by Fungus and Aphid

Aphids are highly polymorphic, small (1.5–3.5 mm), soft-bodied, pear-shaped, sap-sucking insects in the family Aphididae that often occur in clustered colonies and feed exclusively on sap from the vascular tissues of plants, mainly the phloem. Several aphid species attack peach such as peach leaf curl aphid, *Brachycaudus helichrysi*, peach black aphid, *Pterochlorus persicae*, and green peach aphid, *Myzus persicae*. Green peach aphid also transmits *Plum pox virus*. Peach leaf curl aphid, *B. helichrysi* is one of the most important pests of peach trees and is reported from all around the world. The pest is highly polyphagous in India infesting about 175 species belonging to 115 genera and 49 plant families. Besides peach, it also infests nectarines, almonds, apricots, prune, and plum. Damage is severe because the aphid colonizes young shoots, buds, grafts and young plants. Both nymphs and adults are very destructive. They suck cell sap from the leaves, petioles, blossoms, and fruits leading to curling and withering of leaves (Figure 13.2) and blossom, reduced fruit set, and slow development of fruit due to lack of nutrition.

Figure 13.2 Peach Leaf Curl Caused by Aphid. In aphid induced leaf curls presence of aphids, sticky honeydew and black sooty mold can be noticed.

The infested leaves turn pale or yellow. Premature fruit and leaf drop also occurs. Sticky honeydew secretion by aphids leads to the development of black sooty mold (fungus) on leaves and twigs that reduce the surface area for photosynthesis. Black sooty mold also develops on fruits. From March to May is the peak activity period of peach leaf curl aphid. Clustered colonies of aphid can be noticed on infested plant parts especially on lower surface of the leaves. The difference between peach leaf curl induced by fungus and aphid is given in Table 13.1.

Table 13.1 Difference between peach leaf curl caused by fungus and aphid

Features	Fungus	Aphid
Causal organism	*Taphrina deformans* (Ascomycetes)	Peach leaf curl aphid, *Brachycaudus helichrysi* (Aphididae: Hemiptera)
Attack	Leaves, flowers, fruit, stem, young green twigs, and shoots.	Leaves, petioles, blossoms and fruits.
Symptoms	Leaf curling, puckering, deformation, swelling; later the leaf give a grayish white, powdery or velvetlike appearance due sporulation in infected tissues.	Leaf curling and withering; sticky leaves due to honeydew excreted by the aphids. Swelling is absent. Wilting of terminal shoots. Black sooty mold grows on the honeydew staining the leaves and fruit. Leaves, blossoms and young fruits drop pre maturely. Reduced tree vigour.

(*Contd.*)

Features	Fungus	Aphid
Color development	Disease characterized by coloration; generally yellow-reddish brown color develop on leaves, fruits and stem.	Not prominent; occasionally infested leaf turn pale or yellow.
Size of leaves	Generally increases (abnormal enlargement due to hypertrophy).	Generally reduced
Gummosis	Present in infected plant parts.	Absent
Fruit symptoms	Cracking, swelling and gummosis.	Reduced fruit set and slow development of fruit.
Management	Use of fungicide such as lime sulphur, Fytolon or Copper oxychloride, Ferbam, etc.	Use of insecticide such as Imidacloprid, Thiamethoxam, Dimethoate, etc.

Causal Organism and Etiology

The common leaf curl of peach is caused by *Taphrina deformans*. It belongs to the sub-division Ascomycotina. The mycelium is multicellular, and the cells are typically binucleate. The hyphae are intercellular or intracellular. Asexual reproduction takes place by small, oval, or spherical, uninucleate, yeast-like conidia called blastospores, which develop by budding of ascospores while they are still within the ascus. The pathogen has a yeast-like phase on host surfaces and a parasitic phase (mycelial or filamentous) inside the vegetative growth and fruit. Asci are produced in a palisade on or from within the epidermis of the infected host tissue; there are no specialized fruiting bodies. The asci are exposed on the infected leaf surface; splits open at its tip, and expel a cloud of infectious spores. The blastospores either may bud and produce more blastospores or may germinate by producing a germ tube which develops into mycelium and then infects the host. No specific sex organs develop in *Taphrina*.

Disease Cycle and Epidemiology

The pathogen, *Taphrina deformans* survives as an asexual, yeast-like bud-conidia stage (blastospores) in crevices of the bark, twigs, under the bud scales, and old infected leaves. Infection occurs through bud scales in mid-to-late winter just as buds begin to swell and blastospores are formed by budding of the ascospores. Balstospores germinate and infect the young leaves. The ascospores are formed on freshly infected tissues and are the chief source of secondary infection. Intercellular growth of the pathogen causes cell division and cell enlargement (hyperplasia and hypertrophy) that result in thickened, curl to convoluted, or blistered leaves, shoots or fruits. The suitable temperature for infection is 10°C–21°C. Pathogen produces ascospores on naked asci. Rain enhances the infection, disease development, and spreads. The disease can be very severe in wet humid regions where bud burst can be delayed for 2–7 weeks. In temperate regions, young leaves and flowers are infected each spring; asci and ascospores are produced in late spring or early summer. The spores bud during warm, moist weather, giving rise to a yeast phase that can survive saprobically on plant surfaces. Infections occur at temperatures of 10°C–21°C, but ascospores and bud-conidia can survive hot, dry conditions for several months. Periods of cool, wet weather during early bud development favor leaf curl disease. When temperatures at early leaf development are high, infections rarely become established. Emerging leaves are infected when the fungus changes to a mycelial parasitic phase. Naked asci are then produced on the leaf surface and the sexual ascospores are forcibly discharged. These spores germinate and form more bud-conidia that contaminate twig tissues.

Management

Cultural Methods

- Remove the infected and curled leaves from the canopy.
- Reduce drought stress.
- Maintain vigor of infected trees by irrigation and N-fertilization management.
- Reduce stress from crop load by extra thinning.

Chemical Control for Fungus

- Spray lime Sulphur @ 0.8–2 L/acre.
- One spray of lime Sulphur before bud break and a second spray in the fall are effective to eliminate the disease.
- Spray Fytolon (0.3%) or Copper oxychloride (0.25%) or Chlorothalonil also gives protection from the disease.

Note: Apply first leaf curl spray when 50% of the leaves have fallen and again at delayed dormant in late February before floral buds begin to open.

Management of Aphids

- Install yellow sticky traps (@ 20–25 per hectare) for trapping winged aphids.
- Spray Imidacloprid 17.8 SL @ 150–175 ml or Thiamethoxam 25 WG @ 200 g or Dimethoate 30 EC @ 660 ml/ha or Acetamiprid 20 SP @ 75 g/ha in 500–1000 litres of water.

14 Strawberry (*Fragaria* × *ananassa* Duch.)

14.1 LEAF SPOTS

14.1.1 Common Leaf Spot or Mycosphaerella Leaf Spot

It is one of the most common diseases of strawberry that affects both wild and cultivated strawberries throughout the world. It is also known as "Mycosphaerella leaf spot", "bird's-eye spot", "strawberry leaf spot", "Ramularia leaf spot", "gray spotness", and "white spot".

Symptoms

Mycosphaerella leaf spot can infect leaves, petioles, runners, fruit stalks (pedicels), and berry caps or calyxes.

Leaves: The initial symptoms appear as small, circular, reddish to deep purple spots on the upper surface of leaves (Figure 14.1). A definite reddish purple to rusty brown border surrounds the spots. Later, these spots enlarge and the centers turn grayish to white on older leaves and light brown on young leaves. Small, black, fruiting bodies (perithecia) of the fungus may appear in the center of older spots. When numerous, spots may cause a rapid scorching of the leaves and a gradual defoliation of the plant. Leaf and fruit stalks, stolons, and calyxs may show similar spots. Lesions also develop on stems, petioles, and runners.

Fruits: Firm, dark brown, sunken spots are produced on fruit. The spots form on ripe berries around groups of seeds. The spots are usually only one or two spots per fruit. However, some fruits may be more severely infected.

Symptoms of Mycosphaerella leaf spot can be confused with early stages of leaf blight (*Phomopsis obscurans*), leaf scorch (*Diplocarpon earliana*) and leaf blotch (*Gnomonia comari*) although the lesions of these diseases do not enlarge aggressively.

Figure 14.1 Common Leaf Spot of Strawberry. Source: Clemson University-USDA Cooperative Extension Slide Series, Bugwood.org.

Causal Organism and Etiology

Leaf spot of strawberry is caused by the fungus, *Mycosphaerella fragariae* (teleomorph). The pathogen belongs to Kingdom–Fungi, Phylum–Ascomycota, Class–Dothideomycetes, Order–Capnodiales, Family–Mycosphaerellaceae. It reproduces by producing perithecia that are black, partially embedded, globose, and minutely ostiolate. Within each perithecium, asci that are cylindrical to clavate, have short stalks, and contain 8 spores, are formed in small clusters. Ascospores are hyaline and two-celled with a median septum with each cell generally containing two oil drops. The anamorphic state (*Ramularia brunnea*) of the fungus produces elliptical to cylindrical hyaline, and 0–4 septate conidia.

Disease Cycle and Epidemiology

The fungus, *Mycosphaerella fragariae* overwinters as microsclerotia and spores in the lesions of leaves, diseased ratoon crops, or undecomposed infected strawberry residues in the soil. The fungus may also be introduced in the field through infected runners. The fungus infects the plants and reproduces in spots on the upper and lower leaf surface. The symptoms can appear on the upper leaf surface after 10–14 days depending on weather conditions. Spores produced in fruiting bodies are spread by splashing rain, and wind during wet weather. Eye spot is usually more severe in low-lying areas or where plants are shaded. Cool conditions favor the disease. Temperatures between 18°C–24°C, with periods of high rainfall and humidity, are optimal for the growth of fungus and lesion development. If frequent rains occur a few infection sites can start an epidemic. Large numbers of conidia can cause secondary infections during prolonged periods of damp to wet, moderately warm weather. Temperatures between 18°C–24°C are optimal for the growth of fungus and for lesion development. Infections can continue to occur throughout the growing season.

Management

- Select healthy runners for planting.
- Remove diseased leaves from less-affected plants before planting.

- Do not plant strawberries in low-lying or shady areas.
- Use drip irrigation.
- Remove infected leaves.

Host Resistance

- Grow resistant cultivars of strawberries.

Chemical Control

- Spray Copper oxychloride @ 2 g/L or Carbendazim + Mancozeb @ 1 g/L of water.

14.1.2 Black Spot or Anthracnose

Black spot is considered a major strawberry disease that can cause serious crop losses.

Symptoms

The symptoms are most commonly produced on runners, stalks, crowns, flowers, and fruits. On mature fruits, circular, firm, whitish-tan, and sunken lesions develop that later turn into black spots. In the center of the spots, masses of slimy, cream- to salmon-colored fungal spores develop. A white fungal growth can also develop on the fruit. Lesions turn dark brown and enlarge until the entire fruit is rotted. Eventually, infected fruits become dried, shriveled, and mummified. Within the crown tissue, red-brown discoloration and necrosis of the fruits or even the wilting of infected plants may occur. Under favorable conditions, extensive infection occurs that causes wilting and death of plants.

Causal Organism and Etiology

Black spot or anthracnose of strawberry is caused by *Colletotrichum* spp. (*Colletotrichum fragariae*, *C. gloeosporioides,* and *Colletotrichum acutatum*). The fungus *Colletotrichum acutatum* is most commonly associated with anthracnose fruit rot.

Disease Cycle and Epidemiology

The fungi can persist in infected plants, plant debris, and mummified fruit as dormant spores or other fungal structures. The disease is spread from infected plants and fruit by rain splash, overhead irrigation, and on the hands of fruit pickers. Rain splash or movement of equipment spread the spores to a short distance in the field. Long-distance spread can occur with the movement of strawberry transplants from the nursery to the grower. To colonize the host, these fungi develop many specialized infection structures, including germ tubes, appressoria, intracellular hyphae, and secondary necrotrophic hyphae. The infection begins when the conidia adhere to plant surfaces, produce germ tubes, and then continue forming appressoria, which penetrate the cuticle directly. The pathogen grows beneath the cuticle by forming a subcuticular intramural network of hyphae before spreading throughout the tissue. In the case of *C. gloeosporioides*, the process continues with the formation of secondary necrotrophic hyphae, which kill plant cells by using cellular debris as nutrients. The disease develops and spreads very rapidly during warm, rainy, and humid weather.

Management

Cultural Methods

- Plant only disease-free transplants.
- Wash all soil from plant crowns prior to planting.
- Select a planting site with full sun, good air circulation, and soil drainage.
- Avoid shady locations.
- Plant new beds periodically in different sites.
- Reduce the overall density of the strawberry beds.
- Avoid the use of excess nitrogen fertilizer.
- Soil solarization.
- Crop rotation with non-host crops.
- Remove weed regularly.

Host Resistance

- Grow resistant/tolerant cultivars to reduce disease pressure.

Chemical Control

- Spray Copper oxychloride @ 2 g/L or Carbendazim + Mancozeb @ 1 g/L of water.
- Fumigation of soil.

14.1.3 Angular Leaf Spot

Angular leaf spot or bacterial angular leaf spot is the most economically important bacterial disease of strawberry. It was first reported from Minnesota, USA in 1960.

Symptoms

The symptoms appear on almost all plant tissues, including leaves, stems, crowns, sepals, and calyxes. The symptoms first appear as minute, water-soaked lesions (1 to 4 mm in diameter) on the lower surfaces of leaves which enlarge to form dark green or translucent angular spots. Lesions may coalesce to form reddish spots with a chlorotic halo on the upper side of leaves that eventually become larger necrotic spots before the plants suffer from vascular collapse. Under high humidity, bacterial exudates appear on infected tissues. Dead tissues drop out of the leaf, leaving a raggy appearance. Flowers, but not fruit, are affected. Calyx infections occasionally extend into the pedicel, causing blight of the associated flower tissues.

Causal Organism and Etiology

Angular leaf spot of strawberry is caused *Xanthomonas fragariae*. It is an aerobic, gram-negative, non-spore former, non-capsulated, rod-shaped bacterium, measuring 0.4 × 1.3 µm, with a single polar flagellum. *X. fragariae* is considered quarantine organism by the European and Mediterranean Plant Protection Organization (OEPP/EPPO 1986).

Disease Cycle and Epidemiology

The bacterium, *Xanthomonas fragariae*, is not free-living in soil but is able to survive on infected plant debris, in between the growing seasons. The bacterium is resistant to desiccation and can survive in dry leaf debris or buried leaves in soil for long periods. Crowns of systemically infected living plants are another source for overwintering inoculum. The bacterium is generally introduced into a planting on symptomless systemically infected nursery plants. Commercial strawberry runners used for planting may spread the bacterium over short and long distances. The young leaves are infected by the bacterial cells present in the infected crop residue at the beginning of the growing season. Succulent leaves and calyx tissue are most susceptible to infection. The disease begins with *X. fragariae* invading the plant through natural openings, such as stomata, hydathodes, or wounds. During the growing season, multiple cycles of secondary infections may occur. The bacterial cells exuded from primary lesions on the lower surface of leaves serve as the source of secondary spread in plantings. The bacterial cells are spread by rain, wind-splashed rain, and overhead/sprinkler irrigation. Systemic infections may arise under damp nursery conditions. The bacterium may cause systemic infection of crown pockets during epiphytotic when environmental conditions are favorable for exudation and spread. The favorable conditions for infection are high relative humidity, cool to moderate daytime temperature (18°C–21°C), and cold night temperature near freezing. Long periods of precipitation, sprinkler irrigation, and heavy dew in the spring also favor disease development.

Management

Cultural Methods

- Use only certified and disease-free planting material.
- Remove and destroy infected plant parts.
- Follow crop rotation.
- Avoid overhead-irrigation.
- Avoid moving equipment or harvesting when plants are wet to minimize the spread of angular leaf spots within infected fields.
- Avoid wounding plants.

Host Resistance

- Grow resistant cultivars of strawberry.

Chemical Control

- Spray Copper oxychloride @ 2 g/L.

14.2 BOTRYTIS GREY MOLD

The Botrytis fruit rot (BFR) or Botrytis grey mold (BGM) is one of the most important, common, and destructive diseases of strawberry worldwide. This pathogen affects fruit in the field, storage, transport, and market. The disease affects all aerial parts of the plant but flower blight and fruit rot phases of the disease are the most damaging. Infected fruit may rot in the field prior to harvest or in transit or storage after harvest. Under favorable conditions yield losses may exceed 50%. Under wet conditions, more than 80% of strawberry flowers and fruits can be lost if plants are not sprayed with fungicides.

Symptoms

The disease attacks shoots, leaves, flowers, and fruit, and is more destructive on mature or senescent tissue. The pathogen initially infects the leaf or flower tissues damaged by insects, frost or other causes. Commonly blossoms turn brown and die. Initial symptoms on fruits appear as pale to light brown, water-soaked areas. Eventually, the sign and symptoms of the disease appear as a fuzzy, greyish-colored, and soft growth on strawberry fruits. Mushy, grey-colored, soft spots may also be evident on flowers, stems, and leaves. Under high humid conditions, the spots are covered with a grey mycelial growth, and sporulation. Completely decayed fruits covered with a grey, dusty powder (spores of the fungus), become firm, turn darker brown, shrivel, dry, mummified and later black sclerotia develop on these rotted berries (Figure 14.2). Berries touching soil or infected plant parts are most commonly infected. Sometimes, the mycelial growths with sporulation on infected tissues are confused with *Rhizopus* sp., which has a black growth. Berries with Rhizopus rot also often show "leak", whereas Botrytis causes dry rot.

Figure 14.2 Grey Mold of Strawberry.

Causal Organism and Epidemiology

Grey mold of strawberry is caused by the ascomycete fungus *Botrytis cinerea* (teleomorph: *Botryotinia fuckeliana*). It is a necrotroph, i.e. it prefers to infect and grow damaged or senescing tissues, eventually causing tissue death. *B. cinerea* usually produces greyish mycelia and conidia as well as black-colored sclerotia due to the accumulation of melanin. It produces abundant hyaline to slightly colored, smooth, unicellular, ovoid to ellipsoid, multinucleate (3–6 nuclei) conidia (macroconidia), measuring 10–12 × 8–10 µm. Conidia are produced on short sterigmata on the swollen tips of aerial, free, branched conidiophores. Microconidia (male spermatia) rarely occur in nature; they are small, globose, unicellular, and uninucleate that scarcely germinates and never infects the plants. Apothecia of *B. cinerea* are rare in the field but are more common in other *Botrytis* spp. Chlamydospores (survival structure) has been described in *B. cinerea*. The fungus also produces black, melanized, elongated, or spherical sclerotia measuring 3–5 mm in length. The melanized rind and *β*-glucans encasing the internal mycelium protect sclerotia from desiccation, UV radiation and microbial attack over long periods.

Host Range: *B. cinerea* has no apparent host specificity and can infect more than 1000 plant species and causes disease in many flower, fruit, and leafy vegetable crops.

Disease Cycle and Epidemiology

Botrytis cinerea, with a wide host range, is an opportunistic pathogen that readily colonizes dead or dying tissues where it sporulates profusely. It can survive in soil, living and dead host tissues as mycelium, or conidia, or sclerotia (dark-colored resting bodies). Sclerotia commonly germinate by producing conidia, occasionally by producing hyphae, and rarely by producing apothecia and ascospores. However, during the summer in high temperatures, the pathogen is unable to survive and commonly introduced into the fields, or new areas through infected transplants. Foliage from nursery transplants is a known source of inoculum of *Botrytis cinerea*. Old infected leaves gradually die after transplanting becoming a source of inoculum (mycelium or conidia) responsible for the primary infection of emerging leaves and flowers. The grey mold fungus can sporulate profusely, on dead and decaying plant material and in infected neighboring plants. *B. cinerea* mostly initiates infection through airborne spores (conidia) landing on healthy plants throughout the growing season. *B. cinerea* mainly enters the host through natural openings or wounds. They germinate and form appressoria for host tissue penetration. In primary infections, *B. cinerea* infects flower organs of strawberries during, or after flowering, and or in secondary infections by penetrating fruit receptacle tissues. The stigmatic fluid in flowers of the 'wet stigma' type serves as a nutrient medium for airborne conidia. Primary infections of fruits can be facilitated by infected senescent petals, stamens, and calyxes. Following infection of the unripe receptacle by *B. cinerea*, fungal growth is usually arrested and a symptomless quiescent phase occurs. As grey mold develops on infected berries, these become sources of secondary inoculum or infections of adjacent berries. *B. cinerea* is a polycyclic pathogen. The fungus continuously produces abundant conidia which is ubiquitous and available throughout the season. The conidia are dispersed through the air, water splash, agricultural operations, and direct contact with infected leaves, flowers, stems, and fruits. Irrespective of the type of infection, the pathogen generally enters a short latent or asymptomatic (quiescence), biotrophic phase in the beginning of the disease cycle. An aggressive necrotrophic phase commonly succeeds the quiescent or asymptomatic phase once plant organs start to senesce or ripen (fruits), during which *B. cinerea* causes rapid decay of the infected tissues. *B. cinerea* produces cell-wall-degrading enzymes, toxins, and oxalic acid. Strawberry plants are susceptible to *B. cinerea* during bloom and as the fruits ripen. *B. cinerea* growth is facilitated by cool, humid and wet weather temperatures. Heavy rain or overhead irrigation also favours the growth of the fungus. A combination of weather conditions such as frequent rains, high relative humidity, and moderate temperatures (15°C–25°C), and a continuous leaf or surface wetness for at least 13 hours during flowering are conducive to the development of this disease. Infection rate of *B. cinerea* approaches 90% when strawberry fruit or flowers are wet for 24 hours or longer.

Management

Cultural Methods

- Select healthy plants from the nursery.
- Use varieties that are suited to the growing area with necked fruit or reflexed calyx.

- Select the sites for planting that have good air circulation.
- Avoid planting in shade areas and where plants are subject to frost injury.
- Remove diseased berries and dead leaves from the crops and burn them.
- The removal of infected leaves, fruits, and infected plant parts, on a regular basis especially during the fruit development stage reduces the inoculum.
- Remove blossoms in the planting year.
- Grow plants on polythene mulch that prevents fruit from touching the soil.
- Avoid dense planting to improve air circulation between the plants so that the foliage and developing fruit can dry more rapidly.
- Work with strawberry plants in the field or polyhouse when they are dry.
- Avoid overhead irrigations, overwatering, and waterlogging.
- Irrigate the plants in the early morning hours when temperatures are increasing and foliage has time to dry before nightfall.
- Disinfest the tools, especially pruning tools.
- Use drip irrigation to reduce free water on the plants.
- Do not overfertilize the plants. Avoid excessive use of nitrogenous fertilizer. High nitrogen levels promote excessive leaf growth and available surfaces for infection.
- Shading of thick foliage reduces the rapid drying of berries.
- Harvest berries regularly and place them in a cool room as soon as possible.
- Growing strawberries in plastic tunnels have proven to effectively reduce the incidence of *Botrytis* infections.
- Often, after harvesting, strawberries (fruits) are stored in modified atmospheres i.e. generally low oxygen and high carbon dioxide to slow down metabolic processes, senescence, and fungal decay. During storage, usually high relative humidity (about 85%–90%) is kept to prevent fruit dehydration, but it limits the fungal growth.
- To prevent *B. cinerea* infections during storage novel postharvest treatments of strawberries have been suggested such as edible fruit coatings of silk fibroin or methylcellulose, chitosan that prevent dehydration and can act as an antifungal compound.

Physical Control

- Heat treatment of transplants prior to planting may be useful in some situations. Preheat strawberry plants at 37°C for an hour then move to 44°C for 4 hours.
- To reduce or inhibit the reactivation of *B. cinerea* quiescent infections is rapid and constant cooling of strawberry fruits at temperatures below 2.5°C is another critical strategy to avoid the post-harvest losses.
- Rapid and constant cooling of strawberry fruits at temperatures below 2.5°C reduces or inhibits reactivation of *B. cinerea* quiescent infections and the post-harvest losses.

Host Resistance

- Grow resistant cultivars.

Biological Control

- Most bio-fungicides for edibles are preventive (applied prior to the infection). Some bio-fungicides such as *Bacillus subtilis* QST 713 (Cease, Serenade), *Streptomyces griseoviridis* (Mycostop), and *Streptomyces lydicus* WYEC 108 (Actinovate AG) can be used against the grey mold of strawberry.

Chemical Methods

- No fungicide is recommended when conditions are not suitable for infection or disease development.
- Spray fungicide, if necessary. Use the recommended protectant fungicide such as Copper oxychloride, Chlorothalonil, Captan, Mancozeb, Iprodione, Thiram, Sulfur, Ziram, and Potassium bicarbonate @ 2 g/L from the onset of flowering until harvest. Apply a fungicide at 5%–10% bloom and at full bloom. During wet seasons, fungicide spray may be repeated every 7–10 days.

15

Potato (*Solanum tuberosum* L.)

15.1 EARLY BLIGHT

Early blight is one of the most common diseases of potato wherever it is grown in the world. In India, it is more common than late blight and may cause up to 40% loss in tuber yield. However, in tomato, the losses may reach up to 78%. J.B. Ellis and G.B. Martin in 1882 first described the causal organism of early blight as *Macrosporium solani*. The first reference to the fungus as a parasite and its association with potato leaf blight was by B.T. Galloway in 1891 in Australia. L.R. Jones was the first to suggest the name 'early' blight to distinguish the disease from 'late' blight in 1893. The name originated from the fact that early blight attacks early-maturing cultivars more severely than medium or late-maturing ones, whereas late blight is more severe on medium or late-maturing cultivars.

Symptoms

Symptoms of early blight occur on the fruit, stem, and foliage of tomatoes and foliage and tubers of potatoes. The first symptoms of early blight appear as small, circular, or irregular, dark-brown to black spots on the lower leaves (Figure 15.1). These spots enlarge and gradually may become angular-shaped. The spots subsequently get covered with grayish fungal growth. The tissues turn necrotic, and concentric rings of raised and depressed necrotic zones are formed, giving the typical "target board" or "bull's eye" appearance, often with a chlorotic halo surrounding the lesion which is highly characteristic of early blight. Multiple lesions on the same leaf also may coalesce, or grow together, to form one mass. As lesions coalesce, chlorosis may be visible due to clusters of infection. Elongated, brown-to-black lesions also may develop on stems and petioles of infected plants. On potato tubers, the symptoms are characterized by sunken, irregular lesions, which are often surrounded by a raised purple border. Beneath the surface of the lesion, the tuber tissue is leathery or corky with a brown discoloration.

Figure 15.1 Early Blight of Potato.

Causal Organism and Etiology

Early blight of potato is caused by *Alternaria* spp. (*A. solani* and *A. alternata*). The genus stands distinct from other genera by it's transversely and longitudinally-septate (muriform) conidia. Two major features of *Alternaria* species are the production of melanin, especially in spores, and the production of host-specific toxins in the case of pathogenic species. The inter- and intra-cellular mycelium of *A. solani* consist of septate and branched, light brown hyphae which become darker with age. The spores of *Alternaria solani* are tan to olive-brown that are normally produced singly and are 8–10 celled, ellipsoid to oblong, with a long, narrowing terminal cell or beak. The spores of *A. alternata* are smaller, formed in chains, and do not have a long beak. *A. alternata* can cause brown spotting on the leaves and black pitting in tuber tissue.

Disease Cycle and Epidemiology

The fungus survives as mycelium and conidia in crop debris, infected tubers, other Solanaceous crops, and weeds, as well as in the soil, and cause primary infection. The mycelium remains viable in dry infected leaves for a year or more. The pigmentation of the mycelium increases its resistance to lysis and extends its survival for years. Chlamydospores are also formed, but infrequently. Contaminated tubers with conidia or mycelium during harvesting are another source of primary inoculum. Infection of lower leaves first takes place, during periods of warm, rainy, humid weather, through conidia formed on the soil. Alternating wet and dry periods favor spore formation and dissemination, which peaks during the morning drying periods. Spores germinate and penetrate leaf tissues directly through the cuticle, natural openings (e.g. stomata), or wounds. Secondary infection of the disease occurs through conidia produced on infected tissues (primary spots). Early blight is considered a polycyclic with repeating cycles of a new infection. These conidia are disseminated mainly by wind, splashing rains, insects, and overhead irrigation. Abrasions and skinning of tubers during harvesting increase vulnerability to infections. Young or wounded tubers are more susceptible to infections. In presence of free moisture, the spores germinate and penetrate lenticels and wound sites. Viable spores can also be carried on tuber surfaces into storage prior to germination and infection. Lesions continue to expand slowly in storage causing dry rot. Warm temperature (24°C–30°C) and relative humidity of about 95% favor infection and the disease. Intermittent dry and wet weather is more conducive for early blight.

Management

Cultural Methods

- Use pathogen-free seeds (tubers) and healthy transplants for planting.
- Avoid overhead irrigation to reduce periods of leaf wetness. If only overhead irrigation is available, it should not be applied at night as this practice will increase the spread and infection.
- Avoid irrigation in cool, cloudy periods or late in the evening.
- Always avoid the cultivation of other Solanaceous crops (chilli, brinjal, and tomato), beans, *Brassica* spp., some weeds like datura, etc. nearby potato fields.
- Select fields with good drainage.
- Remove and burn infected plant debris and weed hosts of the fungus to reduce primary inoculum.
- Crop rotation is the best way to avoid soilborne inoculum.
- Field sanitation practices should be followed.
- Removal and destruction of infected plant debris should be done because the spores lying in the soil are the primary source of infection.
- Collect the dead haulms and burn them immediately after crop harvest.
- Apply balanced doses of fertilizers, especially nitrogen.
- Kill off the foliage at least 2–3 weeks before harvesting to prevent tuber infection
- Avoid harvesting too early or under wet conditions and do not bruise the tubers during harvesting and sorting.
- Avoid injuring potato tubers during harvesting to prevent tuber infections.
- Store the tubers under conditions that heal the wounds through suberization. A temperature of about 13°C, relative humidity of 90%, and good aeration help the suberization of the tubers.

Host Resistance

- Grow resistant/tolerant varieties such as Kufri Naveen, K. Sinduri, K. Jeevan, etc. The variety Kufri Sindhuri possesses a fair degree of resistance.

Chemical Control

- Spray any one or combination of the following fungicide(s).

Fungicides	Rate of Application
Aureofungin 46.15% w/v SP	0.005% in 300 litres of water/acre.
Captan 50% WG	600 g in 200 litres of water/acre (second spray after 5 days intervals).
Captan 75% WP	666 g in 400 litres of water/acre (second spray after 8 days intervals).

(*Contd.*)

Fungicides	Rate of Application
Captan 50% WP	1 kg in 300–400 litres of water/acre.
Chlorothalonil 75% WP	350–500 g in 240–320 litres of water/acre (second spray after 14 days interval).
Copper oxychloride 50% WP	1 kg in 300–400 litres of water/acre.
Mancozeb 35% SC	0.5% or 500 g in 100 litres of water as required depending upon crop stage and equipment used.
Mancozeb 75% WP	600–800 g in 300 litres of water/acre.
Hexaconazole 2% SC	1.2 litres in 200 litres of water/acre (second spray after 21 days intervals).
Propineb 70% WP	0.30% or 300 g in 100 litres of water as required depending upon crop stage and plant protection equipment used (second spray after 15 days intervals).
Captan 70% + Hexaconazole 5% WP	200–400 g in 200 litres of water/acre (second spray after 21 days intervals).
Carbendazim 12% + Mancozeb 63%	200 g in 200 litres of water/acre.

15.2 LATE BLIGHT

Late blight is one of the most devastating diseases of potato and remains the most costly potato pathogen to manage worldwide. It affects both potato foliage in the field and tuber in the storage, causing either a dry or wet rot, which can absolutely destroy a crop and cause 100% crop loss. Late blight is the disease that caused the Irish potato famine of the 1840s and mass emigration in Ireland. Approximately one million people left Ireland for the United States and Canada in one of the largest human migrations in modern times, and approximately 2 million people died of starvation. The pathogen was first described by M.J. Berkeley and subsequently named *Phytophthora infestans* by Anton de Bary ("father of plant pathology") in the 1870s. de Bary's conclusive studies disproved the theory of spontaneous generation or the wrath of God. Thus, late blight signifies the official beginning of the science of plant pathology. These early studies also contributed to Louis Pasteur's germ theory which was proposed 15 years later. It is presumed that in India the disease was first introduced into Nilgiri hills between 1870 and 1880 from Britain.

Symptoms

Late blight lesions can occur on all plant parts. Late blight disease of potato produces blackish/ brown lesions on leaves and stems. Severely infected fields with late blight have a distinct fetid odor.

Leaves: The symptoms initially are small, pale to dark green, circular to irregular shaped, water-soaked lesions, and or have a pale green or chlorotic border but expand rapidly into large, dark brown or black spots and the entire leaf becomes necrotic (Figure 15.2). In dry weather, the water-soaked halo disappears from the lesions and the dead tissue becomes brown and dry. These lesions usually appear at the tips of the leaves or on actively growing tissue. During cool and humid weather, sporangia and sporangiophores emerge through stomata on

Figure 15.2 Late Blight of Potato. (A) Water-soaked lesions on upper surface of leaves and (B) White mycelial growth on lower surface of leaves and on stem.

the lower surface of infected tissue and result in a glistening white sporulation at the border of lesions on lower surface of leaves. As several lesions coalesce, the entire plant can be destroyed. Under favorable conditions, late blight spread rapidly from leaves to petioles and stems and infects an entire field in only a few days.

Stems: The lesions on the stems are often black strips along the length of the stem. Severely infected stem leads to wilting and death of leaves and floral parts above the point of infection.

Tubers: Infected tuber tissues shows small to large, irregular shaped, copper brown, reddish, or purplish colored areas. Sporulation may occur on the surface of infected tubers in storage or on discarded cull piles. The tuber may show wet or dry rot. In dry rot, granular, brownish-red patches appear on the tubers that extend about half an inch below the skin. The tubers decay very fast and are accompanied by a foul smell (caused by secondary infection by soft rot bacteria).

Causal Organism and Etiology

Late blight disease of potato and tomato is caused by an Oomycetes (water mold), *Phytophthora infestans* (Kingdom–Chromista, Phylum–Oomycota, Class–Oomycetes, Order–Pythiales, Family–Pythiaceae). In Greeks, *Phytophthora* means plant destroyer (phyto=plant, phtora=destroyer). Oomycetes are not true fungi but are more closely related to brown algae. The mycelium of Oomycetes consists of hyaline, branched and coenocytic (few septa) hyphae and the nuclei are diploid which are intracellular with single or double club-shaped haustoria. Most of the true fungi are haploid. Eva Sansome (a plant geneticist) described first time the diploid life cycle of *Phytophthora*. The sporangiophores emerge through stomata on leaves and lenticels on the tubers. They are slender, hyaline, branched, and indeterminate. The sporangia are multinucleate (7–30 nuclei), thin-walled, oval, or pear-shaped with a definite papilla at the apex. The method of germination of sporangia is temperature dependant. Sporangia germinate by producing zoospores at a lower temperature (15°C or below) and by direct germ tubes at high temperature (16°C–24°C or above). The zoospores are biflagellate, one flagellum possesses fine hairs (tinsel type) while the other does not (whiplash type). Sexual reproduction is not common. In presence of both mating types (A1 and A2), the formation of oospores may be quite common. The antheridia are amphigynous and the oogonium is spherical. Oospores germinate by a germ tube.

Disease Cycle and Epidemiology

In absence of the oospores, *P. infestans* survives as mycelium in infected tubers of the previous season lying in the field or storage and serves as the primary inoculum. Under cool and moist conditions, sporangia and sporangiophores develop on infected tubers or new volunteer sprouts. Under favorable conditions, 100,000 sporangia per lesion are produced. The sporangia are dispersed to healthy potato foliage by wind or soil water. The sporangia germinate directly by germ tube at warm temperature (20°C or above) and low moisture. Sporangia germinate by producing zoospores at low temperature 12°C–13°C and when enough moisture is available. The zoospores, after swimming for some time, lose their flagella and form a cyst. A germ tube emerges from encysted zoospores and infects the host either through stomata or directly. The germ tube formed by the sporangium grows and forms an appressorium at the tip which forms infection hypha that is pushed into the leaf by force generated by the turgor pressure in the appressorium. Potato tubers can be infected in the field when sporangia are washed from lesions on the foliage and enter into the soil. Infections generally begin in tuber cracks, eyes or lenticels. The pathogen colonizes the intercellular space and grows on the nutrients from dead cells. Zoosporangia cause secondary infections, directly or through zoospores till plants are available. High humidity (above 90%), cool temperature (10°C to 20°C) and moisture (dew or rain) favor the disease development in epidemic form. On the basis of pathogen–moisture–temperature relation, the forecasting of late blight has been done in the temperate countries using four conditions: (i) night temperature below dew point for at least 4 hours, (ii) minimum temperature of 10°C or slightly above, (iii) clouds on the next day, and (iv) rainfall during the next 24 hours of at least 0.1 mm.

Management

Cultural Methods

- Use of disease-free and certified seed.
- Remove all the infected tubers and buried in the soil
- Select the seed lots carefully to avoid bringing in late blight on the seed.
- Plants should be adequately hilled to avoid tuber infection.
- Destroy the cull piles/potatoes through composting, burial, chopping, freezing, or feeding to livestock prior to new crop emergence.
- Monitor the cull disposal sites.
- Remove the infected tubers before storage.
- Do not overfertilize with nitrogen.
- Make sure plants are adequately hilled.
- Harvest only when vines are dead.
- Avoid harvesting under wet conditions.
- Maintain good air circulation in storage.

Host Resistance

- Grow resistant/tolerant varieties such as Kufri Badshah, K. Naveen, K. Sinduri, K. Kuber, K. Jeevan, K. Jyoti, etc.

Chemical Control

- Treat the emerging sprouts with fungicides.
- Apply protectant fungicides before late blight appears such as Mancozeb @ 2.5 g/L, with the onset of disease at 10 days interval.
- Apply systemic fungicide mixtures (1 or 2 sprays) such as Metalaxyl + Mancozeb or Carbendazim + Mancozeb or Cymoxanil + Mancozeb, etc. at the rate of 2 g/L of water in areas where blight occurs regularly in an epiphytotic form.

15.3 RHIZOCTONIA STEM CANKER AND BLACK SCURF

Black scurf of potato is a serious disease commonly observed in most potato-producing areas of the world. In India, it was first reported in 1912–1915. It is now a wide spread disease in the country. The major damage (upto 34%) is caused by the stem canker phase which reduces the yields of tubers. The black scurf causes qualitative losses as it decreases the marketablity of the tubers. Black scurf is more common than stem canker in India. Black scurf is found in most of the potato growing areas throughout the world.

Symptoms

The symptoms of the disease are found in both above and belowground portions of the plant. There are two distinct phases of the disease: (i) stem canker and blight phase, and (ii) black scurf phase on tubers.

Stem canker and blight phase: This is most damaging phase than black scurf phase. Stem canker occurs on underground portions of plants and often goes unnoticed. Early in the season, pre-emergence killing of the growing tips of sprouting tubers occurs. On growing plants, sunken or shallow-brown cankers develop on stems. If cankers are severe, they may girdle the stem, interfering with the normal movement of water and carbohydrates throughout the plant. Due to damage to the stem, the normal translocation of carbohydrates downward to tubers is disturbed and it accumulates in the tops. This causes stunting, rosetting, enhanced purpling of the skin due to excessive anthocyanin formation, and the development of aerial tubers (small and purple to green) (Figure 15.3 B).

Black scurf phase: It is the most obvious sign of *Rhizoctonia* disease. In this phase, the fungus forms irregular, dark brown to black, hard masses (sclerotia), 1–10 mm, on the surface of the tuber (Figure 15.3 A). These sclerotia appear as a black crust on the skin of tubers. These sclerotia, also known as black scurf, are difficult to remove by washing and brushing. Although these structures adhere tightly to the tuber skin, they are superficial and do not penetrate or damage the tuber, even in storage. The black scurf stage may often produce tubers that are misshapen, cracked, and discolored by the presence of sclerotia on the tuber surface.

Causal Organism and Etiology

Black scurf of potato is caused by *Rhizoctonia solani* AG-3 (mycelial and sclerotial stage-anamorph) (teleomorph-basidial stage: *Thanatephorus cucumeris*). *Rhizoctonia solani* (Deuteromycetes, Mycelia Sterilia) is a species complex of 13 anastomosis groups (AGs), which are categorized according

Figure 15.3 Black Scurf of Potato. (A) Black sclerotia on the surface of potato tubers and (B) Formation small aerial tubers occurs when underground stems are severely infected.

to the ability of their hyphae to anastomose (fuse) with one another. *Rhizoctonia solani* AG3 is most commonly associated with potato. The septa are typically of dolipore type. The lateral branches from the main hypha are invariably constricted at the point origin and a septum occurs in the branch near the junction with the main axis. The sclerotia are usually irregularly shaped and range from small, flat, barely visible blotches to large, raised lumps. The mycelium also produces monilioid or barrel-shaped cells (chlamydospores) in many isolates. The basidial stage is in fact inconspicuous, mostly saprophytic stage of the pathogen. The basidia are borne on small imperfectly symmetrical cymes. They are barrel-shaped, obpyriform or clavate, and bear 4 sterigmata which arise as blunt-knobs and later become horn-shaped. Basidiophores are ellipsoid or oblong-ellipsoid, flattened on one side, hyaline, and truncate. These spores are capable of producing secondary basidiospores (the same shape) by repetition on germination.

Disease Cycle and Epidemiology

The black scurf disease of potato can be initiated by seedborne or soilborne inoculum. The fungus (*Rhizoctonia solani*) overwinters as mycelium and sclerotia on infected tubers, plant residue, or in infested soils. It can also survive on underground parts of weed plants growing in potato fields symptomless. Sclerotia in soil or on seed tubers germinate, and the resulting mycelium colonizes plant surfaces. The early steps of infection are initiated by successive branching of runner hyphae resulting in the formation of short swollen cells giving rise to infection cushions which is a prerequisite for inducing stem and stolon lesions. Hymenia of the teleomorph may form near the soil surface on aerial stems. The hymenia do not cause damage to the plant but basidiospores enclosed in them may serve as a source of subsequent infections. After planting, sclerotia germinate and mycelium invades roots, sprouts, and emerging stems particularly when cool, moist soils slow plant growth. Seedborne inoculum is particularly effective in causing stolon damage because it is so close to developing sprouts. The fungus penetrates young, susceptible tissue, causing cankers that slow or stop the expansion of the infected stem or stolon. Cankers can sever the stolon or shoot from the plant or kill the growing point. Sclerotia form on tubers and in soil, providing inoculum for other growing seasons. When infected seed tubers are planted, the fungus grows from the seed surface to the developing sprout, and infection of root primordia, stolon primordia, and leaf primordia occurs. Seedborne inoculum is particularly effective in causing disease because of its close proximity to developing sprouts and stolons. Soilborne inoculum can be as damaging as seedborne inoculum, especially when tubers are planted in infested soil. However, the infection can only occur when the sprouts develop in proximity to the *Rhizoctonia* mycelia. Infection may occur anytime during the growing season, although most infections probably occur early in the growing season because the plant's resistance to stem and stolon infection increases after emergence, eventually limiting the expansion of lesions. The disease is favored by the capacity of fungus to survive in soil as sclerotia and mycelium in plant debris for long periods, and environmental conditions of low soil temperature and high soil moisture. Sclerotia can germinate at 8°C–30°C with the optimum temperature at 23°C. The optimum temperature for germination of basidiospores is 21°C–25°C. However, the optimum temperature for the infection and lesion development on potato is 18°C.

Management

Cultural Methods

- Use disease-free seed.
- Use seed treatments of registered fungicides to reduce some infections, especially from an infected seed piece.
- Warm the seed prior to planting.
- Any practice that promotes rapid emergence will reduce attack by *Rhizoctonia.*
- Use proper crop rotation, preferably grasses or cereals.

Host Resistance

- Grow resistant or tolerant varieties.

Biological Control

- Seed tuber treatment with *Trichoderma harzianum* or *Gliocladium virens* has been shown to reduce disease incidence in the field by 50% and 55% respectively.
- *Verticilium biguttatum* is *Rhizoctonia*-specific bioagent that suppress the sclerotia in soil. It is compatible with fungicides such as Pencycuron and Flutalonil.

Chemical Control

- Dip potato tubers in 1.75% Sulphuric acid or 3% Boric acid for 20 minutes to suppress sclerotia production on the tuber surface. Boric acid can also be applied as a spray with a hand spray on the seed tubers.
- Pre-plant seed tuber treatment with a combination of Sodium hypochlorite (@ 500 ppm for 8 minutes followed by Thiophanate methyl (50 g a.i./100 kg of seed) is highly effective in managing seedborne black scurf and common scab.
- Pencycuron 250 SC and Pencycuron 25 WP provide almost complete control of black scurf when used as seed tuber dip treatment at 0.5% and 0.75% concentration. Pencycuron, a phenylurea-based fungicide, is a non-systemic contact fungicide.

15.4 WART DISEASE OR BLACK WART

Potato wart is known by various names including cauliflower disease, black wart, black scab, potato tumor, potato cancer, potato canker, wart, and warty disease. It is a serious disease of potato in a temperate climate. Potato wart was first observed in Germany in 1908 by Spieckermann in Westphalia. Hilli (1932) described potato warts as a "social disease". Wart disease was first reported in India from Darjeeling (Das and Gupta, 1953) and restricted to Darjeeling and adjoining areas only. The potato wart disease can cause 50%–100% yield losses. *S. endobioticum* is considered an A2 pest by the EPPO.

Symptoms

Buds on underground stems, tubers, and stolen are the major sites of infection and origin of the warts. Dark-brown to black cauliflower-like outgrowths (warts) appear on potato tuber eyes, stolon buds, and the base of the stems. On the tubers, the warts are more typical and conspicuous, and they vary in shape and size and large masses may cover the entire tuber (Figure 15.4). Initially, warts are white or green, and darken and decay as they age. The cells adjoining the infected site undergo hyperplasia and hypertrophy, and form the tumors.

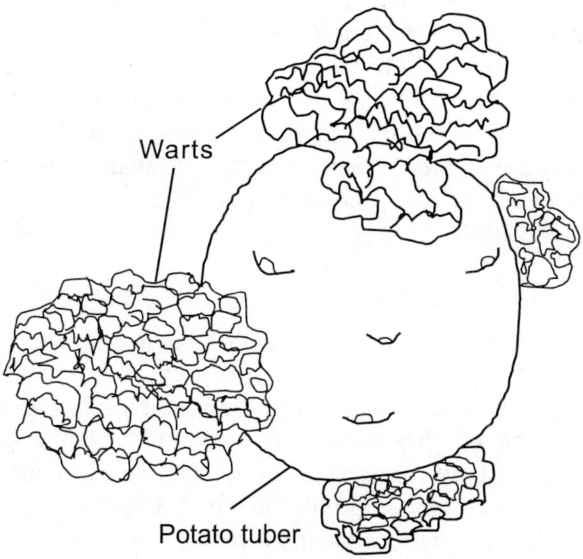

Figure 15.4 Wart Disease of Potato. *Synchytrium endobioticum* causes rough galls (warts) on potato tubers.

Morphologically the wart consists of distorted, proliferated branched structures grown together into a mass of hyperplastic tissue. In advanced stages, the warts are darker in color, sometimes black, and undergo decaying due to the attack of saprophytes. The infected plants, sometimes, show symptoms on the above-ground parts, like the formation of warts at the base of the stem near the soil and stunted growth. Potato roots are not infected. Potato wart symptoms can be confused with other potato diseases. See the difference in powdery scab, potato smut, and false wart or pseudowart in Table 15.1.

Table 15.1 Difference in powdery scab, potato smut, and flase smut of potato.

Disease	Powdery scab	Potato smut	False Wart or Pseudo-Wart
Causal organism	*Spongospora subterranea*	*Thecaphora solani*	Proliferation of eyes or chemical contaminant
Nature	Soilborne	Soilborne	–
Diagnostic features	It affects stems and stolons and causes wart-like galls on tubers.	It is characterized by warty swellings on the surface of potato tubers.	Simultaneous germination of all buds in one eye results in wart-like outgrowths ('pseudo-wart') very similar to those caused by potato wart disease.

Causal Organism and Etiology

Potato wart is a quarantined disease which is caused by the obligate biotrophic, soil-borne fungus *Synchytrium endobioticum*. It is a member of the Kingdom: Fungi, Phylum: Chytridiomycota, Class: Chytridiomycetes, Order: Chytridials, and Family: Synchytriaceae. It is a primitive fungus characterized by the lack of hyphae, the formation of zoospores, and the development of long-lived resting sporangia. It is an obligate, holocarpic, endobiotic parasite. The cells within these warty tissues contain thick-walled resting spores. Each large-sized wart has the potential to release 200,000 resting spores. Two forms of sporangia exist, so-called summer and winter sporangia (resting spores), which contain 200–300 motile zoospores. The summer sporangia are polyhedral, ovoid to almost round, light golden brown, aseptate, transparent, smooth, and thin-walled. Resting sporangia are thick-walled structures, usually spherical to ovoid in shape and golden brown in color. This fungus serves as a vector of *Potato virus X* (PVX) and *Potato mop top virus*.

Disease Cycle and Epidemiology

Synchytrium endobioticum depends on living hosts to complete its life cycle. The disease is soilborne and tuberborne. The tubers may have incipient warts that go unnoticed. Under adverse conditions (including water shortages) galls decay and release resting sporangia in the soil. As these galls decay and disintegrate, they release the thick-walled resting sporangia into the soil environment. Resting sporangia are endogenously dormant and can remain viable for 40 to 50 years at depths of up to 50 cm in the soil profile. The resting sporangia are primarily spread in infected seed tubers, which may have incipient warts that pass undetected, or in infested soil adhering to tubers, equipment, and by other carriers of contaminated soil. Resting sporangia survive passage through the digestive system of animals fed infected potatoes, and contaminated manure also can disperse inoculum. Earthworms have also been found to serve as means of inoculum dispersal, and resting sporangia can be dispersed by wind-blown soil or by flowing surface water. The pathogen has two reproductive cycles (asexual and sexual). Both cycles produce sporangia that germinate into infectious motile spores. Each reproductive cycle takes about 10–12 days to complete. The resting sporangia produced through sexual cycle can survive up to 50 years at depths up to 50 cm (20 inches) in the soil. However, the sporangia produced through the asexual cycle are short-lived and responsible for the secondary disease cycle. As long as the conditions are suitable, including adequate water, the disease cycle continues with the production of more sporangia. Resting sporangia in decaying warts and soil, germinate to release posteriorly, uniflagellate (whiplash type), haploid (uninucleate) zoospores. These zoospores migrate in soil water for a limited distance (50 mm or less). When zoospores come in contact with the host epidermal cells of meristematic tissues of growing points, buds, stolon tips, or young leaf primordial, they encyst and loose flagellum and then infect susceptible host tissue. After infection, the host cells greatly enlarge and haploid sori form inside the host cells while neighboring host cells begin to proliferate, resulting in the characteristic warty galls. Each sorus contains 1 to 9 summer sporangia, which in turn germinate to produce new haploid zoospores which reinfect susceptible tissue (i.e., a secondary disease cycle). These rapidly repeating secondary disease cycles ultimately result in an extensive invasion of host cells and rapid onset of gall formation. Under stress conditions, such as water shortage, zoospores also may conjugate (fuse) in pairs to form uninucleate, diploid, and biflagellate zygotes which infect the host tissue to form resting sporangia. Zoospore infection is mainly responsible for hypertrophy (increase in the size of infected and adjacent cells) while the zygote infection mainly causes hyperplasia (repeated cell division). Potato wart is favored by cool, wet soils during tuber development. A soil temperature of at least 8°C and water are required for the germination of both winter and summer sporangia and the dispersal of zoospores. Cool summers with average temperatures of 18°C or less, winters of approximately 160 days at or below 5°C, and annual precipitation of 70 cm are important for the development of the disease, and disease is generally limited to environments with these characteristics. Temperatures of 12°C–24°C favor infection.

Management

Cultural Methods

- Use disease-free seed tubers.
- Avoid potato cultivation on infested soils.
- Use crushed crab leg as a soil amendment.

Regulatory Measure

- The introduction of the pathogen in a field or locality must be effectively checked by practicing quarantine measures. In India, through the application of domestic quarantine regulation, this disease has so far been confined to Darjeeling hills.

Host Resistance

- Grow resistant/tolerant cultivars such as Kufri Muthu, Kufri Sheetman, Kufri Sherpa, Kufri Khasi Garo, Kufri Bahar, Kufri Kumar, and Kufri Kanchan.

15.5 VIRAL DISEASES

Almost 30 viruses are now reported on potato in India but only about 9 are responsible for significant losses. These include *Potato leaf roll viruses* (PLRV) and *Potato mosaic viruses* such as Potato virus X, Y, A, PVM, PVS and certain complexes such as crinkle (PVX + PVA), rugose (PVX + PVY), and aucuba (PVF + PVG). The mosaic symptoms caused by different viruses are often overlapping, depending on host cultivar, virus strains, temperature, light, etc. The symptoms such as crinckle, rugose, and rosette are natural combinations of potato viruses.

Potato leaf roll virus (PLRV) and *Potato virus Y* (PVY) are the two most serious viruses of potato worldwide. The severe strains of PVY and PLRV may cause 60%–75% loss in tuber yield while the mild ones such as *Potato virus* X, S, A, and M may suppress yield by 10%–30%. PVA is distributed worldwide and may reduce yield up to 30%–40%. PVA occurs less frequently than PVY and it causes severe disease in combination with PVX and PVY. Some important viruses of potato are listed in Table 15.2.

Table 15.2 Common viral diseases of potatoes.

Virus	Disease symptoms
(1) *Potato virus X*	Mild mosaic
(2) *Potato virus M*	Symptomless or para crinkling of leaves
(3) *Potato virus Y*	Rugose mosaic
(4) *Potato virus A*	Mild mosaic
(5) *Potato virus S*	Symptomless or mild rugosity
(6) *Potato acuba mosaic virus*	Yellow flecking
(7) *Potato mop-top virus*	Bright yellow blotches and rings
(8) *Potato leaf roll virus*	Leaf rolling
(9) *Alfalfa mosaic virus*	Yellow mottling
(10) *Andean potato latent virus*	Chlorosis and mild mosaic

15.5.1 Leaf Roll

Potato leaf roll disease is also known as potato leaf curl and potato phloem necrosis or net necrosis. Leaf roll of potato causes both qualitative and quantitative damage. It can cause severe yield loss (up to 90%) and, in some cultivars a quality reduction due to internal damage to tubers (net necrosis). The disease is common throughout India and causes up to 50% tuber yield losses. Genetically engineered resistance to PLRV was first reported by Kawchuk et al. (1990) who transformed potato with the PLRV coat protein (CP) gene.

Symptoms

The name "leaf roll" is partly descriptive of foliar symptoms. Inward rolling of leaves is the diagnostic symptom. The rolled leaves are stiff, thick, leathery and brittle, and sometimes have a brown or pink margin (Figure 15.5). Plant also shows dwarfing or stunting of the entire plant when raised from infected tubers. Plants display an erect growth habit. In primary infection (infection occurring during the season), the rolling of leaves starts in the upper part (youngest leaves) of the plant and progresses downward. In secondary infection (plant raised from infected tubers), this rolling of leaves starts in the lower leaves and progresses upward throughout the plants. Internal necrosis of the tuber is usually associated with primary infection but can also develop in tuber of plants with secondary infection. Two types of necrosis may occur: stem end browning and net necrosis (brown ring in the vascular system).

Figure 15.5 Potato Leaf Roll.

Causal Agent and Etiology

Potato leaf roll is caused by the *Potato leaf roll virus* (PLRV) and affects only potato. PLRV is also known as potato phloem necrosis virus. *Potato leaf roll virus* belongs to the family Luteoviridae and genus Polerovirus. PLRV particles are isometric (24 nm in diameter) with a monopartite (+)–ssRNA genome. It occurs in extremely low concentrations and is confined to the phloem cells. Infected potato foliage contains about 1–2 mg PLRV particles per gram of leaf tissue.

Disease Cycle and Epidemiology

The infected tubers, volunteers, and weed hosts serve as reservoirs of the primary inoculum. The virus remains confined to phloem tissue and circulates to other parts of the plants. Replication probably occurs in phloem companion cells and the virus moves long-distance within the phloem sieve elements. PLRV is not sap transmissible. In nature, the transmission occurs through infected tubers and by several species of aphid vectors in a circulative, persistent, non-propagative manner. PLRV is the only virus infecting potato that is persistent in the vector. The aphids feeding on the phloem transmit the virus to other plants in the fields and thus cause secondary infections. Green peach aphid, *Myzus persicae* is the main aphid vector and responsible for 95% of secondary infections. Other aphids are *Myzus circumflexus*, *M. convolvuli*, *M. solonafolii*, and *Aphis gossypii*. The aphid has to feed on the phloem for about 30 minutes to acquire the virus and it becomes infective after 24 hours after feeding. The vector may transmit the virus to over long distances. The epidemiology of the virus is largely associated with the ecology of its aphid vector, *Myzus persicae*. Moderate temperature (18°C–25°C) and dry weather favor vector build-up and consequently virus spread.

Management

Cultural Methods

- Use disease-free seed.
- Plant only the certified or virus-tested tubers.
- Remove and destroy infected plants.

- Remove all self-sown potatoes and solanaceous weeds.
- Rogue (remove) infected plants showing virus symptoms from seed potato crops.
- Dehaulming or de-topping of the foliage in third to fourth week of December to first week of January or as soon as aphid population exceeds critical limits of 20 aphids per 100 compound leaves.
- Avoid planting new and seed potato crops near old, ware, or processing crops.
- Regular inspections of plantings for any aphid and leaf roll virus symptoms.

Host Resistance

- Grow resistant/tolerant varieties such as Kufri kuber, etc.
- At least seven species are known to contain resistance to PLRV including *Solanum brevidens*, *S. etuberosum*, *S. chacoense*, and *S. raphanifolium* which have very high levels of resistance.
- Diploid genotypes of potato are highly resistant to PLRV and some other viruses.
- Hybrids developed by somatic fusion between diploid *Solanum verrucosum* and dihaploid *S. tuberosum* have been shown to be highly resistant to PLRV.

Biological Control

- Spray neem seed kernel extract (NSKE) @ 5%.
- Biological control of the aphids by predatory insects, such as ladybird beetles, lacewings, minute pirate bugs, damsel bugs, and seed bugs is much more effective.

Chemical Control

- Spray Imidacloprid or Methimedophos @ 0.2% or Methyl demetone 25 EC @ 1.2 ml/L or Monocrotophos 36 SL @ 0.4 g/L regularly to check vectors if needed.
- Apply Carbofuran 3% CG @ 6.64 kg/acre or Thiamethoxam 25% WG @ 40 g in 200 L of water/acre or soil drenching of Thiamethoxam 25% WG @ 80 g in 200 L water/acre for aphid control.

15.5.2 Mosaic Diseases

The mosaic diseases of potato have been categorized according to symptoms (Table 15.3).

Table 15.3 Mosaic symptoms and potato viruses.

Potato Mosaics	Viruses
Latent or faint mosaic	PVX, PVS, PVT
Mild mosaic	PVM, PVA
Severe green mosaic	PVY
Severe yellow mosaic	*Potato mop top virus*, *Potato yellow dwarf virus*, *Tobacco rattle virus*, and *Alfalfa mosaic virus* (calico disease)

15.5.2.1 Potato Virus A (PVA)

PVA is confined to members of the Solanaceae. The virus is widespread and can cause up to 40% yield loss. PVA is considered less important than PVY.

Symptoms

Haulm: *Potato virus A* (PVA) causes mild mosaic symptoms not dissimilar to those caused by *Potato virus X* (PVX). Differences can be difficult to detect visually, but PVA mottles may appear on the veins, and infected leaves look shiny. Infected plants may have a more open habit.

Roots, stolons, and tubers: Typical symptoms are not normally visible.

Causal Agent and Etiology

The PVA genome is composed of a (+)-ssRNA, flexuous filamentous particles measuring 730×11 nm. *Potato virus A* (PVA) belongs to *Potyviridae* (genus *Potyvirus*), the largest virus family infecting plants.

Disease Cycle and Epidemiology

At least seven aphid species including *Aphis frangulae*, *Macrosiphum euphorbiae*, and *Myzus persicae* are capable of transmitting *Plant virus A*. The virus is non-persistent and is lost from aphids as they go through their life cycle. As with PVY, the virus can be acquired rapidly from an infected plant (less than 1 min) and transmitted equally rapidly.

15.5.2.2 Potato virus X (PVX)

Potato virus X causes mild or latent mosaic. The disease and pathogen are worldwide in distribution. It alone can cause 15%–20% yield loss but mix infection with other potato viruses (PVA and PVY) cause severe losses.

Symptoms

Mild mosaics and mottles, but some strains cause severe or rugose mosaics (Figure 15.6). Tuber necrosis occurs in some cultivars. Symptoms masked in vigorously growing plants above 21°C. No symptoms development and virus multiplication occur at temperatures above 30°C.

Causal Agent and Etiology

Potato virus X is also known as *Potato mild mosaic virus*, *Potato latent virus*, and *Potato mottle virus* and belongs to potex virus group. It is a filamentous particle (515 × 13 nm) with monopartite ssRNA and constitutes 8% of the particle weight. The particles are diffusely distributed in dense X-bodies which may fill a greater part of the cell. Particles and inclusion bodies are seen in plastids, dictyosomes, and endoplasmic reticulum.

Host Range: Potato, tomato, tobacco, eggplant, pepper, *S. nigrum*, *Petunia* spp., *D. stramonium*, *Gomphrena globosa*, *Capsicum pendulum*, and *Chenopodium amaranticolor*.

Figure 15.6 Mosaic Disease of Potato.

Disease Cycle and Epidemiology

PVX is highly contagious, spread in field by contact or by cutting of seed tubers. Some stains of the virus can persist on surfaces of equipment used in the field. *Potato virus X* is easily transmitted by sap inoculation, cutting, tuber-core grafting and sap inoculation from infected to fresh potato plant stems and grasshoppers and other chewing insects, dodder (*Cuscuta compestris*), zoospores of *Synchytrium endobioticum*, and contact of plant parts in the field. No insect vectors are known except the grasshoppers that may mechanically transmit the virus. The natural host range of PVX appears to be limited almost entirely to potato and tomato.

15.5.2.3 Potato virus Y (PVY)

Potato virus Y causes severe mosaic (vein banding)/rugose mosaic of potato. High levels of PVY have been responsible for the rejection of many seed lots as certified seeds. Yield losses may reach upto 10%–80%. In combination with *Potato virus X*, it causes an even more destructive disease known as rugose mosaic. *Potato virus Y* (PVY) contains strain groups that cause different symptoms in potato and other Solanaceous crops, thus varying in importance according to the crop. Status of the disease PVY^{O} occurrence is worldwide. However, PVY^{N} and PVY^{NTN} strains are limited to certain countries; therefore, these are of quarantine significance. PVY^{O} is one of the most damaging potato viruses in terms of yield loss.

Symptoms

Leaves: Symptoms vary with the strain; environment and the host variety involved, from mild to severe mosaic to foliage necrosis, leaf drop streak or acropetal necrosis (see the Table 15.4). PVY^{N} induces transient mild mottle. Primary symptoms are necrotic or chlorotic rings or spots on the first infected leaves, followed by thickening of veins and green/pale yellow mottling of the top leaves (Figure 15.7). Symptoms of secondary infection are apparent early in the growing season.

Figure 15.7 Mosaic Disease of Potato Caused by *Potato virus Y.*

Haulm: Primary symptoms of common strain (PVY^{O}) and stipple-streak strain (PVY^{C}) are necrosis, mottling, yellowing of leaflets, leaf drops, and premature death of plants. Necrosis of foliage is a hypersensitive-type of reaction and generally starts at a veinal point on foliage and may result in spots or rings. Affected leaves may drop (leaf drop streaks) or remain clinging to the stem. These symptoms may be restricted to a few leaves or to a single shoot. Secondary symptoms result in a dwarf plant with mottled or crinkled foliage. Necrosis of foliage may also accompany these symptoms in certain cultivars. Some PVY^{O} (rugose strain) symptoms resemble those of early blight. Co-infection with PVY + PLRV induced symptoms that were characterized by bright blotching and necrotic leaf margins with purpling of the leaf tips, and leaf margins, stuntedness and leaf distortions.

Roots and stolons: Distinct symptoms are not normally seen.

Tubers: While primary symptoms of tobacco veinal necrosis strain (PVY^{N}) in potato foliage are vague and lack tuber symptoms, there are some strains that cause tuber ring spot necrosis (PVY^{NTN}). The PVY^{NTN} strains cause mild mosaic to brightly chlorotic foliage and clearly visible raised necrotic ring spots on the tuber surface. The rings may remain isolated or coalesce together to cover the whole tuber. Under the surface of necrotic rings the tissues may be thick in contrast to adjacent watery flesh. Rings may become sunken and skin cracked. Necrosis often increases during storage.

Table 15.4 Variability of symptoms caused by different strains of PVY.

Strains of *Potato virus Y*	Symptoms induced
PVY°(Common strain; rugose strain)	Crinkling (mottling) and, in some varieties, stipple streak necrosis. The first infected top leaf may wither and drop from the plant or it may remain clinging to the stem. Crinkling due to secondary infection with PVY^{o} strain is usually accompanied by foliage necrosis.

(*Contd.*)

Strains of *Potato virus Y*	Symptoms induced
PVY^C (Stipple-streak strain)	Stipple streak symptoms. The primary symptoms are necrotic rings or spots and bands along veins, thickening of veins and mosaic-like mottling. Secondary symptoms consist of necrotic spots and dots in the leaves, mosaic symptoms and necrosis of the petioles and stems, resulting in plant death. Stipple streaks in the foliage can lead to necrosis in tubers, often in the eyes. The infected eyes usually die.
PVY^N (Tobacco veinal necrotic strain)	Only mild mosaic in leaves and no symptoms in tubers.
PVY^{NTN} (Tuber necrotic strain of PVY^N)	Rapid systemic necrosis in potato stems and tubers in many varieties. Severe superficial tuber necrosis (potato tuber necrotic ringspot disease) and may also induce necrotic foliar symptoms.

Causal Agent and Etiology

The particles of *Potato virus Y* (potyvirus group) are elongated, flexuous, helically constructed rods, non-enveloped, measuring 730–740 × 11–15 nm. Its genome is made of a unipartite single-stranded, positive sense RNA. Cylindrical inclusion bodies are seen in the epidermal cells. *Potato virus Y* was also known as *Potato acropetal necrosis virus*, *Potato severe mosaic virus*, and *Tobacco vein banding virus*.

Host range: PVY has a wide host range, naturally infecting plants in more than nine families, including 14 genera of the Solanaceae, such as pepper, tomato, eggplant, and tobacco.

Disease Cycle and Epidemiology

The main sources of PVY inoculum are infected seed tubers. Aphids feeding on plants emerging from infected tubers acquire PVY within a few seconds and also inoculate the virus to healthy plants within seconds. Thus, aphids probing on the potato plants are potential vectors of PVY. The peach potato aphid, *Myzus persicae* is the most efficient vector in many areas (Figure 15.8). Virus spread mainly takes place by winged aphids. In most areas, infected tubers are the initial source of virus, although in tropical countries Solanaceous weeds may also serve as a reservoir of virus for vectors. PVY is carried in infected tubers, is transmitted locally non-persistently by aphids, and is occasionally transmitted by plant-to-plant contact. In potato varieties susceptible to *Potato virus Y* strain, PVY^{NTN}, growth and storage temperature of 20°C results in tuber necrosis even at the time of harvesting.

Figure 15.8 Aphid Vector Feeding on Lower Surface of Potato Leaf.

Management of Mosaic Diseases

Cultural Methods

- Planting certified seed free from virus.
- Eliminate sources of the virus by rouging seed crops and destroying volunteers.
- Harvest early seed crops.
- Detop to remove the foliage from the plants in the third or fourth week of December when the aphid population starts building up and then leave tubers in the soil to mature.
- Straw mulching between rows reduces aphid infestation and PVY incidence.

Host Resistance

- Grow resistance varieties.
- The variety Kufri Mega is resistant to PVX and PVS.
- Using microtubers in potato for the production of virus-free planting material are a promising future technique against potato viruses. Microtubers are miniature tubers developed under tuber-inducing conditions in vitro using the tissue culture technique. These small, dormant tubers can be easily handled and stored at low temperature for several months.
- The source of resistance to potato viruses has been found in wild tuber-bearing *Solanum* species. *Solanum stoloniferum* shows resistance through the expression of a hypersensitive response to all potyviruses.
- A clone DW.84-1457 was described to be highly resistant to PLRV, PVX, and PVM. Another clone G8107(1) has been found to show high resistance in graft and vector-transmission tests.

Biological Control

- Spray 1% or 5% neem oil @ 5 ml/L inhibits aphid transmission of PVY.
- Antiviral activity of extracts of a number of marine algae belonging to the phylum Heteroknotophyta has been shown to reduce 80% infectivity of PVX when tested on a local lesion host. The extract caused aggregation of virus particles.
- In areas where virus inoculum and aphid pressure are high, oil sprays prior to the arrival of aphids can be helpful. Spray of horticultural mineral oil, refined rapeseed oil, and soya oil can cause over 80% mortality of aphids. Repeated sprays of oil emulsion (10ml/liter of water) on plants raised from Imidacloprid treated seed tubers are reported to reduce the number of PVY infected-plants by 40% (rapeseed oil) and 60% (mineral oil).

Chemical Control

- The spread of a non-persistently transmitted viruse like PVY cannot be reduced by controlling aphids with insecticides. In fact, insecticide sprays stimulate aphids to probe more profusely, which may result in more virus spread.
- Spray Imidacloprid or Thiamethoxam or Pymetrozine @ 0.2% to reduce aphid population.

16

Cucurbits

16.1 DOWNY MILDEW

Downy mildew is one of the most common and destructive diseases affecting all cucurbit crops. Downy mildew of cucurbits was first documented by Berkeley and Curtis in 1868 from Cuba. The disease is common in North India.

Symptoms

Downy mildew affects mostly the leaves; occasionally stems and fruits. Symptoms vary in different cucurbit species and varieties, specifically in terms of lesion development, shape and size. On the upper surface of leaves, initially small, numerous pale-yellow, oily, angular lesions (restricted by the leaf veins), develop which later turn brown (Figure 16.1 A). On the lower side of the leaves gray-brown to purplish-black, fluffy downy growth appears under humid and moist weather (Figure 16.1 B). This is actually the sporulation (sporangia and sporangiophores) of the pathogens extruding from the stomata. Sporulation can occur also on the upper leaf surface during very severe infections. As the disease progresses, these lesions coalesce and form larger lesions, and may eventually cover the entire leaf. In severe infection, leaves become necrotic; curl upwards, and finally, the whole plant dies quickly (within 4–10 days from first symptoms) (Figure 16.1 C). Premature defoliation caused by downy mildew will reduce fruit size and predispose fruit to sunscald injury. Fruits are not directly attacked. But, in case of severe infection, a few smaller size fruits develop and are often misshapen.

Causal Organism and Etiology

Pseudoperonospora cubensis is the causal agent of downy mildew of plants belonging to Cucurbitaceae (gourd family), including cucumbers, melons, squashes, gourds, and watermelons. It is a biotroph or obligate parasite (requires living host tissue for growth and reproduction) and

Figure 16.1 Downey Mildew of Cucumber. (A) Young, pale-yellow, oily, angular lesions on upper leaf surface, (B) Gray-brown to purplish, fluffy downy growth on lower side of leaf, and (C) Severely infected leaf with necrosis.

belongs to Kingdom–Chromista, Phylum–Omycota, Class–Omycetes, Order–Peronosporales, Family–Peronosporaceae. It is an oomycete, is not a true fungus, and is often referred to as a "water mold" due to the fact that it thrives in wet or very humid conditions. The mycelium is branched, hyaline, coenocytic, and intracellular with small, ovate haustoria which sometimes develop finger-like branches. The sporangiophores are branched dichotomously or intermediate between dichotomous and trichotomous branching habits. The sterigmata are subacute. The sporangia are grayish to olivaceous purple, void to elliptical, thin-walled, and with a papilla at the distal end. The germination of sporangia occurs by the production of biflagellate zoospores. The oospores are spherical, rarely obovoid to ellipsoid, light yellow, and smooth-walled. However, oospores are not common in nature.

Disease Cycle and Epidemiology

Pseudoperonospora cubensis cannot overwinter in geographical locations with killing frosts. It perpetuates in the form of active mycelium and sporangia on self-grown or cultivated sponge gourd growing in sheltered places such as greenhouses during severe winters and also in the open during mild winters. Sporangia are the source of primary inoculum. Rarely oospores (thick-walled resting structures), play an important role in survival (in the absence of a living host) and in the disease cycle as primary inoculum. As an obligate biotroph, *P. cubensis* requires living host tissue for reproduction and dispersal. Sporangia are dispersed from infected plants, locally or over long distances, by the wind. The sporangia may also be dispersed locally by rain splash, contact between diseased and healthy leaves, and physical transfer on equipment within a field. Low humidity favors the dispersal of sporangia. Spotted cucumber beetles also transport the sporangia. After deposition on the leaf surface of the host plant, sporangia require water (e.g., rain or dew) to germinate. Germination is indirect: the multinucleate protoplast differentiates into 5–15 biflagellate zoospores. Sporangia germinate via cytoplasmic cleavage, resulting in the release of 2–15 motile, biflagellate zoospores. The zoospores actively swim towards the open stomata where they settle, lose their flagella and encyst. Germ tubes emerge from encysted zoospores and produce appressoria. A penetration hypha develops from the appressorium and enters through the stomatal aperture into the leaf tissue. The penetration via stomata is the most frequent mechanism of penetration of *P. cubensis*. Rarely, a direct (epidermal) penetration occurs. Once the host tissue is infected, intercellular hyphae form clavate-branched haustoria within plant cells (mesophyll cells) which provide nutrients for survival and asexual reproduction. Under favorable conditions, the colonization of the parasite in the tissue of a susceptible host proceeds relatively quickly and sporangiophores emerge from stomata within 5 to 7 days, mainly on the lower side of the leaves. On susceptible hosts, a new infection cycle takes place once in 7 to 14 days. Symptomatic plants with yellow lesions have the greatest sporulating capacity. *P. cubensis* causes a polycyclic disease. Optimal temperature for sporulation is 15°C with 6–12 hours of available moisture. The optimum temperature for the germination of sporangia is 10°C–20°C. Although the liberation and dispersal of sporangia occur under conditions of low leaf moisture, leaf wetness is necessary for the pathogen to successfully germinate, and infect the host plant. At 15°C, the optimum temperature for infection, at least 2 hours of leaf wetness are required for infection when high levels of inoculum. Sporangiophores emerge from stomatal openings only when relative humidity is 90% or above.

Management

Cultural Methods

- Burn the crop debris to reduce the inoculum.
- Do early sowing of the crop.
- Downy mildew severity can be decreased by encouraging airflow and reducing leaf wetness.
- Trellising cucurbits and growing in passive or traditional greenhouses can help to reduce relative humidity and leaf wetness.
- Avoid overhead irrigation or irrigate only in the late morning hours to reduce leaf wetness duration.
- The crop should be grown with wide spacing in well-drained soil.
- The bower system of cropping reduces the disease incidence.
- Seed production should be preferably carried out in the summer season because summer crop is often free from disease.
- Remove and destroy weed hosts such as golden creeper, wild cucumber, and volunteer cucumbers in neighboring fence rows and field edges.
- Apply a balanced dose of fertilizers (N, P, and K).

Host Resistance

- Grow resistant/tolerant verities. For example cucumber lines like Summer Prolific. In Punjab, two lines of muskmelon (LC-8 and PPDMR-4) have been identified as resistant to downy mildew.

Chemical Control

- Spray protective fungicide such as Mancozeb @ 0.25% at seven days intervals.
- If the disease is severe give one spray of Metalaxyl + Mancozeb @ 0.2%.

16.2 POWDERY MILDEW

Powdery mildew is diagnosed by white, powdery mold on plant tissues. Powdery mildew is the most common disease of cucurbits and a serious threat in many countries. A powdery mildew infection acts as a sink for plant photosynthesis causing reductions in plant growth, premature foliage loss, and consequently a reduction in yield. The yield loss is proportional to the severity of the disease and the length of time that plants have been infected. In India, the disease was reported first time by Butler in 1918 on various hosts from Uttar Pradesh and Bihar on the basis of conidial characteristics of the pathogens.

Symptoms

Powdery mildew disease appears on all foliar parts including leaves, petioles, and stems. Initial symptoms appear as tiny, white to dirty grey spots or sometimes with reddish brown tinge on leaves, and stems. As the disease progress, the superficial white powdery mass quickly covers most of the leaf surface and leads to a heavy reduction in photosynthetic area (Figure 16.2). The transpiration rate is very fast from infected leaves. Black pin-point bodies (cleistothecia)

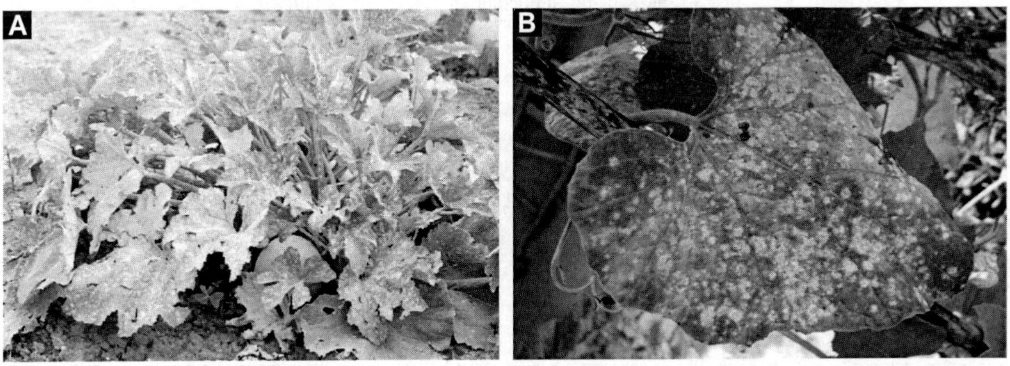

Figure 16.2 Powdery Mildew of Cucurbits. (A) Powdery mildew of summer squash and (B) Close up of bottle gourd leaf infected with powdery mildew disease.

appear rarely in the late season. They have been found only during the winter months in India. Severely infected plants may defoliate prematurely, wither and die. The fruits remain undersized and often deformed.

Causal Organism and Etiology

Powdery mildew of cucurbits is caused by two fungal pathogens, *Sphaerotheca fuligena* (syn. *Podosphaera xanthii*) and *Erysiphe cichoracearum* (syn. *Golovinomyces cichoracearum*) in India, and elsewhere. *S. fuliginea* is more commonly reported worldwide. Both of these fungi belong to Kingdom-Fungi, Phylum-Ascomycota, Subdivision-Ascomycotina, Class-Pyrenomycetes, Order-Erysiphales, and Family-Erysiphaceae. *E. cichoracearum* is reported to attack potato and tomato seedlings, lettuce, sunflower, mango, caster, and many other plant species. Some cucurbit crops (such as cucumber) are also susceptible to a third powdery mildew fungus, *Leveillula taurica* (anamorph: *Oidiopsis taurica*). Cleistothecia are not common. The conidia of these two fungi are very similar and difficult to distinguish from each other, but the presence of fibrosin bodies in the conidia of *S. fuliginea* enables a firm identification of the fungus. The other distinguishing features between these fungi, *Sphaerotheca fuligena* and *Erysiphe cichoracearumi* are given in Table 16.1.

Table 16.1 The difference between *Sphaerotheca fuligena* and *E. cichoracearum.*

Features	*Sphaerotheca fuligena*	*E. cichoracearum*
Recent name	*Podosphaera xanthii*	*Golovinomyces cichoracearum*
Host range	Broad	Broad
Distribution	Worldwide	Worldwide
Weather preference	Warmer weather	Cooler weather
Mycelium	Hyaline	Well developed, evanescent but sometimes persistent and effused
Conidia	Ellipsoid to barrel-shaped, hyaline; produced in long chains; conidia germinate by producing forked germ tubes but no appressoria.	Ellipsoidal or barrel-shaped, present in long chains; conidia germinate with straight germ tubes and unlobed appressoria.
Fibrosin bodies in conidia	Present (very distinct)	Absent
Cleistothecia	Cleistothecia are rare and globose, have numerous hypha-like appendages.	Cleistothecia are not commonly observed but are gregarious or scattered, globose, becoming depressed, or irregular. Appendages are numerous, basally inserted, myceloid, interwoven with the mycelium, and hyaline to dark brown.
Asci	Broadly elliptical to subglobose.	Ovate to broadly ovate, rarely subglobose, more or less stalked
Number of asci per ascocarp	Single, unitunicate ascus	10–25, unitunicate asci.
Ascospore	Asci contain eight ascospores which are ellipsoid to nearly spherical.	Asci contain 2 or rarely 3 ascospores that are ovoid to ellipsoid.

Disease Cycle and Epidemiology

The fungi have a wide host range and, therefore, are always available on one or the other collateral host. As obligate pathogens, all powdery mildew species overwinter on volunteer cucurbit hosts or weed host plants and serve as sources of the initial inoculum. Dormant mycelium or cleistothecia in infected plant debris of the previous year left in the field or conidia from collateral hosts/alternate hosts serve as primary inoculum. Cucurbits under protected cultivation may also be sources of the initial inoculum. The disease cycle is initiated by windblown conidia. Conidia are spread by wind, thrips, and other insects. Conidia are dispersed over a long distance by wind. Penetration is direct and confined to the epidermal cells. When conidia land and establish on the leaf surface of a susceptible host, one or two germ tubes develop and penetrate one or two epidermal cells. Powdery mildew develops quickly under favorable conditions. Symptoms appear in 3–7 days after infection and large numbers of conidia are produced in a short time. Disease development is optimum at temperatures of 20°C–27°C. High relative humidity (95%) favors infection, though conidia can germinate and infect plants at 50% or lower relative humidity. High relative humidity also favors conidial survival. *E. cichoracearum* is more active at lower temperatures than *P. fuliginea*. Dry conditions are favorable for colonization, sporulation, and dispersal of the pathogen. Rain and free water on the plant surfaces are unfavorable for infection. However, the disease development occurs in the presence or absence of dew. Dense plant growth and low light intensity also favor disease development. Susceptibility of leaves is greatest 16–23 days after unfolding.

Management

Cultural Methods

- Destroy crop debris.
- Avoid dense growth of crops.
- Avoid using high doses of nitrogen fertilizers.
- Increase air movement inside the canopy.
- Avoid planting new crops next to those that are already infected with powdery mildew.
- Remove old and heavily diseased leaves to improve air circulation and reduce inoculum.

Host Resistance

- Grow resistant/tolerant line. Sources of powdery mildew disease resistance are listed below:

Crops	Varieties
In USA	
Melon	PMR 45 (race-1 resistant), Seminole (race-2 resistant), Georgia-47 (moderately resistant), Planter's Jumbo (highly resistant to race-1 and race-2), Campo (resistant to an unidentified race), PI 124111, PI 79376, PI 134198, and PMR 450,
Muskmelon	SR-91 and V-1 are two important cultivars of muskmelon resistant to sulphur injury in California
Cucumber	PI 200815, PI 197087, PI 200818, *Cucumis sativus* var. *hardwickii*, wise 2757, Palmetto, Ashley, Poinsett, and Cherokee.
In India	
Muskmelon	PMR 5, PMR 6, Arka Rajhans, Diguria, and Haragola
Watermelon	Arka Manik is one of the highly resistant varieties of watermelon under Bangalore condition

Biological Control

Several biofungicides have been registered for the management of powdery mildew in cucurbits.

- AQ10 (Ecogen, Inc.) was developed specifically for powdery mildew. Its active ingredient is fungal spores of *Ampelomyces quisqualis*, which act to parasitize and destroy the powdery mildew fungi.
- Serenade® (AgraQuest, Inc.) has a bacterium, *B. subtilis*, as an ingredient and prevents the powdery mildew from infecting the cucurbit plant.
- Recently, a yeast-like fungus *Sporothrix flocculosa* (syn. *Pseudozyma flocculosa*) has been tested for control of powdery mildew in greenhouse-grown cucumbers with promising results. It has been formulated as a wettable powder (Sporodex®) for use against powdery mildew on greenhouse crops.

Chemical Control

- Apply protectant fungicides such as wettable sulfur @ 0.2%.
- Foliar sprays of Penconazole @ 0.05% or Tridemorph @ 0.1% or Carbendazim @ 0.1%, give very good control of the disease.
- Spray Benomyl 50% WP @ 80 g in 200 litres of water/acre or Thiophanate methyl 70% WP @ 572 g in 400 litres of water/acre.
- Dusting with 200-mesh Sulphur @ 25–30 kg/ha is quite effective to control the disease.
- Sprays with other fungicides like Calexin (0.1 or 0.05%), Copper sulphate, and Karathane are also effective.

16.3 WILTS

16.3.1 Bacterial Wilt

Bacterial wilt impacts all cucurbit crops except watermelon. Watermelon is generally resistant. Cucumbers and cantaloupe are most susceptible, followed by squash, pumpkin, and gourd. Bacterial wilt is among the most devastating diseases for cucurbits that may cause up to 80% of yield loss in susceptible varieties. *Erwinia tracheiphila*, the causal agent of bacterial wilt of cucurbits, was one of the first bacterial plant pathogens ever described. Bacterial wilt of cucurbits was first studied by Erwin F. Smith. Herbivory by leaf beetles can include exposure to *E. tracheiphila*, a relationship first recognized by E.F. Smith and his students.

Symptoms

Initial symptoms of the disease appear as drooping and wilting of a few to several leaves and stem. The infected plants appear dull green in patches. As the disease progress, plants exhibit chlorosis, rapid, and extensive wilting (Figure 16.3 A). Eventually, foliar necrosis occurs, and the entire plant collapses, shrivels, and dies. Wilted leaves shrivel and dry up and infected stems turn pale and become soft. The wilting can occur on individual runners or stems, or throughout an entire plant's foliage. In a cross-section of a symptomatic stem, bacterial oozing can be seen from multiple blocked xylem vessels (Figure 16.3 B). A fine, sticky, thread-like strand or slimy cloudy mass of bacterial ooze can be observed when the cut ends of infected stems are slowly pulled apart (Figure 16.3 C) or placed into the water. Small, irregular water-soaked areas appear on the fruit surface. Slime rot of squash fruit can occur in storage conditions.

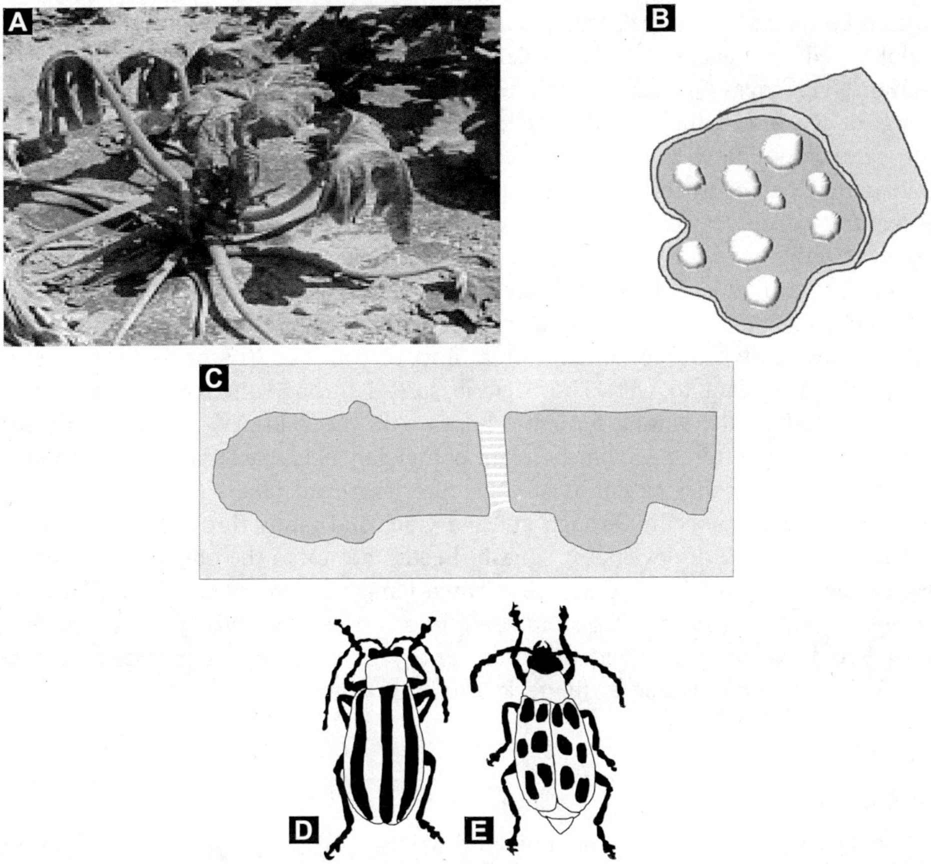

Figure 16.3 Bacterial Wilt of Cucurbits. (A) Wilting symptoms on cucurbits, (B) Illustration of bacterial ooze from a cross section of infected stem, (C) Thread-like strands can be seen when cut ends of infected stems are slowly pulled apart, (D) Striped (*Acalymma vittatum*), and (E) Spotted (*Diabrotica undecimpunctata*) cucumber beetle which transmit the bacterium *Erwinia tracheiphila*.

Causal Organism and Etiology

Bacterial wilt of cucurbits is caused by *Erwinia tracheiphila*. It is a gram-negative, motile by peritrichous flagella, xylem-limited, facultative anaerobic bacterium.

Disease Cycle and Epidemiology

The bacterium, *Erwinia tracheiphila*, is not a seedborne or soilborne, and survives in plant debris for only about one month. The bacterium survives in the digestive tract (gut) of hibernating adults of striped cucumber beetle (*Acalymma vittatum*) and spotted cucumber beetle (*Diabrotica undecimpunctata howardi*) (Figure 16.3 D–E) or in weeds. It is believed that the relationship between insects and bacterium is symbiotic. Infected wild cucurbits can also serve as primary inoculum and from these plants, the spread occurs through beetles. Cucumber beetles prefer wilted plants for feeding resulting in an increase in the number of contaminated beetles. Each contaminated beetle can inoculate several healthy plants, after one feeding only on an infected plant. Adults feed on foliage, flowers, pollens, and roots, cause deep wounds and transmit the bacterium. Transmission occurs when adults feed on plants through their contaminated mouth parts or when inoculum from infested frass enters freshly wounded tissues on leaves, stems, or nectaries of flowers where the beetles aggregate. *E. tracheiphila* is an obligately vector-transmitted pathogen that replicates only in the xylem of susceptible host plants or the digestive tract of insect vectors. The bacteria enter the xylem vessels exposed by wounds and become systemic within the plant. Once inside the xylem, the bacterium, multiplies, and extracellular polysaccharides (slime), and block xylem vessels. The accumulation of bacterial cells, polysaccharides, and gums, along with the formation of tyloses results in the plugging of the vascular system which blocks the normal flow of water and nutrients to the rest of the plant and causes the plant to wilt. The bacteria spread through the xylem vessels to other parts of the plant. The first wilting symptom appears on the field at 6–7 days after infection, and another 7 days is required to wilt the plant completely. The bacteria present in the xylem vessels of infected plants die within 30–60 days after the dead plants dry up. Fruit infection of squash plants generally occurs through infected vines and rarely through beetles feeding on the blossoms and the rind of developing squash. Beetles attract to the odors of infected plants. In spring, overwintering adults become active when temperature exceeds 12°C. The spread of the pathogen is fast under normal moisture conditions because the activity of beetles decreases in prolonged cool and rainy weather. In extremely dry weather, the bacteria cannot penetrate the exposed vascular systems due to the lack of film of water.

Management

Cultural Methods

- Use floating row covers (spunbond or woven plastic, polyester or polypropylene material), fine netting, cones, perimeter trap crops (e.g. buttercup squash, *Cucurbita maxima*), and straw or reflective plastic mulches or other barriers to prevent the feeding of cucumber beetles on vines. Aluminum-coated or reflective mulches repel cucumber beetles with variable results, whereas black plastic mulch can reduce the survival of striped cucumber beetle larvae.
- Soil management practices and the deployment of plastic mulches can reduce cucumber beetle populations in muskmelon, cucumber, and squash.
- Place yellow sticky cups after every 20 feet along the rows to trap the beetles.
- Remove and destroy the diseased plants to prevent the further spread of the disease, to reduce sites for beetle hibernation, and to minimize the survival of other cucurbit pathogens.
- Do not plant cucurbits near weedy woods or brush where the cucumber beetles are likely to overwinter.

- Crop rotation (for a 3-year cycle) or intercropping with non-cucurbit crops to reduce cucumber beetle population. Avoid susceptible or highly susceptible cucurbits 2 out of 3 years during this rotation.
- Avoid direct seeding of plants.
- Remove all the weeds from the field.

Host Resistance

- Use resistance varieties.

Biological Control

- Tachinid flies (*Celatoria setosa*) and braconid wasps (*Censtistes diabroticae*) have been reported to parasitize striped cucumber beetles.
- Foliar application of Kaolin clay (Surround WP), insecticidal soap containing neem oil, and rotenone 1% dust.
- Introducing entomopathogenic nematodes through drip irrigation of cucumbers grown in black plastic mulch can also control cucumber beetle larvae.
- Application of Plant growth-promoting rhizobacteria (PGPR) to cucumber seeds can significantly reduce bacterial wilt incidence.
- Use of Kaolin clay, Pyrethrum, Spinosad, and plant extracts such as Neem oil and inducers of systemic acquired resistance for the management of the disease or the beetle vector.

Chemical Control

- Use any of the fast-killing insecticides like Methoxychlor, Diazinon, and Carbryl. The Carbryl should be sprayed late in the day to prevent honeybee mortality.

16.3.2 Fusarium Wilt

Fusarium wilt is one of the most, problematic, destructive, and economically important diseases that affect most cucurbits. Fusarium wilt of watermelon is one of the oldest described Fusarium wilt diseases and the most economically important disease of watermelon worldwide. This disease was first described by Erwin F. Smith in 1894 in the southeastern region of the United States. In India, Fusarium wilt was first reported from Maharashtra in 1955 and now it occurs in almost all the states. The disease is rated as one of the most serious diseases of muskmelon that is capable of causing up to 90% or more mortality.

Symptoms

Symptoms of fusarium wilt are similar in all cucurbits. In the field, the disease symptoms often appear in patches. The symptoms of Fusarium wilt are characterized by loss of turgor pressure of the vines. Vines may recover during the evening but eventually, wilt permanently. Initial symptoms often include a dull, gray-green appearance of leaves that precedes a loss of turgor pressure and wilting. Wilting is followed by a yellowing of the leaves and finally necrosis. The primary diagnostic symptom of fusarium wilt is a discoloration of the vascular system (xylem), which can be observed readily in longitudinal or cross-section of roots or stems. In watermelon and muskmelon, a brown necrotic streak may be visible externally on the lower stem and extending along the length of the vine. A white to pink fungal growth develops on the surface of infected stems during wet weather.

Causal Organism and Etiology

The fusarium wilt pathogen is very host-specific. Fusarium wilt of cucumber (*Cucumis sativus* L.) is caused by *Fusarium oxysporum* f. sp. *cucumerinum*, watermelon (*Citrullus lanatus*) by *F. oxysporum* f. sp. *niveum*, and muskmelon (*Cucumis melo* L.) by *Fusarium oxysporum* f. sp. *melonis*. On cucumber *Fusarium oxysporum* f. sp. *cucumerinum* causes pre- and post-emergence damping-off of seedlings and wilting of mature plants. Morphologically, these species produce three types of spores: microconidia, macroconidia, and chlamydospores. Chlamydospores are formed in older mycelium. Single-celled, hyaline, ovoid to ellipsoid microconidia are produced in abundance. Relatively less in numbers of fusiform, mostly 2–3 septate and hyaline macroconidia are produced. Several *Fusarium* species and physiological races have been identified on cucurbits from different parts of the world.

Disease Cycle and Epidemiology

The disease cycle for all cucurbit Fusarium wilts is similar. Fusarium wilt of cucurbits is a seed and soilborne disease. The *Fusarium oxysporum*, persists for long periods in the soil or in association with plant debris as durable spores (chlamydospores). The pathogens are transmitted through soil, contaminated seeds, farming tools and vehicles, humans and animals, and water. The pathogen is both externally and internally seedborne. If the seeds are infected, the fungus can be transported directly into the xylem from infected inner tissues. If the fungus is in the soil, it germinates and penetrates the fibrous root system at the root tips or at any openings. From inside the root tissue, the fungus enters the xylem and produces conidia that block water flow in the host. One-sided wilt symptoms appear after 3–4 weeks of planting. Since the pathogen is soilborne, the diseased plants often occur in random patches. The pathogens remain in dead plant debris or in the soil in the form of mycelia, macroconidia, and chlamydospores, until a new susceptible host is available. The pathogens may attack the plant at any stage, although, outbreaks often follow periods of crop stress, including hot, dry weather conditions. Soil temperatures of 18°C–25°C and 50%–65% relative humidity support rapid disease progression. At high soil temperatures, plants become infected but may not wilt; rather they develop severe stunting. This disease can be quite severe in fields with light, sandy soils with pH 5.5–6.5, less than 25% soil moisture, and high nitrogen (especially NH_4–nitrogen) content.

Management

Cultural Methods

- Plant clean and pathogen-free seeds.
- Remove and destroy the infected plants and plant debris.
- Avoid water stagnation and maintain proper drainage.
- Plant on land not previously cropped with the cucurbit species.
- Apply lime to bring the soil pH to 6.5–7.0 which can reduce disease.

- Do not move soil from infested fields.
- Do not replant the same cucurbit species for 5–7 years. This can help to manage watermelon wilt but is considered ineffective for melon and cucumber wilt.
- Follow crop rotations with crops other than cucurbits.
- Deep plowing of severely infested fields.
- Add cellulosic materials (straw) to the soil to reduce the survival of the pathogen in the soil.

Physical Control

- Treat the seed with hot water treatment at 55°C for 15 minutes to reduce seedborne inoculum.

Host Resistance

- Resistant varieties offer the most reliable control of this disease. Because there are races of wilt forms, therefore it is necessary to know which races are present in a particular area before choosing a resistance variety. Some important resistant varieties of muskmelon are Durgapura Madhu and Punjab Sunehri; watermelons are Charleston Gray, Iowa Belle, Garrisonian, Crimson Sweet, Sweet Princess, Jubilee, etc.
- **Grafting for Fusarium resistance**: The common rootstocks used for controlling Fusarium wilt of watermelon are *Cucurbita moschata*, bottle gourd and squash interspecific hybrids (*Cucurbita maxima* × *Cucurbita moschata*). Besides, the rootstocks of *Citrullus* spp., *Cucumis* spp. and *Cucurbita* spp. are commonly used for preventing Fusarium wilt of cucumber, melon and bitter gourd, in addition to watermelon. Grafting melon onto squash interspecific hybrids can provide resistance to *F. oxysporum* f.sp. *melonis* race 1, 2.

Biological Control

- The use of antibiotic-producing soil fungi and bacteria, i.e. *Gliocladium* spp., *Trichoderma* spp., and *Pseudomonas* spp.
- Treat the seed with *Trichoderma harzianum* @ 10 g/kg seeds.
- Soil application of *Trichoderma viride* formulation @ 2 kg mixed with 50 kg of farm yard manure per acre, can effectively manage the disease.
- Use of non-pathogenic strains of *F. oxysporum* that compete with pathogenic forms for root colonization (i.e. *Fusarium oxysporum* strain 47).
- The use of other formae speciales and races of *F. oxysporum* to induce resistance in plants to pathogenic forms.
- Natural or induced soil suppressiveness. The use of cover crops, such as hairy vetch, as soil amendments ("green manures") may reduce fusarium wilt in some locations, but the magnitude of the wilt reduction may vary by location.

Chemical Control

- Treat the seed with Benomyl or Carbendazim 50% WP @ 1–2 g/kg seeds.
- Drench the soil with Captan 50% WP (@ 0.2%–0.3%), and repeat it twice or thrice.

16.4 GUMMY STEM BLIGHT

Gummy stem blight was first observed in 1891, on cucumber (*Cucumis sativus* L.) by Fautrey and Roumeguere in France. It affects leaves, petioles, vines, stems, tendrils, pedicels, flowers, peduncles, fruit, and the seed of all cucurbits such as cucumber, watermelon, melon, cantaloupe, squash, and pumpkin. The fruit rot phase is often called black rot. The susceptibility of cucurbit crops to black rot differs from susceptibility to foliar blight and crown cankers. The pathogen is most common in the southern United States and in subtropical and tropical areas of the world. The disease appears both in field and storage condition and causes significant crop losses. Gummy stem blight causes enormous losses to melon (*Cucumis melo* L.) production worldwide.

Symptoms

Gummy stem blight disease of cucurbits produces various types of symptoms such as leaf spot, stem canker, vine wilt, and black fruit rot. Infection and symptoms can occur on all plant parts except roots. The main diagnostic features of the disease are gummy exudate on stem and fruit lesions, and the abundant pycnidia (asexual fruiting bodies) followed by pseudothecia (sexual fruiting bodies).

Seedlings: Light to dark brown spots appears on the cotyledons. Infection of the cotyledons or hypocotyl causes a large, water-soaked, brown discoloration of the tissues, followed by tissue desiccation and collapse. Seedlings die rapidly after infection of either the hypocotyl or cotyledons. If these plants are transplanted in the field and survive, cankers may form on crowns or vines. Damping-off occurs when the fungus is carried in the seed and kills the seedlings.

Leaves: On older plants, initial symptoms appear as a small, circular gray-green to tan spots sometimes surrounded by a yellow halo. Such spots enlarge rapidly, become irregular in shape, and turn tan to dark brown. When the lesions coalesce, the entire leaf becomes blighted. The spots dry, crack and tear, giving leaves a tattered appearance. Infection often begins at the leaf margins.

Stem and crown: Stem infections consist of oblong water-soaked lesions that later turn brown to dark brown. As the disease progress, stem lesions enlarge and turn into brown cankers that can girdle the entire stem and result in foliage wilting and death of affected areas. A red or brown, sticky exudate gummy fluid often oozes from stem cankers. The most important form of the disease is crown rot, which may kill plants. At first, pale brown, then bleached areas develop, and a reddish gum oozes from the cracks. Tiny, dark brown to black, closely spaced in the groups, pimple-like fruiting bodies (pycnidia) develop within the infected tissues. The pycnidia (asexual structure) and pseudothecia (sexual structures) of the pathogen develop in leaf spots and in lesions on other above-ground plant parts, including petioles, vines, stems, tendrils, and pedicels of flowers and fruit. Sometimes pycnidia are arranged in rings on the fruit or leaf surface and are visible with the necked eye.

Fruits: Fruit develop water-soaked, small oval to circular spots that are a greasy green or yellowish and turn dark brown as the spots enlarge. Amber gummy exudate and black fruiting bodies may develop on the spots. Eventually, spots may turn black. On fruit, dark cracked sunken lesions form, beneath which extensive rot is found.

Difference between gummy stem blight and anthracnose of cucurbits

Gummy stem blight can be distinguished from cucurbit anthracnose by the larger leaf spots, larger lesions that encircle petioles, and crown cankers. On watermelon leaves, the shape of anthracnose spots is distinctly angular and irregular, while leaf spots of gummy stem blight generally are round. The centers of anthracnose leaf spots are more likely to dissolve than the centers of leaf spots of gummy stem blight.

Causal Organism and Etiology

This disease is caused by three genetically distinct species of *Stagonosporopsis*; *S. cucurbitacearum* (syn. *Didymella bryoniae*), *S. citrulli*, and *S. caricae*. The three species are pathogenic to cucurbits, but *S. caricae* also causes leaf spots and stem and fruit rot in papaya (*Carica papaya*). The fungus *Didymella bryoniae* is a necrotrophic pathogen that belongs to the Domain–Eukaryota, Kingdom–Fungi, Phylum–Ascomycota, Subphylum–Pezizomycotina, Class–Dothideomycetes, Subclass–Pleosporomycetidae, Order–Pleosporales, Family–Didymellaceae, Genus–*Didymella*, and Species–*bryoniae*. The sexual stages of the fungus, *Didymella bryoniae*, produce pseudothecia and the asexual stage, *Stagonosporopsis cucurbitacearum*, produces pycnidia. Pycnidia are tan when young and become brown as they age. They usually are slightly larger than pseudothecia. The conidia are hyaline, cylindrical with rounded ends, non- or monoseptate. Pseudothecia are dark brown to black and usually are slightly immersed in the host tissue. The numerous bitunicate asci usually produce eight ascospores per ascus. The ascospores are hyaline, monoseptate with a constriction at the septum, and rounded on the ends. The upper cell is usually wider than the lower cell. Hyaline pseudoparaphyses are present in pseudothecia.

Disease Cycle and Epidemiology

Didymella bryoniae can survive as chlamydospores in infected seeds, weeds, and plant debris from previously infected cucurbit crops. These are the primary sources of inoculum. Conidia act as secondary inoculum and spread through wind and dripping or splashing water. Beetles, mainly striped cucumber beetles, serve as a vector of the disease, which also spread conidia from diseased plants to others. Leaves are penetrated either directly through the cuticle or intercellular spaces around the bases of trichomes. Stems are penetrated through wounds or by the extension of leaf lesions. Fruits are penetrated either through wounds or through flower scars at the time of pollination. Cotyledons and young leaves of watermelon and melon are very susceptible, but those of cucumber and certain squash are resistant and become susceptible only as they mature. The optimum temperature for infection is 20°C in melon and 24°C–25°C in watermelon and cucumber. Leaf wetness or high humidity is essential for the release of conidia of the pathogen and its germination. For the production of pseudothecia require warm and high humid weather condition. Peak ascospore dispersal occurs after rain and during dew periods at night. Free moisture on leaves for at least one hour is necessary for infection, and further continuous leaf wetness is required for lesion expansion.

Management

Cultural Methods

- Use healthy, disease-free seeds.
- Disease-free seeds should be produced in arid locations.

- Infected seeds may be symptomless, and seed treatment may be necessary if not previously treated.
- Inspect seedlings regularly for signs and symptoms of infection as the disease is common at the seedling stage.
- Follow crop rotation with non-cucurbit crops (such as maize and sorghum) for 3–4 years to reduce soilborne inoculum.
- Remove and eradicate weeds from the field before planting cucurbit crops.
- Soil amendment with poultry refuses @ 6–8 kg/pit 15 days before seed sowing.
- Inspect the water-melon field at least twice a week during the growing season.
- Remove infected plant parts from the field and destroy them.
- Crop refuse should be ploughed deeply immediately after harvest to reduce fungus survival.
- Avoid wounding the fruits during harvest and store the harvested fruit at 7°C–10°C to prevent postharvest black rot.
- Maintain proper drainage and aeration in the field.
- Avoid exotic hybrids and varieties due to the high degree of susceptibility.
- Summer ploughing and green manuring followed by *Trichoderma* application.
- Avoid injury near the collar region.

Biological Control

- *Bacillus subtilis* is available for use as a bioagent against *D. bryoniae*, but its success is dependent on environmental conditions at the time of application.

Host Resistance

- Resistant cultivars are currently not available.

Chemical Control

- Seed treatment with Carbendazim @ 0.25%.
- One drenching of Carbendazim @ 0.1% near collar region.
- Apply Bordeaux paste (1:1:10) on the stem up to one foot from the soil at the early stage.
- Spray Mancozeb or Copper oxychloride @ 2 g/litre water, three times in 12 days intervals.

17 Onion (*Allium cepa* L.) and Garlic (*Allium sativum* L.)

17.1 PURPLE BLOTCH

The name "purple blotch" for this disease was proposed by Nolla (1927). Purple blotch is the most destructive disease of *Allium* spp. (onions, garlic, shallots, leeks, scallions, and chives). It causes heavy yield losses ranging from 2.5%–97% in the bulb and seed crop of onion and garlic worldwide.

Symptoms

The disease often begins on older leaves or seed stalks as small, sunken, water-soaked lesions that rapidly develop white centers. As the disease progresses, lesions enlarge, become oval-shaped or elliptical or oblong in shape, and turn purple to brown, often with yellow rings that create a distinctive zonate (target spot) or "bull's-eye pattern" (Figure 17.1 and 17.2). In moist weather, the surface of the lesion may be covered with brown to dark grey masses of spores. Leaves of severely infected plants wilt and die. Sometimes, purple blotch and Stemphylium leaf blight disease occur on the same leaf or plant therefore the microscopic study is necessary to distinguish these two fungi. Onion bulbs may be infected at harvest or in storage through the neck or wounds in the bulb scales. Bulb rot symptoms begin as soft, water-soaked areas; eventually, bulbs turn dark reddish-purple, then brown or black.

Causal Organism and Etiology

Purple blotch is caused by *Alternaria porri* (a dematiaceous hyphomycete). Conidia of *A. Porri* are brown to golden-brown, normally ovoid, obclavate, ellipsoid, and subcylindrical with 8–12 transverse septa and usually a few longitudinal septa. Conidia have a long, tapering beak and are borne singly. The pathogenicity of *Alternaria* spp. is due to the production of host-specific or nonspecific toxins (secondary metabolites) that may induce disease.

Figure 17.1 Purple Blotch of Garlic.

Figure 17.2 Purple Blotch of Onion.

Disease Cycle and Epidemiology

Alternaria porri overwinters as dormant mycelium in crop residue, on or near the soil surface. The pathogen is seedborne in onions. The spores are spread by wind and splashing rain or irrigation. The pathogen is introduced into onion and garlic fields by windblown spores from nearby plants. Spores infect the leaves through stomata or directly through the epidermis. The fungus can also enter the plant through wounds caused by thrips feeding or hail. Spores produce on the lesions throughout the growing season, and disperse to adjacent leaves and plants. Spores are produced at night and released in the morning as humidity decreases. These spores infect new plants during periods of warm (25°C–30°C) humid weather. Free moisture, in the form of rain, persistent fog, or dew, is required for infection and spore production. Typical lesions are produced when the leaf wetness duration is more than 16 hours while flecking may appear at 12 hours. Purple blotch disease is severe in high humidity (80%–90%) and moderate temperature (25°C–30°C). The older leaves are more susceptible to infection than the young leaves, although damage by thrips may increase the susceptibility of young leaves. Under favorable conditions, epidemics develop quickly, as the symptoms develop in 1–4 days and new spores are produced after only 5 days.

Management

Cultural Methods

- Use healthy and quality seed or planting materials from recognized institutes or organizations or seed agencies for the better and good quality bulb and seed production.
- Follow crop rotation to reduce the amount of inoculum present and reduce disease incidence for at least 2–3 years with non-host crops such as small grains, millets, corn, and sorghum.
- Summer ploughing has been found to reduce the incidence of purple blotch.
- Good drainage.
- Use recommended doses of nitrogenous and other fertilizers.
- Apply adequate but not excessive nitrogen fertilizer before bulb initiation.
- Avoid dense plantings of late-maturing varieties.
- Avoid planting during wet conditions which favors disease development.
- Remove weeds, especially those of the *Allium* family such as wild onion during land preparation.
- Remove infected plant parts and dispose by burning or burying them deeply.
- Do not irrigate the crop as harvesting time approaches.
- Sort and cure the bulbs to separate infected and noninfected onions.
- Reduce over-wintering and survival of the pathogens by eliminating culls, onion debris, and volunteers.
- Undercut onion roots and harvest at full maturity when intact necks top over naturally, ideally during dry weather.
- Check plants in the nursery and remove any seedlings that show leaf spots before transfer to the field.
- Harvest the crop when plants are mature, and preferably during dry weather, and treat the bulbs before storage.
- Remove plant debris after harvest or incorporate it into the soil so that it decomposes quickly.
- Bury deeply infected plant material.
- Manage onion thrips, as plants weakened by thrips infestation are more susceptible to this disease.
- Harvest the crop in dry weather and avoid injury to the necks.
- Allow onions to cure properly before leaf removal.
- Store garlic and onion bulb at 1°C–4°C and humidity 65%–70% in dry and well-aerated stores.

Physical Methods

- Hot water soaking (50°C for 20 minutes) of onion seed proved to be the best treatment for reducing purple blotch pathogen.

- Heat treatment of onion bulbs at 35°C for 8 hours before planting and prophylatctic spray of Metalzxyl + Mancozeb @ 0.25% gave good control in the onion seed crop.

Biological Control

- Treat seeds with *Trichoderma viride* @ 4–5 g/kg seed before nursery sowing.
- Spray *Trichoderma viride* @ 10 g/L at weekly intervals.

Host Resistance

- Grow some tolerant cultivars of garlic such as G-1, G-50, Bombay Red, Red Creole, or Red Comet F1, and G-323. Onion varieties Agri found Light Red, Agrifound Dark Red, Red Globe, and VL Piyaz 3 are reported to be moderately resistant.

Chemical Control

- Treat seeds with Captan @ 2.5 g/kg seed or Thiram @ 2.5 g/kg seed before nursery sowing.
- Treat the seed with the chemical Thiophanate-methyl (@ 2.5 g/kg seed) or Imidacloprid + Tebuconazole (@ 0.4 ml into 30 ml of water or 1 kg of seed) to save the new crop from pathogen attack.
- Spray effective fungicides like Dithane M-45 (Mancozeb) @ 2.5 g/L or Chlorothalonil (Kavach) @ 2.0 g/L or Antracol (Propineb) @ 2.0 g/L or Copper oxychloride (Blitox-50) @ 3.0 g/L or Iprodione @ 2.5 g/L at 15 days of intervals.
- Apply a combination of Mancozeb @ 0.25% + Abamectin @ 0.05% for effective management of thrips and purple blotch.
- Spray Propineb 70% + Cymoxanil 6% based products such as Milraz WP 76 @ 40–50 ml in 20 L).
- Spray of other fungicides such as Metalaxyl + Mancozeb (e.g. Redomil Gold 75 WP @ 200 gm/100 litres of water/acre) can be done when the first blotch symptoms appear, at biweekly intervals, and/or when conditions are favorable for disease development.

17.2 STEMPHYLLIUM BLIGHT

Stemphylium leaf blight or Stemphylium leaf blight and stalk rot is more common than purple blotch. In Ethiopia, farmers locally call it "Yeshinkurt Ebola", meaning "Ebola of onion". The disease can drastically reduce garlic yield, as much as 70% and 60% of onion. Stemphylium leaf blight of onion and garlic was first identified from India by Rao and Pavgi in 1975.

Symptoms

On garlic, two types of leaf lesions can be observed: (i) small, dirty white to light grey spots and (ii) progressing purple spots (with sunken tissues) (Figure 17.3). These small white lesions soon develop into larger, elongated white spots that eventually become purple and water soaked. If the infection is severe, necrosis occurs in all aerial plant organs and the plant desiccates. On onion, initial infections on the leaves produce small, light-yellow to brown, water-soaked lesions. As the lesions expand, they coalesce, causing extensive blight of the leaves. The center

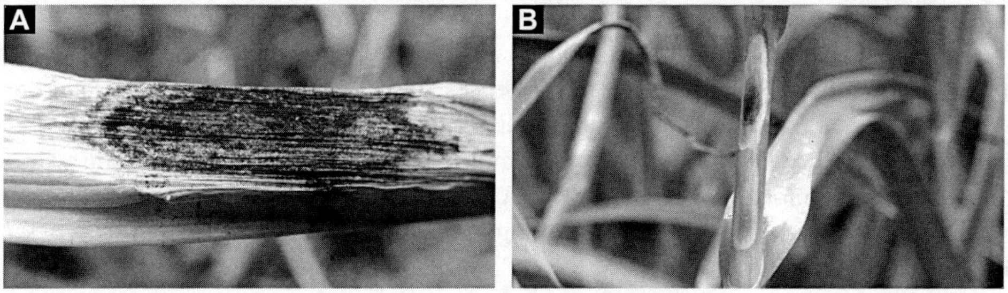

Figure 17.3 Stemphylium Blight of Garlic. (A) Leaf lesion and (B) Stem lesion.

of the lesions of onion and garlic plants, turn brown and finally dark-brown to black as the fungus sporulates. In advanced stages, the lesions may girdle and kill the leaves and seed stems. Whitish fruiting structures, pseudothecia (sexual fruiting bodies) can be seen on leaf lesions. The symptoms of Stemphylium leaf blight can be confused with purple blotch, which is caused by *Alternaria porri*. Both Stemphylium leaf blight and purple blotch lesions may occur on the same plant, and spores of each may occur on the same lesion. The cause of symptoms cannot be determined easily until the spore phases of the two pathogens occur. Infection is usually limited to leaves, and does not extend down to the scales of the bulb.

Causal Organism and Etiology

Stemphyllium blight is caused by *Stemphylium vesicarium* (Teleomorph: *Pleospora allii*). This fungus produces olive-brown to a golden brown, oblong to oval conidia that have up to four longitudinal septa and varying numbers of transverse septa. Conidia have a length-to-width ratio of 1.5–3.0. Conidia are borne singly on conidiophores that have a distinctly swollen tip. Conidia are olive-brown, oval to ovoid, and are borne on conidiophores that are pale to brown with dark edges and bands. The conidia have 1–5 transverse septa and are constricted at 1–3 transverse septa. The conidia also have 1–2 complete longitudinal septa. In some regions, the pseudothecia and ascospores of the sexual stage, *Pleospora allii*, can be found associated with crop debris. The pseudothecia are black and bear several cylindrical asci. Ascospores are yellowish brown and ellipsoidal, with the upper half narrowly tapered. Matured ascospores have 5–7 complete transverse septa and zero to several incomplete longitudinal septa.

Host Range: *Stemphylium vesicarium* is a pathogen of a wide range of hosts including: garlic (*Allium sativum*); onion (*Allium cepa*); asparagus (*Asparagus ocinalis*); European pear (*Pyrus communis*); lucerne (*Medicago sativa*); soybean (*Glycine max*) and mango (*Mangifera indica*).

Disease Cycle and Epidemiology

The disease cycle of *Stemphylium vesicarium* is characterized by sexual and asexual phases, and the fungal spores are blown into onion and garlic fields from nearby plants. In sexual phase, the *Pleospora allii* produces pseudothecia which contain asci and ascospores. Pseudothecia develops on both diseased and symptomless tissues of the host and non-host plants. Pseudothecia development is restricted to the leaves and inflorescence and requires 3–6 months for the production of asci and ascospores. The pseudothecia are able to persist on crop residues and serve as the primary inoculum in the following season. In the asexual cycle, numerous conidia (asexual spores) are produced and released from leaf lesions resulting from primary infections. Infection of onion and garlic leaves by conidia can occur by various mechanisms such as through stomatal openings, wounds caused by other diseases, insect pest feeding, or physical damage such as wind, hail, and herbicide damage, and direct infection of dead or dying leaves. Conidia disseminated by wind and water splash, and multiple infection cycles can occur within the cropping season, resulting in polycyclic epidemics. *S. vesicarium* could attack via functioning tissues or dead tissue with the help of developing appressoria. *S. vesicarium* is a weak pathogen, so it often infects onion and garlic plants secondarily, after other diseases develop or when the plant is otherwise stressed. This fungus normally invades dead and dying onion tissues. The rate of photosynthesis in onion plants is reduced because of leaf dieback. The maturation of pseudothecia and release of ascospores are closely associated with temperature and high relative humidity mild. In temperate regions, the development of pseudothecia occurs in the winter at temperatures between 5–15°C and high relative humidity. At these low temperatures, ascospores maturation takes 1–6 months. Pseudothecia release ascospores in the spring, coinciding with rainfall events. Pseudothecia are commonly observed in onion leaves, but ascospores have not been observed within these structures. Warm (18°C–25°C) humid conditions and long periods of leaf wetness (16 hours or more) favor severe disease development.

Management

Cultural Methods

- Remove or bury or destroy the infected plant debris.
- Rows of plants should follow the direction of the prevailing wind to allow adequate airflow and avoid long periods of leaf wetness.
- Reduce plant density.
- Good field drainage.
- Field sanitation: Collect and burn crop residues
- Follow long rotations with non-host crops.
- Follow proper field drainage and reduce plant density.

Physical Control

- Hot water soaking of onion seed (50°C for 20 minutes).

Host Resistance

- Grow resistant varieties such as Welsh onion (*Allium fistulosum*) lines, which are moderately resistant to Stemphylium leaf blight.

Chemical Control

- Spray Mancozeb 75% WP @ 600–800 g in 300 litres of water per acre or Zineb 75% WP @ 600–800 g in 300–400 litres of water per acre.

18

Chilli (*Capsicum annuum* L.)

18.1 ANTHRACNOSE AND FRUIT ROT

Anthracnose, ripe fruit rot and dieback is the most economically important disease fungal disease of chilli crop worldwide, affecting both fruit and seed quality. Anthracnose of chilli was first reported from New Jersey, USA, by Halsted in 1890. In India, it causes a calculated yield loss of 10%–54%, however, 10%–80% has been reported worldwide due to anthracnose disease. Anthracnose causes extensive pre- and postharvest damage to chilli fruits causing anthracnose lesions.

Symptoms

The word "anthracnose" is derived from Greek word meaning "coal", is the common name for a plant disease, characterized by very dark, sunken lesions and containing fungal spores. Anthracnose disease can occur on leaves, stems, and both pre-harvest and post-harvest fruits. Besides fruit rot, it also causes leaf spots, dieback on the stem, seedling blight, or damping off. The disease causes two major types of symptoms in chilli plants: (i) fruit rot and (ii) dieback.

Fruit rot: The characteristic symptoms appear on the ripe fruit of chilli, which appears as dark circular or angular lesions, sunken necrotic tissue with concentric rings of acervuli (asexual fruiting bodies) that are often wet and produce pink to orange conidial masses (Figure 18.1). Often several lesions coalesce and cause severe fruit rot. When a diseased fruit is cut open, the under surface of the skin is found covered with minute, elevated, spherical black stromatic masses or micro-sclerotia of the fungus. In the advance stage, the seeds are covered with a mat of fungal mycelium. Such seeds turn rusty in color. The disease affects the fruit quality, reduces dry weight of fruit, and the quantity of capsaicin and oleoresin.

Figure 18.1 Chilli Anthracnose.

Dieback: The disease causes necrosis of tender twigs from the tip to backward. The entire branches or top of the plants wither away. In advanced stage, the dead twigs are water-soaked to brown, becoming grayish-white to straw-colored. Numerous black dots (acervuli) are found scattered all over the necrotic surface of the affected twigs. On stems and leaves, the lesions appear as small sunken grayish-brown spots with dark margins, further on which acervuli develop in concentric rings.

Causal Organism and Etiology

Anthracnose and fruit rot is caused by a complex of *Colletotrichum* species (*Colletotrichum capsici, Colletotrichum truncatum*, *C. acutatum*, and *C. gleosporoides*). In India, primarily three important species, namely, *Colletotrichum truncatum*, *C. acutatum*, and *C. gleosporoides*, are responsible for the chilli anthracnose. *C. acutatum* especially causes anthracnose of unripe, green fruits of chilli. *Colletotrichum* species belongs to the Kingdom–Fungi, Phylum–Ascomycota, Class–Sordariomycetes, Order–Phyllachorales, and Family–Phyllachoraceae. The anamorphs are *Glomerella* species. *Colletotrichum* species causing anthracnose of chili in Asia were previously identified as *C. acutatum* (straight conidia with acute ends), *C. gloeosporioides* (straight conidia with obtuse ends), and *C. truncatum* (falcate conidia).

Disease Cycle and Epidemiology

The pathogens survive under unfavorable conditions as mycelia, acervuli and micro-sclerotia in plant debris, rotten fruits, seed and soil, which serves as primary inoculum. Fungi can overwinter on alternative hosts such as other solanaceous or legume crops in the field. During warm and wet periods, conidia from acervuli (asexual fruiting bodies) and micro-sclerotia are splashed by rain or irrigation water from diseased to healthy fruit and foliage. Diseased fruit acts as a source of inoculum, allowing the disease to spread from plant to plant within the field. Secondary spread of pathogen takes place through conidia formed in acevruli on leaves, stem, branches, and fruits and dispersed by water-splashes and wind. The conidia become attached to the leaf and fruit surfaces of the host. Under favorable conditions of high leaf and fruit moisture, the conidia germinate, and form appressoria normally within 24 hours. Appressoria then produce infection hyphae. Initial infection by *Colletotrichum* species involves a series of processes including the attachment of conidia to plant surfaces, germination of conidia, production of adhesive appressoria, penetration of plant epidermis, growth and colonization of plant tissue and production of acervuli and sporulation. *Colletotrichum* species have different complicated lifestyles including intramural necrotrophic, hemibiotrophic, latent or quiescent and endophytic lifestyles. The most common lifestyle for the major *Colletotrichum* species causing anthracnose of chilli is hemibiotrophy. Infection of green (unripe) fruit can occur but the pathogen remains in a quiescent state in the form of appressorial structures in tissues of unripe fruits and causes infection after fruit ripening or maturing. Chilli anthracnose can cause a significant postharvest disease due to the pathogen's ability to develop a latent or quiescent lifestyle in infected fruit. As the pathogens are seed, soil, water, and airborne, and hence may lead to damage at the seedling stage or on the aerial parts of the plants. Generally, infection occurs during warm, wet weather. Temperatures around 27°C and high humidity (90%) are optimum for disease development. The dieback phase if the disease is closely dependent on heavy and prolonged dew deposit after the rainy season.

Management

Cultural Methods

- Use disease-free healthy seeds.
- Production of pathogen-free planting materials is the key control measure used to manage the disease.
- Early removal of affected plants will control the spread of the diseases.
- Remove and destroy the solanaceous weed hosts and infected plant debris.
- Follow the crop rotation with non-solanaceous crops for two to three years.
- Choose fields that drain well.
- Minimize fruit wounds by controlling insects.
- Intercropping reduces splash dispersal of conidia.
- Stagnation of water should not be allowed in nursery beds and fields in order to avoid fungal infection.

Host Resistance

- Grow resistant/tolerant varieties such as Hisar Shakti, Hisar Vijay, TNAU Chilli Hybrid Co 1, etc.
- Pepper varieties resistant to anthracnose are K. Surkh, CH-107, Chamatkar, Saten Yellow, G-4, Bengal green, H-1, H-4, H-6, S20-1, and Lorai.

Biological Control

- Seed dressing with *Psedudomonas fluorescens* and or *Tricoderma harzianum* @ 10 g/kg of seed is highly effective.
- Stem extract of plant, *Catalpa avata* @ 1.25 mg/L, can give upto 95% control of fruit anthracnose.

Chemical Control

- Treat the seed with Benzimidazole fungicides (Benlate, Delsene M) and strobilurin fungicide (Azoxystrobin), Brassico, Bisdithane, and soak the chilli seeds in 0.2% Thiram (a dithiocarbamate) solution for 12 hours to eliminate the seedborne inoculum.
- Give three sprays of Captan @ 1.5% or Mancozeb @ 0.25%. Just before flowering, at the fruit formation stage, and 15 days after the second spray.
- For anthracnose and fruit rot apply Copper sulphate 2.62% SC @ 400 ml in 200 litres of water/acre or Copper hydroxide 77% WP @ 500 ml in 200 litres of water/acre or Captan 50% WG @ 600 g in 200 litres of water/acre or Captan 70% + Hexaconazole 5% WP @ 200–400 g in 200 litres of water/acre.
- For dieback and fruit rot spray Azoxystrobin 23% SC @ 200 ml in 200–300 litres of water/acre or Benomyl 50% WP @ 80 g in 240 litres of water/acre or Copper oxychloride (Blitox-50) 50 % WP @ 1000 g in 300–400 litres of water/acre or Captan 75% WP @ 800 g in 400 litres of water/acre or Difenoconazole 25% EC @ 50 ml in 200 litres of water/acre or Tebuconazole 25% WG @ 200–300 g in 200 litres of water/acre or Zineb 75% WP @ 600–800 g in 300–400 litres of water/acre or Chlorothalonil @ 75% WP @ 320 g in 300 litres of water/acre or Mancozeb 75% WP @ 600–800 g in 300 litres of water or Hexaconazole 2% SC @ 1200 ml in 200 litres of water/acre or Propineb 70% WP @ 200 g in 200–300 litres of water/acre.

18.2 WILT

Fusarium wilt causes significant yield losses (20%–30%). Leonian (1919) first time reported the wilt disease of chilli caused by *Fusarium* spp.

Symptoms

The symptoms appear as upward and inward rolling of the leaves, epinasty, stunting, vein clearing in younger leaflets, yellowing of older leaves, wilting of plants, and ultimately the death of the plant. Brown discoloration of the vascular system appears, particularly in the infected basal stem and roots. Generally wilting appears in localized areas of the field where a high percentage of the plants wilt and die, although scattered wilted plants may also occur.

Causal Organism and Etiology

Fusarium wilt of chilli is caused by *Fusarium oxysporum* f.sp. *capsici*. *Fusarium oxysporum* and In India, fusarium wilt is commonly caused by two species of fusarium, *F. oxysporum* and *F. solani*. However, recently *Fusarium equiseti* has been reported to cause chilli wilt from Kashmir (Northern Himalayas). Mycelium is grayish-white. Microconidia are formed singly, hyaline and cylindrical. Macroconidia are cylindrical to falcate. Chlamydospores are globose to oval and rough-walled. *Fusarium oxysporum* f.sp. *capsici* is the specific pathogen of pepper Fusarium wilt and causes a significant reduction in pepper yield. Its narrow host specificity has led to the concept of formae speciales. F. oxysporum is the most common species of this genus that causes wilt disease in numerous crops, i.e., chilies, tomatoes, and other members of the solanaceae family. The hyphae are septated, branched, and produce microconidia, macroconidia, and chlamydospores. The microconidia and macroconidia are formed externally on hypha-like conidiophore, which categorizes *Fusarium* as a Hyphomycete. Chlamydospores formed singly or in pairs from normal hypha. Conidia grow from short phialides with a false head. Macroconida are straight fussiform, pedicellate basal cells, 3–5 septate. Microconidia are abundant, ellipsoid or fussiform without, or with 1–2 septa.

Disease Cycle and Epidemiology

The fungi survive as chlamydospores in the soil for several years and are spread by farm equipment, irrigation water, and infected plant debris. The survival of chlamydospores inside the soil initiates the life cycle of *F. oxysporum*, and this stage is referred to as the saprophytic phase. Chlamydospores remain inactive in the residuals of decomposed plants and start germinating after acquiring nutrients from the roots of host plants. The development of the pathogenic strain of *F. oxysporum* chlamydospores often takes place in the hyphae inside the damaged tissue of the host plant, and in an alternative way, they might be developed from macroconidia that emerge from the sporodochia present on lesions at the surface level of the soil. After germination, conidia and chlamydospores emerge from thalli within 8 hours and 3 days respectively, under favorable conditions. Warm soil temperatures (33°C) and high soil moisture generally favor rapid disease development. The survival of chlamydospores inside the soil initiates the life cycle of *F. oxysporum* and this stage is referred to as the saprophytic phase.

Management

Cultural Methods

- Use certified seeds
- Reduce the rate of nitrogen fertilizers.
- Collect and burn any infected plants.
- Maintain the pH level of the soil.
- Flood the field after the emergence of the disease on a few plants.

Host Resistance

- Use of wilt-resistant varieties.

Biological Control

- Apply *Trichoderma viride* to the soil twice per crop season. Apply for the first time at land preparation and the second time at planting.
- Seed treatment with *Trichoderma viride* formulation @ 4 g/kg of seed is effective. Mix 2 kg *T. viride* formulation mixed with 50 kg FYM, sprinkle water, and cover with a thin polythene sheet. When mycelia growth is visible on the heap after 15 days, apply the mixture in rows of chilli in an area of one acre.

Chemical Control

- Treat the seeds with fungicides such as Benomyl (@ 2 g/kg of seed) or Carbendazim (@ 2.5 g/kg of seed) before planting.

- Dip the roots of seedling in 0.2% solution of Dithane M-45 (Mancozeb) before transplanting.
- Metalaxyl + Mancozeb (Ridomil MZ 72) or Mancozeb (e.g. Dithane-M 45) @ 250 g/100 litres of water per acre.
- Apply Fosetyl aluminum @ 100 ml in 100 litres of water per acre.
- Drenching with 1% Bordeaux mixture or Blue copper or Fytolan @ 0.25% may give protection.

18.3 LEAF CURL

Chilli leaf curl, a whitefly-borne-begomovirus disease, is a serious problem and a major limiting factor to chilli and pepper cultivation in India. It causes the frequent epidemic in chilli and pepper and lead to severe yield loss throughout the cropping area. In severe cases, even up to 100% losses of marketable fruits have been reported. Mishra et al. (1963) first time reported the natural occurrence of chilli leaf curl, to be caused by *Tobacco leaf curl virus* (TLCV). Senanayake et al. (2007) reported first time that *Chilli leaf curl virus* affecting the chilli, from Rajasthan (India).

Symptoms

The disease causes typical upward curling (towards midrib), rolling, crinkling, puckering, deformation, yellowing, thickening, and swelling of the veins, blistering of interveinous areas, shortening of internodes and petioles, and crowding of leaves (Figure 18.2). In the advanced stage, the axillary buds produce clusters of leaves with reduced size. The whole plant appears bushy and stunted. Fewer flowers and fruits develop on the diseased plants. Severely affected plants fail to produce fruit.

Figure 18.2 Chilli Leaf Curl.

Causal Agent and Etiology

Chili leaf curl disease is caused by several begomoviruses of which *Chili leaf curl virus* is the most predominant in India. The begomoviruses (Geminviridae) have circular, ssDNA genomes, that are encapsidated in distinctive twinned (geminate) icosahedral particles (approximately 18 × 30 nm in size). The genome of begomoviruses is either bipartite, consisting of two approximately 2.6–2.8 kb, circular, ssDNA components known as DNA A and DNA B, or monopartite, consisting of a single approximately 2.6–2.8 kb, circular, ssDNA component that is a homolog of the DNA A of bipartite viruses.

Disease Cycle and Epidemiology

Leaf curl epidemic results from a complex tripartite interaction between the host, vector, and virus. The virus transmission plays a central role in leaf curl epidemic. Chilli leaf curl disease causing begomoviruses are efficiently transmitted by whitefly (*Bemisia tabaci*), in a persistent manner. Whitefly (*B. tabaci*) is abundant year-round in tropical and subtropical climates where a wide variety of hosts serves as reservoirs. Rainfall prior to seedling may result in the development of an increased vector population due to abundance in the food sources. Temperature in the range of 15°C–35°C plays an important role in an epidemic, as both vector populations and virus transmissions, are influenced by temperature.

Management

Cultural Methods

- Use healthy and disease-free seeds for sowing.
- Raise the nursery in protected structures.
- Uproot and burn or bury the infected plants to avoid further infection.
- Avoid monoculture of chilli crop.
- Cover nursery beds with nylon net or straw to protect the seedlings from viral infection.
- Raise 2–3 rows of maize or sorghum as a border crop to restrict the spread of aphid vectors.
- Remove and destroy virus-infected plants and weed hosts as soon as they are noticed.
- Destroy previous year susceptible crops, particularly solanaceous weeds and volunteer plants.
- Use of yellow sticky traps just above the plants to control insect vectors.
- Use of reflective (silver color) plastic mulch.
- Use of live mulches, border crops, or hedges which are more attractive to the vectors than pepper crops.
- Adjust transplanting dates to avoid the peak season of the vector population.

Biological Control

The following biopesticides or bioagents have been used to control the vector.

- **Parasitoids**—*Eretmocerus emiratus*, *Encarsia adusta*, *Encarsia accenta*, *Eretmocerus eremicus*, etc.
- **Predators**—*Clitostethus arcuatus*, *Coccinella septempunctata*, *Orius* spp., *Sinea confuse*, *Chrysoperla carnea*, and *Chrysopa* spp.
- **Pathogens (Fungi)**—*Paecilomyces farinosus, Paecilomyces fumosoroseus*, and *Verticillium lecanii.*

- **Mycoinsecticides**—Pae-Sin (*Paecilomyces fumosoroseus* ABNPf 303), BotaniGard (*Beauveria bassiana* Strain GHA), Boveril PM (*Beauveria bassiana* CEPA ESALQ PL63), Bea-Sin (*Beauveria bassiana* ABNBb 101), Mycotal (*Lecanicillium muscarium* Ve6) and *Verticillium*.
- **Natural extracts**—Raw milk of cow, neem guard, neem oil, neem seed kernel extract (5%), biosol, repellin, seed extract of *Sapindus trifoliatus* and *Solanum trilobatum*, and leaf extract of *Clerodendrum aculeatum*, and the bark extract of *Terminalia arjuna*.

Chemical Control

- Spray suitable insecticides to reduce the transmission of the virus by insect vectors.
- Soak the seeds in a solution of Trisodium orthriphosphate (@ 150 g/L of water) for 30 minutes to inhibit seedborne inoculum. Wash the treated seed with fresh water and dry it before sowing.
- Apply Carbofuran 3G @ 4–5 kg/acre in the main field to control sucking complex and insect vectors selectively.
- Spray the crop with insecticides such as Dimethoate @ 2 ml/L or Acephate @ 1 g/L of water or Difenthiuron 50 WP @ 0.75 g/L or Monocrotophos @ 0.07% with 0.25% wettable Sulphur or Imidacloprid 17.8 SL @ 0.003% or Imidacloprid @ 0.05% or Acephate (0.1%) or Malathion @ 0.05%.

19

Turmeric (*Curcuma longa* L.)

19.1 LEAF SPOT

Leaf spot is one of the worst diseases of turmeric. This disease was first reported from Coimbatore districts in Tamil Nadu and is now prevalent in all the turmeric growing tracts in India. The disease can cause 50%–60% yield losses.

Symptoms

Elliptical to oblong, brown spots of various sizes, are most prominent on the upper surface of the young leaves. The center of these spots is grayish-white and these spots are surrounded by a yellow halo (Figure 19.1). Later, these spots may coalesce and form an irregular patch covering almost the whole leaf. The center of the spots contains numerous black dots (acervuli) arranged in concentric rings. The acervuli can appear on both upper and lower surfaces of the leaf spots. There is a characteristic black spot on the leaf which represents the acervuli of the fungus in concentric rings on the leaf. The center regions of the spots become thin, papery, and easily tear. Severely affected leaves eventually dry and wilt. The rhizomes develop poorly.

Causal Organism and Etiology

Colletotrichum capsici causes the leaf spot of turmeric.

Disease Cycle and Epidemiology

The pathogen, *Colletotrichum capsici* survives on infected leaf debris for at least one year, which serves as a primary source of infection. The fungus is carried on the scales of rhizomes which are also the source of primary infection. The same pathogen is also reported to cause leaf-spot and fruit rot of chilli where it is transmitted through seedborne infections. If chilli is grown in

Figure 19.1 Leaf Spot of Turmeric.

nearby fields or used in crop rotation with turmeric, the pathogen perpetuates easily, building up inoculum potential for epiphytotic outbreaks. The infection is usually confined to the lamina and may occasionally extend to leaf sheaths. The infection is also seen on turmeric flowers as spots. The pale brownish hyphae accumulate inside the epidermal cells and form the basis of stomatal development. The outer wall of the epidermis is ruptured, and the setae and conidiophores are exposed. The secondary spread is by wind, water, and other physical and biological agents. The disease usually appears in October and November. Relative humidity of 80% and temperatures of 21°C–23°C, favor the primary infection.

Management

Cultural Methods

- Use disease-free seed material.
- Collect and burn infected and dried leaves to reduce the inoculum in the field.
- Field sanitation should be practiced.
- Follow crop rotation with cereal and legume crops.

Host Resistance

- Grow resistant/tolerant varieties like Suguna, Sudarshan, Suroma Roma, Swarna, Nallagatla, Duvvur, Gandikota, Sudarshana, Suguna, Sugandham, Ranga, Rashmi, Rajendra, Sonia, Krishna, Prabha, Pratibha, Alleppey Supreme Sudharshana, TCP 11, TCP 56, and Kedaram.

Biological Control

- Hot water treatment of seed at 50°C for 10 minutes followed by treatment with *Trichoderma viride* in well-decomposed manure @ 1 kg *T. virde*/10 kg of manure.

Chemical Control

- Treat the seed material with Mancozeb @ 3 g/L of water or Carbendazim @ 1 g/L of water, for 30 minutes and shade dry before sowing.
- Spray Mancozeb @ 2.5 g/L of water or Carbendazim @ 1 g/L, 2–3 sprays at fortnightly intervals.
- Spray Blitox-50 or Blue copper @ 3 g/L of water.

20

Coriander (*Coriandrum sativum* L.)

20.1 STEM GALL

Stem gall of coriander is one of the most common, widespread, and serious diseases throughout the world and causes significant yield losses (10%–50%). The disease was first reported by Unger (1833) from Great Britain, and the causal organism was described as *Protomyces macrosporus*. In India, stem gall disease of coriander was reported for the first time by Sydow and Butler (1911) from Pusa (Bihar).

Symptoms

Symptoms appear as tumor-like swellings called galls on leaf veins and veinlets, midrib, leaf stalks, peduncles, stems, and fruits (Figure 20.1). The tumors are at first glossy but later rupture, become rough and hard. The swellings are usually elongated (9–12 mm) and 3–5 mm in diameter. Size of the swellings varies according to the dimensions of the plant parts infected, those on the stem usually being larger than those on the leaf peduncles and flower stalk. These galls develop due to hypertrophy and hyperplasia. The swellings on the veins give a swollen hanging appearance to the leaves. On veins, mid-ribs, and petioles, the gall appears as translucent yellow-white swellings that are often elongated and blister-like. When inflorescence is attacked, the seed formation is considerably reduced. The infected plants produce deformed, hypertrophied and hard seeds which are larger than normal seeds. The diseased seeds do not germinate. Systemic infection provides greater distortion to the plant and plant parts. The inflorescence may show outgrowth on the surface, and uniform invasion of the fruit makes it abnormally large, but partial invasion causes distortion. Severe infection may prevent fruiting and even kill the plant.

Figure 20.1 Stem Gall of Coriander. (A) Stem gall, (B) Leaf galls and deformity, and (C) Swollen seeds and deformity at floral parts.

Causal Organism and Etiology

Stem gall of coriander is caused by an ascomycete fungus, *Protomyces macrosporus*. The genus *Protomyces macrosporus* is an obligate parasite within the Apiaceae family. *P. macrosporus* was the first member of the Protomycetaceae. Presently, *P. macrosporus* has been placed in Kingdom–Fungi, Division–Ascomycota, Class–Archiascomycetes, Order–Taphrinales, Family–Protomycetaceae, Genus–*Protomyces* and species–*macrosporus*. The mycelium of the fungus is found only in the tumors. The fungal hyphae are diploid, intracellular, broad, irregularly branched, and closely septate. The hyphal cells swell into ellipsoid, or globose resting spore initials, which later develop into resting spores (chlamydospores or endospores). The young chlamydospores are yellowish brown, spherical or globose to oval, single-celled, and smooth. As they mature, they become surrounded by thick, hyaline, three-layered walls, and attain a diameter of 50–60 μm. The chlamydospores function as the resting spores of the fungus. These resting chlamydospores germinate in water by rupturing outer walls (exospores) and form sporangia and spores. Longevity of chlamydospores decreases with their age. Ascogenous cells are produced in the galls. Ascogenous cells are spherical to subspherical with thick and smooth walls, formed intercalary in the intercellular mycelium throughout the infected tissue.

Host Range: *Protomyces macrosporus* that forms galls on *Aegopodium podagraria*, *Daucus carota, Angelica sylvestris*, *Anthriscus sylvestris*, and some other members of the family Umbelliferae or Apiaceae, commonly known as umbellifers.

Disease Cycle and Epidemiology

The *Protomyces macrosporus* has a complex life cycle including ascospores and chlamydospores. The disease is seed and soilborne in nature. In soil, the pathogen overwinters in the form of chlamydospores which serve as a primary source of infection. The Chlamydospores (resting spores) are released from the decayed host tissues after the monsoon rain and cause primary infection. The infection takes place at an early stage of seed germination and the symptoms appear in a later stage and deteriorate the seed quality. The chlamydospores germinate in the presence of water by rupturing the outer wall. The inner wall is pushed out to form a vesicle, which appears in continuation with the mouth of the crack. The protoplasm passes into this vesicle and the nucleus divides several times forming 100–200 daughter nuclei. The protoplasm also divides by cleavage into several uninucleate masses. On maturity, these spores separate and collect in the center of the vesicle. The latter bursts and the spores are set free. These spores further multiply by budding in yeast-like fashion and cause infection of the host. High soil moisture and shade predispose the plants to infection. A minimum temperature range of 13°C–15°C and a maximum of 25°C–30°C, along with 45%–75% relative humidity are the most conducive for disease development. Minimum infection occurs at pH 4.6 and maximum at pH 7.4.

Management

Cultural Methods

- Use clean and disease freeseeds, collected from healthy fields.
- Proper field sanitation.
- Follow suitable crop rotation.
- Remove and destroy the diseased plants.
- The disease affects less in the early and late sown crops. Sowing crops around 16 October and 16 November gave minimum yield losses.
- Crop rotation is helpful in reducing the primary inoculum.
- Manage optimum moisture to minimize the disease.
- Apply a balance amount of fertilizers. Potassium and nitrogenous fertilizers reduce stem gall incidence while phosphorus fertilizers increase it.

Host Resistance

- Grow resistant varieties like Rajndra Swati, RCR 41, JD 1, G-5365-91, Pant Haritma, UD 20, Pant-1, CIMAP 2053, Swathi, Karan, and Hisar Bhoomi.

Chemical Control

- Seed treatment with Captan or Thiram @ 4 g/kg of seeds or Bavistin @ 2 g/kg of seeds or Thiram + Bavistin (1:1) @ 3 g/kg.
- Spray Carbendazim @ 0.1% when the symptoms start appearing and repeat the spraying at an interval of 20 days.

Marigold (*Tagetes erecta* L.)

21

21.1 BOTRYTIS BLIGHT

Botrytis blight or grey mold is one of the most destructive diseases of marigolds and several other ornamental plants such as African violet, begonia, chrysanthemum, cyclamen, dahlia, geranium, lily, peony, and rose. The extent and severity depend on weather conditions, and cultural practices.

Symptoms

The symptoms appear as dead blotches on leaves, flowers, and stems. Water-soaked, brown, or grey spots appear on affected flower heads (Figure 21.1). Rotting of stems may cause plants to collapse, flower buds may fail to open and diseased flowers that open become decayed and drop prematurely. A covering of grey fuzzy fungal growth and sporulation (conidia) appears on infected plant tissue.

Figure 21.1 Botrytis Blight of Marigold

Causal Organism and Etiology

Botrytis blight or Grey mold of marigold is caused by *Botrytis cinerea*. The young hyphae are thin, hyaline, and they become brown and septate with age. The conidiophores are light brown, septate, and erect, ramified pseudodichotomically with slightly enlarged tips bearing small pointed sterigmata bearing 1–2-celled, hyaline, oval, or globose conidia forming clusters. Sporodochia forms on the host surface and may turn into hard sclerotial masses. Sclerotia, germinate asexually by producing conidiophores bearing conidia.

Disease Cycle and Epidemiology

Botrytis cinerea persists as mycelium, conidia, or sclerotia on living or dead plant debris, and in infested soil. Sometimes, a seed lot may be contaminated with sclerotia. Upon germination, the sclerotia generally give rise directly to conidia and occasionally to infection hyphae. In some cases, sclerotia of *B. cinerea* germinate by producing apothecia and ascospores although this type of germination is very rare. Spores or conidia produced on germinating sclerotia, on infected plants, or on infected plant debris are dispersed by wind or splashing water to dead, wounded, or extremely soft plant tissues and cause infection. The fungus is capable of causing infection in all growth stages of the host plant and rapidly multiplies in declining foliage. *Botrytis* is often considered a cool weather pathogen with the best growth, sporulation, spore release, germination, and establishment of infection occurring at an optimum of 18°C–23°C; however, it is also active at low temperatures 0°C–10°C. The optimum temperature for conidial germination is 22°C–25°C with free moisture or high relative humidity (90%–100%).

Management

Cultural Methods

- Remove and destroy all infected plant parts.
- Provide proper air circulation between the plants. Wider plant spacing can increase air circulation and promote rapid drying of plants
- Maintain dry foliage and avoid overhead irrigation.
- Avoid fertilizing with excessive amounts of nitrogen. This can cause tender growth that is very susceptible to the fungus.

Host Resistance

- Using disease-resistant cultivars and replacing the susceptible planting material with tolerant cultivars can reduce the disease.

Biological Control

- *Pseudomonas* strain (QBA5) isolated from a healthy tomato plant significantly inhibit the conidial germination and mycelial growth of *B. cinerea*.
- The aqueous extract of *Asteriscus imbricatus* has inhibited completely the growth of *B. cinerea*. Plant extract of *Galenia africana* and *Elyptropappus rhinocerotis* has been found effective against *B. cinerea*.

Chemical Control

- Use fungicides such as Copper, Captan, Chlorothalonil, Mancozeb, Maneb, Sulphur, and Thiophanate methyl and apply them in advance of the disease as a protectant.

22 Rose (*Rosa* sp.)

22.1 DIEBACK

Stem canker or dieback disease is among the most serious diseases of rose and is of complex etiology. In India, rose dieback was first reported by Srivastava (1961) from Delhi. Now, it occurs in all rose-growing areas.

Symptoms

The term "dieback" refers to a progressive death of shoots or branches (Figure 22.1). The disease starts from the pruned surface of the twigs. The stem canker and dieback phase of the disease appears as a yellow to tan or dark brown to a black lesion that enlarges until the stem is girdled. Once the stem is girdled, the foliage above the canker wilts and dies that causeing brown discoloration and death of the terminal areas of the cane. The disease more frequently attacks the woody mature twigs of older plants than the younger herbaceous canes of younger plants. Severe infestation leads to the death of the plant. Tiny black fruiting bodies (pycnidia or perithecia), such as, are sometimes visible on the affected parts of the plant. Dieback also can be an above-ground expression of root damage, where disease, insect, or noninfectious disorders (such as water imbalances) may be involved.

Causal Organism and Etiology

Rose dieback is caused by *Diplodia rosarum* and other fungi including, *Botryodiplodia theobrome*, *Colletotrichum gloeosporoides*, *Coniothyrium fucklelii*, etc. *Diplodia rosarum* produces round, black pycnidia which bear spores. The pycnidiospores are dark-colored and two-celled. Perithecia are immersed in the host tissue and are surrounded by a pseudostroma. Ascospores are ellipsoidal or fusoid, hyaline, two-celled with the septum in or near the middle.

Figure 22.1 Rose Dieback.

Disease Cycle and Epidemiology

The fungi persist in dead twigs and the stalks of the withered blooms. Wet weather and high relative humidity above 80% are predisposing factors for the spread of pathogen. The disease spread faster at temperature 30°C–32°C. Conidia spread through irrigation water. As it is a weak pathogen it gains entry into the plant through wounds, injuries, and pruning cuts. During severe infection, the disease may spread upto to the base of canes and subsequently result in the death of whole plants. When plants are pruned, sometimes, the blackening of twigs starts below the pruned surface which extends further down, killing the whole plant. Older plants are more prone to the attack of the disease.

Management

Cultural Methods

- Field sanitation.
- Prune and destroy dead and diseased plant parts periodically.
- The pruning cut should be oblique to prevent the accumulation of water.
- Disinfect secateurs with 70% alcohol before giving a pruning cut to each cane.
- Overwatering, particularly in coastal areas, should be avoided.
- It is also advisable to procure bud-grafted plantlets from reliable nurseries only.

Host Resistance

- Grow resistant varieties such White christmas, Royal ascot, Crimson glory, Bhim, Blue moon, Jawahar, Pusa sonia, etc.

Chemical Control

- Apply Bordeaux paste (Copper sulphate 1 kg, unslaked lime 1.5 kg, water 10–15 L) or Chaubattia paint (Copper carbonate 4 parts, red lead 4 parts, raw linseed oil 5 parts) to cut ends at the time of pruning of bushes.
- The crop may be sprayed with Copper oxychloride (0.3%) or Mancozeb (0.25%) also at the time of pruning.
- The secateurs should be disinfected with spirit and the cut ends immediately coated with a Chaubatia paste containing 4 parts of Copper carbonate, 4 parts of red lead, and 5 parts of linseed oil.
- Add 0.1% BHC to the paste to check the digger wasp attack.
- Spray wettable Sulphur @ 0.3% or Carbendazim @ 0.1%. Give 2–3 sprays at 15 days intervals.
- Sulphur dust @ 25 kg/ha. However, sulphur is phytotoxic at the higher temperature.

22.2 POWDERY MILDEW

Powdery mildew of roses is a common problem worldwide. Nearly 40% of the fungicide sprayed on roses is to control powdery mildew. It is one of the oldest recorded plant diseases. Powdery mildew of rose is a disease thought to have been first described by Theophrastus in 300 BC.

Symptoms

Powdery mildew disease appears as a white to gray powdery coating (fungal growth) on young leaves, shoots, thorns, peduncles, buds, sepals, and flowers (Figure 22.2). Leaves may distort and drop. Severely affected plants show distorted and stunted growth and are covered with a whitish-gray mildew growth. On leaves, mildew first appears on the upper leaf surface

Figure 22.2 Powdery Mildew of Rose. (A) Symptoms on floral parts and (B) Powdery mildew symptoms on leaves.

as irregular, light green to reddish, slightly raised blister-like areas. Later, the typical dense, powdery white growth (mycelium, conidiophores, and conidia) of the mildew fungus develops. On highly susceptible rose cultivars, buds, young stems (canes), thorns, peduncles, fruit sepals, and even flower petals may become infected and entirely covered with the typically dense, powder-like growth. Flower petals are discolored, dwarfed, fail to open properly, and flowers die early. Severe infections reduce the growth of leaves and flower buds and affect photosynthetic processes.

Causal Organism and Etiology

Powdery mildew is caused by the fungus *Podosphaera pannosa* (syn. *Sphaerotheca pannosa*). This pathogen belongs to the Kingdom–Fungi, Phylum–Ascomycota, Subphylum–Pezizomycotina, Class–Leotiomycetes, Order–Erysiphales, Family–Erisiphaceae, Genus–*Podosphaera* and Species–*pannosa*. No cleistothecia (teleomorphic state of the fungus) were observed. The fungus occurs in an asexual state in crops and is identified as *Oidium leucoconium*. The anamorph state of *Oidium leucoconium* is characterized by white primary mycelium and secondary dense mycelium which form colonies of woolly white to gray appearance. Their primary hyphae are hyaline with a thin, smooth wall. Secondary hyphae are rough, sparsely branched and thick-walled. Hyphal appressoria are almost indistinct, as bumps. Conidiophores are erect with fibrosin bodies and basal cylindrical cells. Conidia (oïdia) are hyaline, ellipsoidal-ovoid to doliform, and developed basipetally in chains of 6–8 conidia. In winter, the fungus forms chasmotecia with mycelloid appendages, immersed in the mycelial layers, more or less gregarious. Asci are hyaline, ellipsoidal to ovoid (subglobose), sessile, containing 4–8 ascospores, ovoid or ellipsoid to doliform.

Disease Cycle and Epidemiology

The fungus *Podosphaera pannosa* overwinters in the fields, primarily as mycelium in plant parts, and occasionally form chasmothecia (small, black sexual fruiting structures) on leaves, petals, twigs, and stems (particularly around the thorns) and plant debris. The pathogen persists only as mycelium and conidia, in roses, grown in a greenhouse. Ascospores from chasmothecia initiate new infections. Ascospores or conidia of the fungus are dessiminated by the wind into plant tissues. The spores germinate at a temperature 21°C and relative humidity 100%. The spore germination decreases if there is a film of water on the leaves. In asexual development, conidia germinate at 20°C with 100% RH in 2–4 hours after being deposited on the host tissue. Then a short germ tube is produced from conidia and within 6 hours an initial appressorium, which hyphae that directly penetrates the cuticle develops and reaches the epidermal cells which form haustoria. The haustoria absorb soluble substances from the host cell, which are translocated to developing mycelium and conidial chains that form on the surface of leaves. Erect and short conidiophores develop over the tissue from mycelium. The complete cycle of infection occurs at temperatures between 15°C and 25°C and RH of 75 to 79%, which favor infection structures of fungus, as mycelium, conidiophores, conidia appressoria, and haustoria. Powdery mildew does not require free water on the plant surfaces to develop and is active during warm, dry summer conditions. Overhead sprinkling (irrigation or washing) during midday may limit the disease by disrupting the daily spore-release cycle.

Management

Cultural Methods

- Collect and burn infected and fallen leaves.
- Remove and destroy diseased canes.
- Prune infected parts and destroy.
- Maintain proper spacing (20 × 20 cm) between plants for good air circulation between plants.
- Avoid overhead irrigation, especially in humid conditions.
- Expose plants to direct sunlight.
- Do not cultivate in shaded areas.
- Do not do hard pruning of roses during the rainy season.
- Use a high-pressure water hose to thoroughly wet all leaf and cane surfaces in the early afternoon so plants dry quickly.

Host Resistance

- Plant cultivars are known to be powdery mildew-resistant varieties such as Ranjana, Kalawati, Dr. Homi Bhava, etc.

Biological Control

- *Ampelomyces quisqualis* (AQ10) and *Bacillus subtilis* (Serenade) have been used as biological agents to control *P. pannosa*.

Chemical Control

- Use fungicides as a preventive measure during humid weather conditions.
- Apply a dormant spray of lime sulfur at one part lime sulfur to nine parts water.
- Foliar applications of potassium salts help to reduce fungal sporulation.
- Spray Thiophanate Methyl (70% WP) @ 1 g/L of water (or 325–600 g/ha) or Chlorothalonil (500 g/L SC) @ 3 ml/L of water (or 950–1800 ml/ha) or Azoxystrobin (250 g/L SL) @ 1ml/L of water (or 500–600 ml/ha).
- Spray Karathane @ 0.05% or Topsin-M @ 0.1% or Bavistin @ 0.1%, or Sulphur 80 WP @ 0.1%.

22.3 BLACK SPOT

Black spot is also known as leaf blotch of roses. It is one of the most common and damaging diseases of roses and causes heavy losses to rose production in India and worldwide. Early descriptions of the disease date back to the early 19th century with the reports of Fries (1815) and Libert (1826).

Symptoms

Circular or irregular shaped, dark brown to black, necrotic spots, frequently with fringed (diffuse or feathery) margins are present on both sides of leaflets (Figure 22.3). The spots

Figure 22.3 Black Spot of Rose. (A–B) Typical black spot with fringed margin and (C) Severe chlorosis on leaf surrounding black spot. On close inspection of black spots reveals the presence of the small black dots (acervuli).

are often surrounded by chlorotic areas. Infected leaves characteristically turn yellow and fall prematurely. In wet weather, numerous spots are produced which coalesce and form large irregular spots covering a third of the leaf surface. On foliar lesions, acervuli (fruiting bodies) which contain conidia of the fungus can be readily seen as small black pustular structures. In cases of severe infection, reddish-purple or black lesions develop on the young stems or canes. Repeated defoliation weakens plants, causes dieback of stems, and reduction in size and number of flowers. Severely affected plants often die if the disease persists over subsequent seasons. Black spot disease may be confused with other fungal diseases such as anthracnose (*Sphaceloma rosarum*) or other minor leaf-spotting fungi. Black spot lesions can be distinguished from others by the diffuse (fringed or irregular or feathered) margin and consistently black color, whereas Cercospora leaf spot (caused by the fungus *Cercospora rosicola*) lesions have more even margins. Anthracnose (caused by the fungus, *Sphaceloma rosarum*, perfect stage: *Elsinoe rosarum*) and Cercospora leaf spot can be separated by season. Anthracnose is a cool-season disease (spring), and Cercospora leaf spot is a hot-season disease (summer).

Causal Organism and Etiology

Black spot of the rose is caused by *Diplocarpon rosae* (teleomorph). Its anamorph is *Marssonina rosae*. It is classified as an ascomycete in the family Dermateaceae and is a hemibiotrophic fungus, restricted to the genus *Rosa*. *Diplocarpon rosae* grows with a septate monokaryotic mycelium within host tissues. The vegetative body of the fungus consists of two parts viz., the subcuticular mycelium and the internal mycelium. The fungus produces acervuli on the central part of the tar spots as blister-like projections. Conidiophores are hyaline short and cylindrical. Conidia are hyaline, two-celled, fusiform or allantoid to obclavate, upper-end round, base narrow, guttulate, 18–25 × 5–6 µm. Fruiting structures (apothecia) of the sexual stage are uncommon. Asci are discoid, sub-epidermal, erumpent and 84–224 µm in diameter.

Disease Cycle and Epidemiology

The fungus, *Marssonina rosae,* overwinters as mycelium and acervuli on living (e.g. leaves, stems, and bud scales) or dead plant tissues (e.g. fallen leaves or in infected canes) left on the soil. Infection is spread by asexually produced two-celled conidia and sexual structures (apothecia) have rarely been found. Spores (conidia) are spread by wind, rain or water splash from infected plants, fallen leaves, prunings from infected plants, and mechanically during cultivation or by contact with insects. It can also be spread by the introduction of infected plants. The conidia require free water to germinate. After landing of conidia on leaf surfaces, it takes at least 9 hours of leaf wetness to infect. Usually, it is the lower leaves that are first infected and then the disease progresses upwards. The infection starts with the germination of conidia, the formation of a germ tube, and penetration of the cuticle via appressoria. During the early biotrophic stage the fungus invades the host tissue with mycelia that develop intracellular haustoria for the extraction of nutrients. This is followed by a mixed biotrophic/necrotrophic phase resulting in some tissue damage. Visual symptoms may develop in 4 days after spore germination and penetration. Once the fungus is established in plant tissues, the acervuli (fruiting bodies) develop on the spots in 11–30 days. A new crop of spores is produced and spread to healthy parts of the plant. The new infection cycle begins within 10–18 days. Leaves infected with black spots produce ethylene (plant hormone). In leaves, high ethylene content leads to leaf drop. As a result, rose plants infected with black spots are defoliated early and look bare. Leaves may drop after only one or two lesions. Fallen leaves around the plant base then serve as a reservoir of fungal spores; re-infect the plant during subsequent rains or irrigation. Warm and wet conditions favor disease development. Rainy periods in any season and temperatures between 10°C–26°C, encourage black spots.

Management

Cultural Methods

- Plant cultivars are known to be resistant.
- Avoid dense plantings and shaded areas.
- Avoid overhead watering and keep foliage as dry as possible.
- Rake up and burn all leaves in a rose planting at the season's end.
- Prune canes back to two buds if canes are infected.
- Remove and destroy diseased canes before the bud breaks.
- Roses should be planted where the sun can quickly dry the night's dew. Space roses far enough apart for good air circulation
- Remove infected canes and burn diseased leaves.
- Mulching can serve as a mechanical barrier between spores formed on old fallen leaves and new growth.
- Avoid over-fertilizing which causes soft growth which is very susceptible to black spot.

Host Resistance

- Grow resistant/tolerant rose varieties such as Martin Frobrisher, Belaya, John Cabot, Carefree Beauty, Arka Parimala, etc.
- *Rosa wichuriana*, *R. rugosa*, *R. laevigata*, *R. multiflora*, *R. virginiana* and *R. roxburghii* were highly resistant to black spot.

Chemical Control

- Spray Mancozeb or Chlorothalonil @ 0.2% or Benomyl @ 0.1% or a Copper dust.

23

Lists of Important Diseases of Crops

23.1 WHEAT (*TRITICUM AESTIVUM* L.)

Disease	Causal Organism
Fungi and Oomycetes	
Leaf rust	: *Puccinia triticina*
Stem rust	: *P. graminis* f. sp. *tritici*
Stripe rust	: *P. striiformis* f. sp. *tritici*
Loose smut	: *Ustilago tritici*
Karnal bunt	: *Tilletia indica* (*Neovossia indica*)
Powdery mildew	: *Blumeria graminis* f. sp. *tritici*
Alternaria blight	: *Alternaria triticina*
Leaf blotch/Septoria leaf blotch	: *Septoria tritici*
Glume blotch	: *Septoria nodorum*
Flag smut	: *Urocystis agropyri*
Bacterium	
Tundu/Yellow year rot or Spike blight	: *Rathayibacter tritici* (*Corynebacteriun tritici*) + *Anguina tritici*
Black chaff	: *Xanthomonas translucens* f. sp. *undulosa*
Nematode	
Ear cockle	: *Anguina tritici*
Molya disease	: *Heterodera avenae*

23.2 SUGARCANE (*SACCHARUM OFFICINARUM* L.)

Disease	Causal Organism/Agent
Fungi and Oomycetes	
Red rot	: *Colletotrichum falcatum* (teleomorph, *Glomerella tucumanensis*)
Smut	: *Ustilago scitaminea* (syn: *Sporisorium scitaminea*)
Wilt	: *Fusarium sacchari* (syn. *Cephalosporium sacchari*).
Eye spot	: *Drecheslera sacchari*
Red sheath	: *Sclerotium rolfsii*
Pokkah boeng	: *Fusarium fujikuroi* species complex (*Fusarium proliferatum, F. moniliformae, F. verticillioides, F. sacchari* and *F. andiyazi*)
Pineapple disease	: *Ceratocystis paradoxa*
Downy mildew	: *Sclerospora sacchari*
Bacteria	
Red stripe	: *Xanthomonas rubrilineans*
Gummosis	: *Xanthomonas vasculorum*
Ratoon stunting	: *Leifsonia xyli* subsp. *xyli*
Leaf scald	: *Xanthomonas albilineans*
Mottled stripe	: *Pseudomonas rubrisubalbicans*
Bacterial leaf blight	: *Acidovorax avenae* subsp. *avenae*
Phytoplasma	
Grassy shoot	: Phytoplasma
Virus	
Mosaic	: *Sugarcane mosaic virus* (Potyvirus group)
Parasitic plants	
Bunga	: *Aeginetia indica* (root parasite)
Striga	: *Striga densiflora*

23.3 SUNFLOWER (*HELIANTHUS ANNUUS* L.)

Disease	Causal Organism/Agent
Fungi and Oomycetes	
Sclerotinia stem rot	: *Sclerotinia sclerotiorum*
Alternaria blight	: *Alternaria helianthi*
Powdery mildew	: *Erysiphe cichoracearum* (*Golovinomyces cichoracearum*), *Podosphaera xanthii* and *Leveillula taurica*
Downy mildew	: *Plasmopara halstedii*
Rust	: *Puccinia helianthi*
Virus	
Stem necrosis	: *Tobacco streak virus* (Ilarvirus)

23.4 MUSTARD (*BRASSICA JUNCEA* L.)

Disease	Causal Organism
Fungi and Oomycetes	
Alternaria blight	: *Alternaria brassicae, Alternaria brassicola, Alternaria raphani*
White rust or blister	: *Albugo candida*
Downy mildew	: *Peronospora brassicae*
Powdery mildew	: *Erysiphe cruciferarum, Erysiphe polygoni*
Sclerotinia stem rot or stem blight	: *Sclerotinia sclerotiorum*
Black leg or dry root	: *Leptosphaerella maculans (Phoma lingam)*
Club root	: *Plasmodiophora brassicae*
Bacteria	
Black rot	: *Xanthomonas campestris* pv. *campestris*

23.5 CHICKPEA (*CICER ARIETINUM* L.)

Disease	Causal Organism
Fungi and Oomycete	
Alternaria blight	: *Alternaria alternata*
Ascochyta blight	: *Ascochyta rabiei* (syn. *Didymella rabiei*)
Black root rot	: *Fusarium solani, F. solani* f. sp. *pisi*
Botrytis gray mold	: *Botrytis cinerea*
Black streak root rot	: *Thielaviopsis basicola*
Collar rot	: *Sclerotium rolfsii* (syns. *Athelia rolfsii*)
Fusarium root rot	: *Fusarium acuminatum, F. arthrosporioides*, F. *avenaceum, F. equiseti, F. solani* f. sp. *eumartii* (syn. *F. eumartii*)
Fusarium wilt	: *Fusarium oxysporum* f. sp. *ciceris*
Ozonium wilt and collar rot	: *Ozonium texanum* var. *parasiticum*
Phoma blight	: *Phoma medicaginis*
Powdery mildews	: *Leveillula taurica* (syn. *Oidiopsis taurica*), *Erysiphe* spp.
Rust	: *Uromyces ciceris-arietini*
Black root rot	: *Fusarium solani, F. solani* f. sp. *pisi*
Black streak root rot	: *Thielaviopsis basicola*
Dry root rot	: *Macrophomina phaseolina* (syn. *Rhizoctonia bataticola*)
Sclerotinia stem rot	: *Sclerotinia sclerotiorum, S. trifoliorum*
Seedling or seed rot	: *Aspergillus flavus, Trichothecium roseum*
Root rot	: *Opercullela padwikii*
Verticillium wilt	: *Verticillium albo-atrum, V. dahliae*

(*contd.*)

Disease	Causal Organism
Wet root rot	: *Rhizoctonia solani* (syn. *Thanatephorus cucumeris*)
Bacteria	
Bacterial blight and damping-off	: *Xanthomonas campestris* pv. *cassiae*
Bacterial root rot	: *Pseudomonas radiciperda*
Bacterial stem canker	: *Pseudomonas andropogonis*
Pink seed	: *Erwinia rhapontici*
Nematodes	
Cyst nematode	: *Heterodera ciceri*
Reniform nematode	: *Rotylenchulus parvus*, *R. reniformis*
Root-knot nematode	: *Meloidogyne arenaria*, *M. artiellia* , *M. incognita*, *M. javanica*
Root lesion nematode	: *Pratylenchus brachyurus*, *P. mediterraneus*, *P. neglectus*, *P. penetrans*, *P. thornei*
Stem and bulb nematode	: *Ditylenchus dipsaci*

23.6 LENTIL (*LENS CULINARIS* L.)

Disease	Causal Organism
Fungi and Oomycetes	
Alternaria blight	: *Alternaria* sp., *Alternaria alternata*
Anthracnose	: *Colletotrichum lindemuthianum*, *Colletotrichum truncatum*
Ascochyta blight	: *Ascochyta fabae* f. sp. *lentis*=*Ascochyta lentis*
Wilt	: *Fusarium oxysporum* f. sp. *lentis*
Dry root rot	: *Macrophomina phaseolina* (= *Rhizoctonia bataticola*)
Powdery mildew	: *Erysiphe pisi* = *Erysiphe polygoni*, *Leveillula taurica*=*Leveillula leguminosarum* f. sp. *lentis* (anamorph: *Oidiopsis taurica*)
Rust	: *Uromyces craccae*, *Uromyces viciae-fabae* (=*Uromyces fabae*)
Wet root rot	: *Rhizoctonia solani* (teleomorph: *Thanatephorus cucumeris*)

23.7 COTTON (*GOSSYPIUM HIRSUTUM* L.)

Disease	Causal Organism/Agent
Fungi and Oomycetes	
Anthracnose	: *Glomerella gossypii* (*Colletotrichum gossypii*)
Verticillium wilt	: *Verticillium dahliae*
Fusarium wilt	: *Fusarium oxysporum* f. sp. *vasinfectum*
Areolate mildew	: *Ramularia gossypii*=*Cercosporella gossypii* (*Mycosphaerella areola*)
Ascochyta blight	: *Ascochyta gossypii*

(*contd.*)

Disease	Causal Organism/Agent
Cotton rust	: *Puccinia schedonnardi*
Southwestern cotton rust	: *Puccinia cacabata*
Tropical cotton rust	: *Phakopsora gossypii*
Bacteria	
Black arm/ Bacterial blight	: *Xanthomonas campestris* pv. *malvacearum*
Crown gall	: *Agrobacterium tumefaciens*
Lint degradation	: *Erwinia herbicola* = *Pantoea agglomerans*
Virus	
Leaf curl	: *Cotton leaf curl virus* (genus Bigeminivirus)

23.8 PEA (*PISUM SATIVUM* L.)

Disease	Causal Organism/Agent
Fungi and Oomycetes	
Downy mildew	: *Peronospora viciae*
Powdery mildew	: *Erysiphe baeumleri*, *E. diffusa* (syn. *E. glycines*), *E. pisi* (syns. *E. pisi* var. *cruchetiana*; *E. pisi* var. *pisi*), *E. trifoliorum* (syns. *E. trifolii*; *E. martii*), *Golovinomyces orontii* (syn. *Erysiphe orontii*)
Ascochyta leaf spot	: *Ascochyta pisi* (syn. *Didymella pisi*)
Rust	: *Uromyces pisi*, *U. viciae-fabae* (syn. *Puccinia fabae*)
Alternaria blight	: *Alternaria alternata*
Anthracnose	: *Colletotrichum lentis*, *C. lindemuthianum*, *C. pisi*, *C. pisicola*, *C. truncatum* (syn. *C. capsici*)
Fusarium wilt	: *Fusarium oxysporum* (syn. *F. oxysporum* f. sp. *pisi*)
White mold	: *Sclerotinia sclerotiorum*
Bacteria	
Bacterial blight	: *Pseudomonas syringae* pv. *pisi*, *P. syringae* pv. *syringae*
Brown spot	: *Pseudomonas syringae* pv. *pisi*, *P. syringae* pv. *syringae*
Pink seed	: *Erwinia rhapontici* (syn. *Pectobacterium rhapontici*)
Virus	
Pea enation mosaic	: Enamovirus, *Pea enation mosaic virus* 1 (PEMV1), Umbravirus, *Pea enation mosaic virus* 2 (PEMV 2)
Pea mosaic	: *Bean yellow mosaic virus* (BYMV)
Pea seedborne mosaic	: *Pea seedborne mosaic virus* (PSbMV)
Pea streak	: *Alfalfa mosaic virus* (AMV), *Pea streak virus* (PeSV)
Pea stunt	: *Red clover vein mosaic virus* (RCVMV)

23.9 MANGO (*MANGIFERA INDICA* L.)

Disease	Pathogen
Fungi and Oomycetes	
Anthracnose	: *Glomerella cingulata* (*Colletotrichum gloeosporioides*)
Malformation	: *Fusarium mangiferae* var. *subglutinans*
Powdery mildew	: *Acrosporium mangiferae* (syn. *Odium erysiphoides* f. sp. *mangiferae*)
Bacteria	
Bacterial blight	: *Xanthomonas axonopodis* pv. *mangiferaeindicae* (Syn. *X. campestris* pv. *mangiferaeindicae*, *X. citri* pv. *mangiferaeindicae*)
Algae	
Red rust	: *Cephaleuros virescens*
Abiotic/disorder	
Black tip	: Smoke of brick-kilns (carbon monoxide and carbondioxide, sulphur dioxide, and acetylene), Boron deficiency

23.10 CITRUS (*CITRUS* SPP.)

Disease	Causal Organism/Agent
Fungi and Oomycetes	
Gummosis	: *Phytophthora citrophthora*, *P. hibernalis*, *P. nicotianae* (syns. *P. nicotianae* var. *parasitica*, *P. parasitica*, *P. palmivora*
Melanose	: *Diaporthe citri* (syn. *Phomopsis citri*)
Anthracnose	: *Colletotrichum gloeosporioides* (syn. *Glomerella cingulata*)
Sooty mold (superficial, not pathogenic)	: *Capnodium citri*, *C. citricola*
Fly speck	: *Schizothyrium pomi*
Mal secco	: *Plenodomus tracheiphilus* (syns. *Deuterophoma tracheiphila*; *Phoma tracheiphila* L.)
Bacteria	
Canker	: *Xanthomonas citri* subsp. *citri* (syns. *X. axonopodis* pv. *citri*, *X. campestris* pv. *citri*)
Yellow dragon or greening or haunglongbing	: '*Candidatus Liberibacter africanus*' '*Ca. L. americanus*' '*Ca. L. asiaticus*'
Stubborn	: *Spiroplasma citri*
Citrus variegated chlorosis	: *Xylella fastidiosa*
Virus	
Tristeza	: *Citrus tristeza virus* (Closterovirus)

(*contd.*)

Disease	Causal Organism/Agent
Viroid	
Bark cracking	: *Citrus bark cracking viroid* (Cocadviroid), *Citrus exocortis viroid* (Pospiviroid)
Exocortis	: *Citrus exocortis viroid*
Cachexia (Xyloporosis)	: *Citrus cachexia viroid* (Hostuviroid)

23.11 GRAPEVINE (*VITIS VINIFERA* L.)

Disease	Causal Organism
Fungi and Oomycetes	
Angular leaf spot	: *Mycosphaerella angulata* (syn. *Cercospora brachypus*)
Downy mildew	: *Plasmopara viticola*
Powdery mildew	: *Uncinula necator*
Gray mold	: *Botrytis cinerea*
Anthracnose	: *Sphaceloma ampelinum* (Teleomorph: *Elsinoe ampelina*)
Bacteria	
Bacterial blight	: *Xylophilus ampelinus* (syn. *Xanthomonas ampelina*)
Crown gall	: *Agrobacterium vitis*
Pierce's disease	: *Xylella fastidiosa*

23.12 APPLE (*MALUS DOMESTICA* BORKH.)

Disease	Causal Organism/Agent
Fungi and Oomycetes	
Scab	: *Venturia inaequalis* (*Spilocaea pomi*)
Powdery mildew	: *Podosphaera leucotricha*
Alternaria blotch	: *Alternaria mali* = *A. alternata*
Nectria canker	: *Nectria galligena, Cylindrocarpon heteronemum*
American hawthorne rust	: *Gymnosporangium globosum*
Cedar apple rust	: *Gymnosporangium juniperi-virginianae*
Japanese apple rust	: *Gymnosporangium yamadae*
Marssonina blotch	: *Marssonina coronaria*
Bacteria	
Blister spot	: *Pseudomonas syringae* pv. *papulans*
Crown gall	: *Agrobacterium tumefaciens*
Fire blight	: *Erwinia amylovora*
Hairy root	: *Agrobacterium rhizogenes*

(*contd.*)

Disease	Causal Organism/Agent
Virus	
Apple stem pitting	: *Apple stem pitting virus*
Apple mosaic	: *Apple mosaic virus* (Ilarvirus)

23.13 PEACH (*PRUNUS PERSICA* L.)

Disease	Causal Organism
Fungi and Oomycetes	
Anthracnose	: *Colletotrichum gloeosporioides* (teleomorph: *Glomerella cingulata*), *C. acutatum*
Armillaria crown and root rot	: *Armillaria mellea*, *A. ostoyae*, *A. bulbosa*, *A. tabescens* = *Clitocybe tabescens*
Brown rot blossom blight and fruit rot	: *Monilinia fructicola*, *M. laxa*
Ceratocystis canker	: *Ceratocystis fimbriata*
Powdery mildew	: *Podosphaera clandestina*, *Podosphaera leucotricha*, *Sphaerotheca pannosa*
Gummosis (blister canker)	: *Botryosphaeria dothidea*= *B. berengeriana*, *Fusicoccum aesculi*, *B. obtuse*, *B. rhodina*, *B. ribis*
Leaf curl	: *Taphrina deformans*
Rust	: *Tranzschelia discolor* f. sp. *persica*
Scab	: *Cladosporium carpophilum*=*Fusicladium carpophilum*, *Venturia carpophila*
Shot hole	: *Wilsonomyces carpophilus*=*Stigmina carpophila*
Bacteria	
Bacterial canker	: *Pseudomonas syringae* pv. *syringae*
Bacterial spot = bacteriosis	: *Xanthomonas arboricola* pv. *pruni*=*X. campestris* pv. *pruni*
Crown gall	: *Agrobacterium tumefaciens*
Phony disease = phony peach	: *Xylella fastidiosa*

23.14 STRAWBERRY (*FRAGARIA* × *ANANASSA* DUCH.)

Disease	Causal Organism
Fungi and Oomycetes	
Black leaf spot	: *Alternaria alternata* f. sp. *fragariae*, *Colletotrichum gloeosporioides* = *C. fragariae*
Common leaf spot or Mycosphaerella leaf spot	: *Mycosphaerella fragariae* (anamorph: *Ramularia tulasnei*)

(*contd.*)

Disease	Causal Organism
Cercospora leaf spot	: *Cercospora fragariae, C. vexans*
Gray mold	: *Botrytis cinerea*
Alternaria fruit rot	: *Alternaria tenuissima*
Anther and pistil blight	: *Rhizoctonia fragariae* (teleomorph: *Ceratobasidium* sp.)
Leaf scorch	: *Marssonina fragariae* (*Diplocarpon earliana*)
Phomopsis leaf blight and soft rot	: *Phomopsis obscurans*
Leaf blotch and stem end rot	: *Zythia fragariae* (*Gnomonia comari*)
Anthracnose and anthracnose fruit rot and black spot	: *Colletotrichum acutatum, C. dematium, C. gloeosporioides=C. fragariae* (teleomorph: *Glomerella cingulata, Gloeosporium* spp.)
Target spot	: *Corynespora cassiicola*
Powdery mildew	: *Podosphaera aphanis* var. *aphanis* (syn: *Sphaerotheca macularis=S. humuli*)
Downy mildew	: *Peronospora potentillae=P. fragariae*
Leaf rust	: *Phragmidium potentillae=Frommea obtusa*
Synchytrium root gall	: *Synchytrium fragariae*
Bacteria	
Angular leaf spot	: *Xanthomonas fragariae*
Bacterial wilt	: *Ralstonia solanacearum*
Cauliflower disease	: *Rhodococcus fascians=Corynebacterium fascians* and *Aphelanchoides fragariae*

23.15 POTATO (*SOLANUM TUBEROSUM* L.)

Disease	Causal Organism/Agent
Fungi and Oomycetes	
Early blight	: *Alternaria solani*
Late blight	: *Phytophthora infestans*
Rhizoctonia canker and black scurf	: *Rhizoctonia solani* (teleomorph: *Thanatephorus cucumeris*)
Wart	: *Synchytrium endobioticum*
Common rust	: *Puccinia pittieriana*
Deforming rust	: *Aecidium cantensis*
Fusarium dry rot	: *Fusarium* spp., *F. sambucinum, Gibberella pulicaris* (teleomorph:=*Fusarium sulphureum*), *F. solani coeruleum*
Gangrene	: *Phoma solanicola* f. sp. *foveata, P. foveata=P. exigua* var. *foveata=P. exigua* f. sp. *foveata, P. exigua* var. *exigua*
Black dot	: *Colletotrichum coccodes=C. atramentarium*
Brown spot and Black pit	: *Alternaria alternata=A. tenuis*

(*contd.*)

Disease	Causal Organism/Agent
Powdery mildew	: *Erysiphe cichoracearum*
Powdery scab	: *Spongospora subterranea* f. sp. *subterranea*
Bacteria	
Blackleg and bacterial soft rot	: *Pectobacterium carotovorum* subsp. *atrosepticum=Erwinia carotovora* subsp. *atroseptica*, *Pectobacterium carotovorum* subsp. *carotovorum=E. carotovora* subsp. *carotovora*, *Pectobacterium chrysanthemi=E. chrysanthemi=Dickeya dadantii*
Bacterial wilt or brown rot	: *Ralstonia solanacearum*
Ring rot	: *Clavibacter michiganensis* subsp. *sepedonicus=Corynebacterium sepedonicum*
Pink eye	: *Pseudomonas fluorescens*
Common scab	: *Streptomyces scabiei=S. scabies*, *Streptomyces acidiscabies*, *Streptomyces turgidiscabies*
Nematode	
Potato cyst/golden nematode	: *Globodera rostochiensis*; *G. pallida*
Virus	
Leaf roll	: *Potato leaf roll virus* (Luteovirus)
Potato mop-top	: *Potato mop-top virus* (Furovirus)
Potato rugose mosaic	: *Potato virus Y* (Potyvirus)
Mosaic	: *Potato virus A*, *Potato virus M*, *Potato virus S*, *Potato virus X* *Potato virus Y*
Viroid	
Potato spindle tuber	: *Potato spindle tuber viroid* (PSTV)
Physiological Disorder	
Black heart	: Oxygen deficiency and high temperature in storage

23.16 CUCURBITS

Disease	Causal Organism/Agent
Fungi and Oomycetes	
Downy mildew	: *Pseudoperonospora cubensis*
Powdery mildew	: *Golovinomyces cichoracearum* (syn. *Erysiphe cichoracearum*); *Podosphaera xanthii* (syns. *Sphaerotheca fuliginea*)
Fusarium wilt	: *Fusarium oxysporum* f. sp. *benincasae*, *F. oxysporum* f. sp. *lagenariae* (bottle gourd, calabash gourd), *Fusarium oxysporum* f. sp. *luffae* (loofah, vegetable sponge), *Fusarium oxysporum* f. sp. *melonis* (melon), *Fusarium oxysporum* f. sp. *momordicae* (bitter melon), *Fusarium oxysporum* f. sp. *niveum*
Gummy stem blight	: *Stagonosporopsis caricae, S. citrulli, S. cucurbitacearum* (syns. *Didymella bryoniae*; *Phoma cucurbitacearum*; *Phyllosticta cucurbitacearum*)

(contd.)

Disease	Causal Organism/Agent
Anthracnose	: *Colletotrichum magnum* (on fruit), *C. orbiculare* (syns. *Colletotrichum lagenaria*; *Glomerella lagenaria*)
Pythium cottony leak (Pythium fruit rot)	: *Pythium aphanidermatum*, *P. oligandrum*, *P. ultimum* (syn. *Globisporangium ultimum*, *Pythium* spp.
Bacteria	
Angular leaf spot	: *Pseudomonas syringae* pv. *lachrymans*
Bacterial brown spot	: *Pantoea ananatis* (syn. *Pantoea ananatis* pv. *ananatis*)
Bacterial fruit blotch	: *Acidovorax citrulli* (syns. *Acidovorax avenae* subsp. *citrulli*; *Pseudomonas avenae* subsp. *citrulli*)
Bacterial rind necrosis	: *Erwinia* spp.
Bacterial soft rot	: *Pectobacterium carotovorum* (syn. *Erwinia carotovora* subsp. *carotovora*)
Bacterial wilt	: *Erwinia tracheiphila*
Xanthomonas leaf spot (formerly bacterial leaf spot)	: *Xanthomonas cucurbitae*
Virus	
Cucumber green mottle mosaic	: *Cucumber green mottle mosaic virus* (CGMMV)
Cucumber leaf spot	: *Cucumber leaf spot virus* (CLSV)
Cucumber mosaic	: *Cucumber mosaic virus* (CMV)
Curly top	: *Beet curly top virus* (BCTV)
Zucchini yellow mosaic	: *Zucchini yellow mosaic virus* (ZYMV)
Tomato ringspot	: *Tomato ringspot virus* (ToRSV)

23.17 ONION (*ALLIUM CEPA* L.) AND GARLIC (*ALLIUM SATIVUM* L.)

Disease	Causal Organism/Agent
Fungi and Oomycetes	
Scape blight	: *Botrytis aclada* (syn. *Botrytis allii*); *Botrytis cinerea* (syn. *Botryotinia fuckeliana*); *Botrytis squamosa* (syn. *Botryotinia squamosa*)
Purple blotch	: *Alternaria porri* (syn. *Macrosporium porri*)
Stemphylium leaf blight and stalk rot	: *Stemphylium vesicarium* (syn. *Pleospora allii*)
Rust	: *Puccinia allii*
Southern blight	: *Athelia rolfsii* (syn. *Sclerotium rolfsii*)
Smudge	: *Colletotrichum circinans* (syns. *C. dematium* f. sp. *circinans*; *Vermicularia circinans*)
Smut	: *Urocystis colchici*, *U. magica*
Downy mildew	: *Peronospora destructor*
Twister	: *Colletotrichum gloeosporioides* (syn. *Glomerella cingulata*)

(*contd.*)

Disease	Causal Organism/Agent
Bacteria	
Bacterial flower stalk and leaf necrosis	: *Pantoea agglomerans* (syns. *Enterobacter agglomerans*; *Erwinia herbicola*), *Pseudomonas marginalis*
Bacterial internal brown rot of bulbs	: *Pseudomonas aeruginosa*
Bacterial leaf spot and flower stalk necrosis	: *Pseudomonas syringae* pv. *porri*, *P. syringae* pv. *syringae*
Bacterial leaf streak and bulb rot	: *Pseudomonas viridiflava*
Bacterial soft rots	: *Dickeya chrysanthemi* (syn. *Erwinia chrysanthemi*); *E. rhapontici*; *Pectobacterium carotovorum* subsp. *carotovorum* (syn. *Erwinia carotovora* subsp. *carotovora*); *Pseudomonas marginalis*
Center rot	: *Pantoea allii*; *P. ananatis* (syn. *Erwinia ananatis*)
Enterobacter bulb decay	: *Enterobacter cloacae*
Slippery skin	: *Burkholderia gladioli* pv. *allicola* (syn. *Pseudomonas gladioli* pv. *alliicola*)
Sour skin	: *Burkholderia cepacia* (syn. *Pseudomonas cepacia*)
Xanthomonas leaf blight	: *Xanthomonas axonopodis* pv. *allii* (syn. *Xanthomonas campestris* pv. *allii*)
Virus	
Yellow dwarf	: *Onion yellow dwarf virus* (Genus Potyvirus)
Iris yellow spot	: *Iris yellow spot virus* (Genus Tospovirus)

23.18 CHILIES (*CAPSICUM ANNUUM* L.)

Disease	Causal Organism/Agent
Fungi and Oomycetes	
Anthracnose and fruit rot	: *Colletotrichum acutatum* (syn. *Glomerella acutata*); *C. coccodes*; *C. gloeosporioides* (syn. *G. cingulata*); *C. truncatum* (syn. *C. capsici*)
Frogeye leaf spot	: *Passalora capsicicola* (syn. *Cercospora capsici*)
Powdery mildew	: *Leveillula taurica* (syns. *Oidiopsis sicula*; *O. taurica*)
Fusarium wilt	: *Fusarium oxysporum* (syn. *F. oxysporum* f. sp. *capsici*)
Phytophthora blight	: *Phytophthora capsici*
Verticillium wilt	: *Verticillium albo-atrum*; *V. dahliae*
Bacteria	
Bacterial canker	: *Clavibacter michiganensis* subsp. *michiganensis* (syn. *Corynebacterium michiganensis*)
Bacterial spot	: *Xanthomonas euvesicatoria*, *X. gardneri*, *X. perforans*, *X. vesicatoria* (syn. *X. campestris* pv. *vesicatoria*)
Bacterial wilt	: *Ralstonia solanacearum*

(*contd.*)

Disease	Causal Organism/Agent
Syringae seedling blight and leaf spot	: *Pseudomonas syringae* pv. *syringae*
Virus	
Leaf curl	: Genus Begomovirus; *Chili leaf curl virus* (syn. *Tobacco leaf curl*; *Tobacco leaf curl virus*)
Physiological Disorder	
Blossom-end rot	: Calcium deficiency

23.19 TURMERIC (*CURCUMA LONGA* L.)

Disease	Causal Organism
Fungi and Oomycetes	
Leaf spot	: *Taphrina maculans*
Leaf blotch/anthracnose	: *Colletotrichum capsici*
Rhizome rot or Root rot	: *Pythium graminicolum*
Leaf spot	: *Phyllosticta zingiberi*
Root rot	: *Pythium aphanidermatum*, *Pythium graminicolum*
Rhizome rot	: *Pythium myriotylum*

23.20 CORIANDER (*CORIANDRUM SATIVUM* L.)

Disease	Causal Organism
Fungi and Oomycetes	
Stem gall	: *Protomyces macrosporus*
Wilt	: *Fusarium oxysporum* f. sp. *corianderii*
Powdery mildew	: *Erysiphe polygoni*
Blight disease complex	: *Alternaria poonensis* & *Colletotrichum* sp.
Stem rot	: *Rhizoctonia solani*
Root rot	: *Pythium sulcatum*
Grain mold	: *Alternaria* sp., *Helminthosporium* sp

23.21 MARIGOLD (*TAGETES ERECTA* L.)

Disease	Causal Organism/Agent
Fungi and Oomycetes	
Alternaria leaf spot	: *Alternaria* spp. (*Alternaria helianthi*, *Alternaria tagetica*, *Alternaria dianthi*)
Flower blight	: *Alternaria zinniae*

(*contd.*)

Disease	Causal Organism/Agent
Flower bud rot	: *Alternaria dianthi*
Botrytis blight	: *Botrytis cinerea*
Fusarium wilt	: *Fusarium oxysporum*
Septoria leaf Spot	: *Septoria tageticola*
Wilt and stem rot	: *Phytophthora cryptogea*
Collar rot	: *Phytophthora* sp., *Pythium* sp., *Rhizoctonia solani*, and *Sclerotium rolfsii*
Damping off	: *Pythium* sp., *Rhizoctoina solani*, *Ceratobasidium* sp.
Cercospora leaf spot	: *Cercospora megalopotamica*
Bacteria	
Bacterial leaf spot	: *Pseudomonas syringae pv. tagetis*
Phytoplasma	
Aster yellows	: *Aster yellows phytoplasma*
Virus	
Mosaic and mottle	: *Cucumber mosaic virus* (CMV), *Marigold mosaic virus* (MMV), and *Marigold mottle virus*
Nematode	
	: Stubby-root nematode (*Trichodorus* spp.), reniform nematode (*Rotylenchulus reniformis* and *R. parvu*), sting (*Belanolaimus* spp.), and awl nematode (*Dolichodorus* spp.)

23.22 ROSE (*ROSA* SP.)

Disease	Causal Organism/Agent
Fungi and Oomycetes	
Dieback	: *Diplodia rosarum*, *Botryodiplodia theobrome*, *Colletotrichum gloeosporoides*, *Coniothyrium fucklelii*, etc.
Powdery mildew	: *Podosphaera pannosa* (syn *Sphaerotheca pannosa*)
Downy mildew	: *Peronospora sparsa*
Black leaf spot	: *Diplocarpon rosae*
Gray mold/Botrytis blight	: *Botrytis cinerea*
Armillaria Root Rot (ARR)	: *Armillaria* sp.
Phytophthora Root Rot	: *Phytophthora* sp.
Stem canker	: *Cryptosporella umbrina*, *Coniothyrium* spp. and *Botrytis cinerea*
Rust	: *Phragmidium mucronatum* and *P. tuberculatum*
Ghost spotting	: *Bipolaris*, *Botrytis*, *Cercospora* and *Cladosporium*
Cercospora Leaf Spot (CLS)	: *Cercospora rosicola*
Anthracnose	: *Sphaceloma rosarum*

(*contd.*)

Disease	Causal Organism/Agent
Bacteria	
Crown gall	: *Agrobacterium tumefaciens*
Virus	
Rose rosette disease (RRD)	: *Rose rosette virus* (RRV)
Rose mosaic	: *Prunus necrotic ringspot virus* (PNRSV), *Apple mosaic virus* (ApMV), and *Arabis mosaic virus* (ArMV)
Nematode	: *Meloidogyne hapla*, *M. incognita*, and *Pratylenchus penetrans*

24 Exercises

24.1 MULTIPLE CHOICE QUESTIONS (MCQs)

1. **Stem rust of wheat is caused by**
   (a) *Ustilago tritici*
   (b) *Puccinia triticina*
   (c) *P. striiformis* f. sp. *tritici*
   (d) *P. graminis* f. sp. *tritici*
2. **Leaf rust of wheat is caused by**
   (a) *Ustilago tritici*
   (b) *P. striiformis* f. sp. *tritici*
   (c) *Puccinia triticina*
   (d) *P. graminis* f. sp. *tritici*
3. **Yellow rust of wheat is caused by**
   (a) *Ustilago tritici*
   (b) *P. striiformis* f. sp. *tritici*
   (c) *Puccinia triticina*
   (d) *P. graminis* f. sp. *tritici*
4. **"*Puccinia Pathway*" in India was discovered by**
   (a) Nagrajan and Joshi (1980)
   (b) Nagrajan and Singh (1975)
   (c) Bushnell and Rowell (1981)
   (d) Cummins and Caldwell (1956)
5. **"*Indian Stem Rust Rule*" was given by**
   (a) S. Nagrajan and L.M. Joshi (1980)
   (b) S. Nagrajan and Hardev Singh (1975)
   (c) S. Nagarajan and D.V. Singh (1990)
   (d) R.S. Singh and Y.L. Nene (1995)
6. **Which of the following statement is correct about the wheat rust pathogens?**
   (a) These are obligate biotrophs, macrocyclic, and heteroecious fungi
   (b) These require graminaceous hosts for the asexual phase

(c) These require major and alternate hosts for the sexual or minor phase
(d) All of these

7. Wheat rusts are distinguished from each other based on
(a) Color of pustules
(b) Arrangement of the pustules
(c) Size of pustules
(d) All of these

8. Who discovered that pycnia are the sexual structure of rust fungi?
(a) J.H. Craigei
(b) K.C. Mehta
(c) S. Nagarajan
(d) Anton de Bary

9. Wrong statement about leaf rust of wheat is
(a) It is heteroecious rust
(b) Wheat is the primary host
(c) *Thalictrum* spp. is alternative host
(d) None of these

10. On a worldwide basis, the primary host of *Puccinia triticina* is
(a) Hexaploid common wheat
(b) Tetraploid durum
(c) Wild emmer wheat
(d) Domesticated emmer wheat

11. The fungus *Erysiphe graminis* f.sp. *tritici* produces
(a) Apothecia
(b) Perithecia
(c) Cleistothecia
(d) Telia

12. The perfect stage of *Colletotrichum falcatum* is
(a) *Glomerella tucumanensis*
(b) *Glomerella cingulata*
(c) *Glomerella lindemuthianum*
(d) None of these

13. Measure applied for the management of sugarcane smut
(a) Roughing out diseased plants
(b) Planting resistant or tolerant cultivars
(c) Hot water treatment of setts
(d) All of these

14. The rust fungi completing their life cycle on a single host are called
(a) Pleomorphic
(b) Autoecious
(c) Heteroecious
(d) None of these

15. The spores which are produced typically by the binucleate mycelium and mainly function as repeating spores for the spread of rust disease in wheat is called
(a) Aeciospores
(b) Basidiospores
(c) Uredospores
(d) Teliospores

16. Promycelium of rust fungi bears
(a) Basidiospores
(b) Uredia
(c) Aeciospores
(d) Telia

17. Dipping the sugarcane setts in which fungicide can completely eradicate the smut infection?
(a) Triademifon (0.1%)
(b) Propiconazole (0.1%)
(c) Both (a) and (b)
(d) None of these

18. Karnal bunt of wheat first time reported in
(a) 1929 (b) 1931
(c) 1941 (d) 1951

19. Bunt disease of rice is caused by
(a) *Urocyctis tritici* (b) *Tilletia foetida*
(c) *Neovossia horrida* (d) None of these

20. The causal organism *Neovossia indica*
(a) Uredospores (b) Zoospores
(c) Oospores (d) Teliospores

21. Karnal bunt of wheat is caused by
(a) *Urocyctis tritici* (b) *Tilletia foetida*
(c) *Neovossia indica* (d) *Puccinia triticina*

22. Spores of *Puccinia graminis tritici* which infect the barberry plants
(a) Teliospores (b) Uredospores
(c) Aeciospores (d) Basidiospores

23. In rust life cycle, the cereal host is infected by
(a) Teliospores (b) Uredospores
(c) Aeciospores (d) Basidiospores

24. The uredospores of *Puccinia graminis tritici* is disseminated by
(a) Bird (b) Wind
(c) Insect (d) Soil

25. Optimum weather conditions for the development of yellow rust of wheat are
(a) 18°C–30°C, free moisture
(b) 15°C–20°C, free moisture
(c) 10°C–15°C, cool and wet weather
(d) 25°C–30°C, free moisture

26. *Barberries* spp. is alternate host of
(a) Yellow rust (b) Brown rust
(c) Black rust (d) All of these

27. *Mahonia* sp. is alternate host of
(a) Yellow rust (b) Brown rust
(c) Black rust (d) Red rust

28. Brown rust disease of wheat is caused by
(a) *Puccinia graminis trittici* (b) *Puccinia striiformis*
(c) *Puccinia hordae* (d) *Puccinia triticina*

29. Which of the following pathogen can survive in the form of cleistothecium in nature?
(a) *Puccinia* (b) *Peronospora*
(c) *Blumeria* (d) *Rhizoctonia*

30. Which spore is known as "*Repeating Spore*" in the disease cycle of wheat rust?
(a) Teliospore (b) Uredospore
(c) Aeciospore (d) Basidiospore

31. Barberry serves as....................for *Puccinia graminis tritici*.
(a) Collateral host (b) Alternate host
(c) Alternative host (d) Primary host

32. Who demonstrated first the heteroecious life cycle of *Puccinia graminis*?
(a) E.J. Butler (b) Anton deBary
(c) K.C. Mehta (d) H.C. Craigie

33. Ug99 is a race of
(a) White rust of mustard (b) Stem rust of wheat
(c) Stripe rust of wheat (d) Red rust of mango

34. Under natural conditions, it is common to find wheat co-infected with
(a) Stripe rust and stem rust
(b) Stem rust and leaf rust
(c) Stripe rust and leaf rust
(d) Stripe rust, stem rust, and leaf rust

35. Stripe rust of wheat produces
(a) Orange brown pustules
(b) Orange-red to dark-red pustules
(c) Orange-yellow pustules
(d) White blister like pustules

36. In a smutted ear of wheat, the black charcoal-like powder is
(a) Conidia (b) Teliospores
(c) Uredospores (d) Basidiospores

37. Wheat variety with a high degree of resistance to stem, leaf, and yellow rust
(a) HD 2967 (b) HS 542 (Pusa Kiran)
(c) PBW 343 (d) HW 1098 (Nilgiri Khapli)

38. Wheat variety resistant to stripe and leaf rust
(a) HD 2967 (b) HS 542 (Pusa Kiran)
(c) PBW 343 (d) Sonalika

39. Which of the following is an alternate host of black rust of wheat?
(a) *Thallictrum* (b) *Isopyron*
(c) *Barberries* (d) All of these

40. Which of the following pathogen can survive in the form of sclerotia in nature?
(a) *Rhizoctonia* (b) *Ralstonia*
(c) Blumeria (d) *Bremia*

41. Which of the following spore causes primary infection in wheat plants?
(a) Teliospore (b) Uredospore
(c) Aeciospore (d) Basidiospore

42. Select the wrong pair
(a) Orange brown pustule—Leaf rust of wheat
(b) Orange-red to dark-red pustule—Stem rust of wheat
(c) Orange-yellow pustules—Stripe rust of wheat
(d) White or creamy yellow blisters—Red rust of mango

43. Loose smut of wheat is caused by
(a) *Ustilago maydis* (b) *Puccinia triticina*
(c) *Uromyces fabae* (d) *Ustilago tritici*

44. Loose smut of wheat is
(a) Externally seedborne disease
(b) Internally seedborne disease
(c) Airborne disease
(d) Soilborne disease

45. Select the wrong pair
(a) Stem rust of wheat—*P. graminis* f. sp. *tritici*
(b) Yellow rust of wheat—*P. striiformis* f. sp. *tritici*
(c) Karnal bunt of wheat—*Tilletia barclayana*
(d) Leaf rust of wheat—*Puccinia triticina*

46. Loose smut of barley is caused by
(a) *Ustilago maydis* (b) *Ustilago tritici*
(c) *Urocystis agropyri* (d) *Ustilago nuda*

47. Loose smut of wheat is
(a) Polycyclic disease (b) Monocyclic disease
(c) Polyetic disease (d) None of these

48. Loose smut of wheat can be effectively managed by
(a) Foliar spray of fungicides
(b) Eradication of seedborne inoculum
(c) Soil solarization
(d) Long-term crop rotation

49. Loose smut of wheat can be effectively managed by
(a) Rouging out of diseased ears
(b) Seed treatment with Carboxin
(c) Solar heat treatment of seeds
(d) All of these

50. Externally seedborne smut disease is
(a) Common bunt of wheat
(b) Covered smut of barley
(c) Loose smut of oat
(d) All of these

51. Generally smut diseases are
(a) Host-specific
(b) Internally or externally seedborne
(c) Managed by rouging
(d) All of these

52. Majority of smut diseases are externally seedborne EXCEPT
(a) Loose smut of wheat and barley
(b) Common bunt and flag smut of wheat

(c) Covered smut and loose smut of oats
(d) Covered smut of barley

53. Flag smut of wheat (*Urocystis tritici*) differs from other cereal smut diseases by exhibiting symptoms in
(a) Stems (b) Heads
(c) Leaves (d) Roots

54. Cereal rust species cause destructive diseases because
(a) They have rapid reproductive cycles
(b) They disseminated over long distances
(c) They have specificity to cultivated hosts
(d) All of these

55. Karnal bunt of wheat was first identified by
(a) L.M. Joshi (b) K.C. Mehta
(c) S. Nagarajan (d) Manoranjan Mitra

56. Karnal bunt of wheat was first identified in
(a) 1928 (b) 1931
(c) 1965 (d) 1971

57. The minimum seed certification standards prescribed for Karnal bunt of wheat for certified seed is
(a) 0.05% (b) 0.25%
(c) 0.01% (d) 0.001%

58. The minimum seed certification standards prescribed for Karnal bunt of wheat for foundation seed is
(a) 0.05% (b) 0.25%
(c) 0.01% (d) 0.001%

59. Wheat containing................bunted grains is unfit for human consumption.
(a) 0.1% (b) 0.25%
(c) 1% (d) 3%

60. A rotten fishy smell in Karnal bunt affected wheat is due to the presence of
(a) Triethylamine (b) Dimethylamine
(c) Trimethylamine (d) Diethylamine

61. Trimethylamine a chemical compound responsible for a rotten fishy odor in Karnal bunt of wheat is secreted by
(a) Teliospores (b) Basidiospores
(c) Mycelium (d) Uredospores

62. Which of the following is a hemibiotrophic and partially systemic pathogen of wheat?
(a) *Tilletia indica* (b) *Ustilago tritici*
(c) *Puccinia triticina* (d) *Alternaria triticina*

63. Fluffy white pustules on wheat leaves, stems, heads, and awns are typical symptoms of
(a) Yellow rust (b) Alternaria blight
(c) Seed gall (d) Powdery mildew

64. "Green-islands" on infected leaves are caused by
(a) Rust pathogens (b) Powdery mildew pathogens
(c) Viral plant pathogens (d) All of these

65. Obligate biotrophs causes "green-islands" on infected leaves which are the site of
(a) Photosynthetically active host tissue
(b) Retarded senescence
(c) Living cells of both host and pathogen
(d) All of these

66. The effect of powdery mildew pathogen on host can be
(a) It can reduce photosynthesis
(b) It can increase respiration
(c) It can increase transpiration rates
(d) All of these

67. The alternative hosts for *Puccinia triticina* (leaf rust) include species of
(a) *Thalictrum* (b) *Isopyrum*
(c) *Anchusa* and *Clematis* (d) All of these

68. Wheat rust can be managed by
(a) Removal of inter-crop 'green bridges'
(b) Eradication of alternative hosts
(c) Spray of Propiconazole
(d) All of these

69. In wheat *Fusarium* spp. causes
(a) Head blight (b) Ear blight
(c) Head scab (d) All of these

70. Wheat scab is caused by
(a) *Fusarium graminearum* (b) *Parastagonospora nodorum*
(c) *Venturia inequalis* (d) *Bipolaris sorokiniana*

71. Spot blotch of wheat is caused by
(a) *Fusarium graminearum* (b) *Microbotryum violaceum*
(c) *Parastagonospora nodorum* (d) *Bipolaris sorokiniana*

72. Powdery mildew of wheat is caused by
(a) *Leveillula* (b) *Blumeria*
(c) *Podosphaera* (d) *Erysiphe*

73. Blister-like lesions arranged in stripes on wheat leaves are caused by
(a) Brown rust (b) Black rust
(c) Yellow rust (d) Powdery mildew

74. Secondary infection of powdery mildew of wheat occurs from
(a) Chlamydospores (b) Conidia
(c) Ascospores (d) Sclerotia

75. Common bunt of wheat is caused by
(a) *Tilletia indica* (b) *Tilletia caries*
(c) *Tilletia barclayana* (d) *Tilletia horrida*

76. Seed gall nematode is
(a) *Meloidogyne incognita* (b) *Radopholus similis*
(c) *Anguina tritici* (d) *Pratylenchus penetrans*

77. Ear cockle of wheat is locally known as
(a) Gegla (b) Sehun
(c) Mamni (d) All of these

78. Seed galls in cereals are also known as
(a) Ear cockle (b) Root knot
(c) Seed lesions (d) Stem gall

79. Ear cockle nematode of wheat can be effectively killed by hot water treatment at
(a) 54°C for 10 minutes (b) 55°C for 20 minutes
(c) 45°C for 10 minutes (d) 60°C for 10 minutes

80. Dusting wheat with Sulphur dust is the best option to manage
(a) Rust (b) Powdery mildew
(c) Ear cockle (d) Alternaria blight

81. Wheat seed galls can be removed by submersion of the seed in
(a) 0.5% brine solution (b) 1% brine solution
(c) 2% brine solution (d) 20% brine solution

82. Seed gall or "ear cockles" of wheat is caused by
(a) *Rathayibacter tritici* (b) *Angunia tritici*
(c) *Clavibactor xyli* (d) *Rathayibacter tritici* + *Angunia tritici*

83. "Tundu" or "Yellow ear rot" of wheat is caused by
(a) *Rathayibacter tritici* (b) *Angunia tritici*
(c) *Clavibactor xyli* (d) *Rathayibacter tritici* + *Angunia tritici*

84. First plant parasitic nematode recorded in
(a) 1665 (b) 1743
(c) 1901 (d) 1888

85. First plant parasitic nematode recorded was by
(a) M.R. Siddiqi (b) John T. Needham
(c) J.R. Christie (d) J.N. Sasser

86. First plant parasitic nematode recorded was
(a) *Xiphinema index* (b) *Angunia tritici*
(c) *Meloidogyne arenaria* (d) *Heterodera schachtii*

87. Foul smell in Karnal bunt affected wheat field is due to
(a) Ethylamine (b) Methylamine
(c) Trimethylamine (d) Triethylamine

88. Each seed gall of wheat infected with *Angunia tritici* may contain
(a) 100–200 (b) 500–800
(c) 1000–1200 (d) 1000–32,000

89. Second-stage juveniles (J_2) of *Angunia tritici* may survive in seed gall upto
(a) 0–1 year (b) 2–3 years
(c) 5–8 years (d) 28–32 years

90. *Angunia*-infected seed galls can be removed from healthy wheat seeds easily by

(a) Winnowing (b) Water floatation
(c) Salt floatation (d) All of these

91. Ear cockle of wheat is caused by

(a) Nemaphos (b) Aldicarb sulfone
(c) Carbofuran (d) All of these

92. Brine floatation method is followed to control

(a) Loose smut of wheat (b) Tundu disease of wheat
(c) Grain smut of sorghum (d) Rice blast

93. In wheat ear, the disease replaces grain with dark masses of fungal spores, giving the heads a black powdery appearance

(a) Powdery mildew (b) Black rust
(c) Common bunt (d) Loose smut

94. Most damaging root rot of wheat is caused by

(a) *Rhizoctonia* (b) *Gaeumannomyces*
(c) *Fusarium* (d) *Pythium*

95. Take-all disease of wheat caused by

(a) *Rhizoctonia solani* (b) *Gaeumannomyces graminis* var. *tritici*
(c) *Zymoseptoria tritici* (d) *Pyricularia graminis-tritici*

96. Which one is the most deadly disease of oilseed Brassica?

(a) Black rot (b) Club root
(c) White rust (d) Alternaria blight

97. Leaf blight of wheat is caused by

(a) *Tilletia indica* (b) *Alternaria triticina*
(c) *Puccinia triticina* (d) *Zymoseptoria tritici*

98. Select the wrong statement about loose smut of wheat

(a) The spikelets entirely transformed into a dry black spore mass
(b) As heads emerge, spores are dispersed by rain and wind
(c) Infected heads emerge later than normal ones
(d) Loose smut of wheat is an internally seedborne disease

99. A severe root and shoot disease of wheat that interrupts plant development

(a) Take-all (b) Loose smut
(c) Rusts (d) Alternaria blight

100. Common bunt of wheat is also referred to as "stinking smut" because of

(a) A fishy odor (b) Its symptoms
(c) Its pathogen (d) Head infection

101. Red rot disease of sugarcane is caused by

(a) Fungus (b) Bacteria
(c) Virus (d) Phytoplasma

102. The disease which is known as "Cancer" of sugarcane

(a) Smut (b) Red rot
(c) Ratoon stunt disease (d) Pokkah boeng

103. Red rot of sugarcane is caused by
(a) *Cephaleuros virescens* (b) *Colletotrichum falcatum*
(c) *Ceratocystis paradoxa* (d) *Sporisorium scitamineum*

104. Red rot of sugarcane was first reported by
(a) Went (1893) (b) Wilson (1914)
(c) Kulkarni (1911) (d) Sartoris (1929)

105. Smut disease of sugarcane was first time recorded from India by
(a) Went (1893) (b) Sydow and Butler (1906)
(c) Kulkarni (1911) (d) Wakker and Went (1896)

106. Smut disease of sugarcane is also referred as
(a) Culmicolus smut (b) Whip smut
(c) The death of sugarcane (d) All of these

107. The symptoms caused by Pokkah boeng disease of sugarcane
(a) Chlorosis (b) Top-rot
(c) Knife-cut (d) All of these

108. Pokkah boeng disease of sugarcane is caused by
(a) *Fusarium* (b) *Xanthomonas*
(c) *Colletotrichum* (d) Phytoplasma

109. Basically, pokkah boeng disease of sugarcane is
(a) Airborne (b) Soilborne
(c) Seedborne (d) Waterborne

110. Secondary infection of sugarcane pokkahboeng disease takes place through
(a) Infected setts and soil (b) Irrigation water
(c) Splashed rains (d) All of these

111. Sugarcane pokkah boeng disease can be managed effectively by
(a) Removing the diseased canes from fields
(b) Follow crop rotation for 3 years
(c) Spray fungicide such as Bavistin or Blitox
(d) All of these

112. Ratoon stunt disease (RSD) of sugarcane is caused by
(a) Fungus (b) Bacteria
(c) Phytoplasma (d) Virus

113. Grassy shoot disease of sugarcane is caused by
(a) Fungus (b) Bacteria
(c) Phytoplasma (d) Virus

114. Ratoon stunt disease (RSD) of sugarcane is caused by
(a) *Xanthomanas* (b) *Clavibacter*
(c) *Ralstonia* (d) *Pseudomonas*

115. Grassy shoot disease of sugarcane is characterized by
(a) Profuse tillering (b) Dwarfing and grassy appearance of the stools
(c) White leaf or albinism (d) All of these

116. Phytoplasma lacks
(a) True cell wall
(b) Sensitivity to penicillin
(c) Regular shape of the cell
(d) All of these

117. Pineapple disease of sugarcane is caused by
(a) *Ceratocystis paradoxa*
(b) *Colletotrichum falcatum*
(c) *Fusarium sacchari*
(d) *Sporisorium scitamineum*

118. Pokkah boeng of sugarcane is caused by
(a) *Ceratocystis paradoxa*
(b) *Colletotrichum falcatum*
(c) *Fusarium* spp.
(d) *Sporisorium scitamineum*

119. Leaf scald disease (LSD) of sugarcane is caused by
(a) *Acidovorax avenae* subsp. *avenae*
(b) *Colletotrichum falcatum*
(c) *Xanthomonas albilineans*
(d) *Clavibacter xyli* subsp. *xyli*

120. Typical symptom of sugarcane smut disease is
(a) Twisted top
(b) Whip-like structure
(c) Grassy shoots
(d) Red discoloration in pith

121. In infected plants, Phytoplasma is found in
(a) Sieve cells
(b) Mesophyll cells
(c) Epidermal cells
(d) Cortex cell

122. Phytoplasma formerly called
(a) Mycoplasma-like organisms (MLOs)
(b) Rickettsia-like organisms (RLOs)
(c) Fungi-like organisms (FLOs)
(d) All of these

123. Primary infection of sugarcane smut takes place by
(a) Soil-borne teliospores
(b) Planting infected setts
(c) Airborne fungal spores
(d) Both (a) and (b)

124. The plants produce a whip like black shoot, often very long and much curved on itself in smut of sugarcane is caused by
(a) *Ustilago nuda*
(b) *Ustilago tritici*
(c) *Ustilago scitamineum*
(d) *Ustilago hordei*

125. Sugarcane wilt is caused by
(a) *Ceratocystis paradoxa*
(b) *Colletotrichum falcatum*
(c) *Fusarium sacchari*
(d) *Sporisorium scitamineum*

126. Red rot of sugarcane is
(a) Airborne pathogens
(b) Sett-borne pathogens
(c) Soilborne pathogens
(d) All of these

127. Grassy shoot of sugarcane is transmitted by
(a) Mealybug
(b) Leafhopper
(c) Thrips
(d) Whiteflies

128. Ratoon stunting of sugarcane is caused by
(a) *Xanthomonas*
(b) *Leifsonia*
(c) *Sugarcane mosaic virus*
(d) Phytoplasma

129. ***Sclerotinia* causes**
(a) Root rot, basal stalk rot, and wilt of sunflower
(b) Sclerotinia stem rot of sunflower
(c) Sclerotinia head rot and midstalk rot of sunflower
(d) All of these

130. Mycelial growth and abundant black sclerotia are present inside (pith) and outside on the infected lesions of the stems in
(a) Sclerotinia stem rot of sunflower
(b) Smut of sugarcane
(c) Rice blast
(d) Black arm of cotton

131. Sclerotinia stem rot of sunflower can be managed by following crop rotation with
(a) Sorghum (b) Corn
(c) Wheat (d) All of these

132. Little leaf of brinjal is caused by
(a) Virus (b) Fungus
(c) Nematode (d) Phytoplasma

133. Sandal spike disease of sandal is caused by
(a) Virus (b) Fungus
(c) Nematode (d) Phytoplasma

134. Which of the following is also a natural host of sugarcane grassy shoot phytoplasma?
(a) Rice (b) Sunflower
(c) Wheat (d) Sorghum

135. Source of primary inoculum of Alternaria leaf spot of sunflower is
(a) Seed (b) Infected plant debris
(c) Both (a) and (b) (d) None of these

136. Alternaria blight of mustard is caused by
(a) *Alternaria brassicae* (b) *Alternaria brassicicola*
(c) *Alternaria raphani* (d) All of these

137. The pathogen of Alternaria blight of mustard survives in the form of
(a) Mycelium (b) Conidia and chlamydospores
(c) Microsclerotia (d) All of these

138. Resistant variety against Alternaria blight of mustard
(a) Karan Rai-4 (b) Taramira-5
(c) Saurabh (d) All of these

139. White rust of mustard is also known as
(a) Stagheads (b) White blister
(c) White blister rust (d) All of these

140. White rust of mustard is caused by
(a) *Cephaleuros mycoids* (b) *Albugo candida*
(c) *Alternaria brassicae* (d) *Puccinia graminis*

141. Viruses are similar to *Albugo* in
(a) Being an obligate parasite (b) Having DNA as genetic material
(c) Being coenocytic (d) Being eukaryotic

142. Clubroot of crucifers is caused by
(a) *Fusarium* (b) *Plasmodiophora*
(c) *Meloidogyne* (d) *Plasmopara*

143. Clubroot disease of crucifers caused by
(a) Soilborne obligate pathogen
(b) Soilborne saprophytic pathogen
(c) Airborne obligate pathogen
(d) Airborne saprophytic pathogen

144. Clubroot is potentially the most serious disease of
(a) Crucifer crops (b) Cereal crops
(c) Pulse crops (d) All of these

145. Haustoria of *Albugo* is
(a) Knob shaped (b) Finger shaped
(c) Digitate (d) Root shaped

146. *Albugo* is
(a) Intracellular, an obligate parasite
(b) Intercellular, an obligate parasite
(c) Intracellular, facultative parasite
(d) Intercellular, facultative saprophyte

147. The order of sporangia development in *Albugo* is
(a) Acropetal (b) Basipetal
(c) Radial (d) Random

148. Typical symptom of white rust of mustard is
(a) White blister on leaves (b) White blister on stems
(c) Distorted floral parts (d) All of these

149. Sporangia of *Albugo candida* collected from mustard are not infective to radish leaves because
(a) Radish leaves secrete some antifungal compounds
(b) Radish leaves have thick cuticles than mustard
(c) *Albugo* is host-specific (biological specialization)
(d) Radish is host crop and mustard is non-host crop

150. The secondary infection of white rust of mustard can be managed by
(a) Crop rotation
(b) Soil fumigation
(c) Seed treatment with Metalaxyl
(d) Foliar spray of Metalaxyl

151. White rust of mustard is favored by
(a) Dry and cool weather (b) Moist and cool weather
(c) Moist and warm weather (d) Dry and warm weather

152. Optimum temperature for the development of white rust of mustard is
(a) 5°C–10°C (b) 10°C–20°C
(c) 20°C–25°C (d) 25°C–30°C

153. The source of primary inoculum of the *Albugo candida* is
(a) Infected plant debris in soil (b) Seed
(c) Perennial weed hosts (d) All of these

154. *Albugo candida* perpetuates as
(a) Sclerotia (b) Oospores
(c) Ascospores (d) Acervuli

155. To cause infection, the germ tube of *Albugo candida* enters through
(a) Stomata (b) Lenticell
(c) Hydathod (d) Cuticle

156. *Albugo* draws nutrition from its host by
(a) Conidiophores (b) Rhizoids
(c) Phagocytosis (d) Haustoria

157. Systemic infection of *Albugo candida* on mustard results in
(a) Abnormal growth and swellings
(b) Twisted and distorted inflorescence
(c) Sterility of flowers
(d) All of these

158. 'Stagheads' in mustard is result of
(a) Systemic infection (b) Hypertrophy
(c) Hyperplasia (d) All of these

159. Which of the following is NOT a fungal disease?
(a) Rust of wheat (b) Black rot of crucifers
(c) Red rot of sugarcane (d) Common smut of maize

160. White rust of mustard differs from downy mildew
(a) White rust produces pustules
(b) Downy mildew produces a powdery coating of sporangia
(c) Downy mildew is caused by *Hyaloperonospora parasitica*
(d) All of these

161. Downy mildew of mustard is caused by
(a) *Hyaloperonospora* (b) *Albugo*
(c) *Plasmopara* (d) *Peronosclerospora*

162. Club-root of crucifers is caused by
(a) *Plasmodiophora brassicae* (b) *Albugo candidia*
(c) *Plasmopara viticola* (d) *Peronosclerospora sorghi*

163. Plant disease which interferes with normal root function
(a) Root rot soybean (b) Clubroot of crucifers
(c) Root knot of okra (d) All of these

164. Abnormal growth of plant parts results from hypertrophy and hyperplasia in
(a) Staghead of mustard (b) Clubroot of crucifers
(c) Root knot of okra (d) All of these

165. Clubroot of crucifers (*Plasmodiophora brassicae*) is most prevalent and severe at pH about
(a) 5.7 (b) 6.5
(c) 7.8 (d) 8.2

166. Club root disease of crucifers can be completely checked at pH
(a) 4.9 (b) 5.7
(c) 6.2 (d) 7.8

167. White cottony rot of mustard is caused by
(a) *Sclerotinia* (b) *Alternaria*
(c) *Macrophomina* (d) *Rhizoctonia*

168. Small cup-like fruiting bodies appear in the mustard fields. The disease most likely to be in mustard crop is
(a) White rust (b) Club root
(c) Stem rot (d) Downy mildew

169. In carpogenic germination, the sclerotia of *Sclerotinia* produce
(a) Mycelium (b) Ascospores
(c) Basidiospores (d) Oospores

170. In India, chickpea wilt was first reported by
(a) U.S. Singh (1991) (b) E.J. Butler (1918)
(c) Y.L. Nene (1978) (d) R.S. Singh (1982)

171. Chickpea wilt is caused by
(a) *Fusarium oxysporum* f. sp. *lycopersici*
(b) *Fusarium oxysporum* f. sp. *vasinfectum*
(c) *Fusarium oxysporum* f. sp. *lentis*
(d) *Fusarium oxysporum* f. sp. *ciceris*

172. Chickpea wilt (*Fusarium oxysporum*) pathogen produces
(a) Macroconidia (b) Microconidia
(c) Chlamydospores (d) All of these

173. Fusarium wilt of chickpea is
(a) Seedborne (b) Soilborne
(c) Airborne (d) Both (a) and (b)

174. In chickpea seed, Fusarium wilt pathogen survives as
(a) Chlamydospores (b) Oospores
(c) Zygospores (d) Basidiospores

175. Chickpea-resistant variety to wilt is
(a) Avrodhi (b) Uday
(c) Vaibhav (d) All of these

176. In India, Ascochyta blight of chickpea was first reported by
(a) U.S. Singh (1991) (b) E.J. Butler (1911)
(c) Y.L. Nene (1978) (d) R.S. Singh (1982)

177. Concentric rings with brown-black dots develop in the lesions of leaves, stem, and pod. The disease most likely to be

(a) Alternaria blight of mustard
(b) Ascochyta blight of chickpea
(c) Rust of beans
(d) White mold of pea

178. Black dots present in concentric circles in Ascochyta blight lesions of chickpea are

(a) Pycnidia (b) Apothecia
(c) Acervuli (d) Cleistothecia

179. Concentric rings of pycnidia are the most important diagnostic features of

(a) Septoria leaf spot of tomato
(b) Ascochyta blight of chickpea
(c) Phomopsis fruit rot of brinjal
(d) All of these

180. *Didymella rabiei* is teleomorph of

(a) *Alternaria solani* (b) *Ascochyta rabiei*
(c) *Septoria tritici* (d) *Phomopsis vexans*

181. *Ascochyta rabiei* can retain its viability in chickpea seed for at least

(a) 2 years (b) 3 years
(c) 4 years (d) 5 years

182. Infected chickpea seeds with Ascochyta blight serve as a major source of

(a) Primary inoculum (b) Long-distance dispersal
(c) Initiation of new infections (d) All of these

183. Resistant variety against Ascochyta blight of chickpea is

(a) Samrat (b) Vardan
(c) Him Chana-1 (d) All of these

184. The first occurrence of Botrytis grey mold on chickpea was reported from India by

(a) Shaw and Ajrekar (1915) (b) Butler and Bisby (1931)
(c) Gossen and Miller (2004) (d) Guar and Singh (1994)

185. Grey mold of chickpea is caused by

(a) *Penicillium expansum* (b) *Sclerotinia sclerotiorum*
(c) *Aspergillus niger* (d) *Botrytis cinerea*

186. The conidia of *Botrytis cinerea* are dispersed by

(a) Fruitfly (*Drosophila melanogaster)*
(b) Grape berry moth (*Lobesia botrana*)
(c) Mediterranean fruit fly (*Ceratitis capitata*)
(d) All of these

187. *Botrytis cinerea* can survive through infected seeds and crop debris in the form of

(a) Mycelium (b) Sclerotia
(c) Conidia (d) All of these

188. "Awarodhi" variety of gram is resistant to

(a) Wilt (b) Drought
(c) Blight (d) Flooding

189. Lentil wilt is caused by

(a) *Fusarium oxysporum* f. sp. *lycopersici*
(b) *Fusarium oxysporum* f. sp. *vasinfectum*
(c) *Fusarium oxysporum* f. sp. *lentis*
(d) *Fusarium oxysporum* f. sp. *ciceris*

190. The multi-septate conidia of *Fusarium oxysporum* f. sp. *lentis* is

(a) Macroconidia (b) Microconidia
(c) Chlamydospores (d) All of these

191. Microconidia of *Fusarium oxysporum* f. sp. *lentis* is

(a) Hyaline (b) Ovoid or kidney-shaped
(c) Single-celled (d) All of these

192. The "forma specialis" concept was first created to distinguish strains of

(a) *Puccinia graminis* (b) *Fusarium oxysporum*
(c) *Blumeria graminis* (d) All of these

193. Pathogen that destroys the vascular system of the plants

(a) *Fusarium* (b) *Ralstonia*
(c) *Verticilium* (d) All of these

194. Wilt can be managed by rotating lentil with

(a) Maize (b) Wheat
(c) Sorghum (d) All of these

195. Lentil rust disease is caused by

(a) *Uromyces* (b) *Puccinia*
(c) *Albugo* (d) *Cephalurus*

196. Development of tyloses in parenchyma cells is associated with

(a) Root rot (b) Wilt
(c) Stem rot (d) Rust

197. Pathogen that can enhance the incidence of Fusarium wilt of chickpea

(a) *Meliodogyne* (b) *Ascochyta*
(c) *Botrytis* (d) *Rhizoctonia*

198. The cotton anthracnose pathogen survives in

(a) Seed (b) Infected crop debris
(c) Several weeds host (d) All of these

199. The fruiting bodies produced by *Colletotrichum gossypii* are

(a) Asexual-Acervuli, Sexual-Perithecia
(b) Asexual-Acervuli, Sexual-Apothecia
(c) Asexual-Pycnidium, Sexual-Perithecia
(d) Asexual-Pycnidium, Sexual-Cleistothecium

200. Pathogen of black arm of cotton survives in
(a) Seed (b) Soil
(c) Diseased crop debris (d) All of these

201. The bacterium remains viable in dry leaves for
(a) 2 years (b) 6 years
(c) 12 years (d) 17 years

202. Finger leaf disease in cotton is caused due to residual effect of
(a) NAA (b) 2, 4-D
(c) Mancozeb (d) Endosulfan

203. Vector transmitting leaf curl virus in vegetable crops
(a) Aphids (b) Honey bee
(c) White fly (d) Thrips

204. The wilt of cotton is
(a) Seedborne (b) Soilborne
(c) Airborne (d) Both (a) and (b)

205. Black arm of cotton is caused by
(a) Phytoplasma (b) Virus
(c) Bacteria (d) Nematode

206. Leaf wilting is a typical symptom of
(a) *Verticillium* (b) *Ralstonia*
(c) *Xanthomonas* (d) *Fusarium*

207. Tiger stripe or tiger claw symptoms on cotton leaves are produced by
(a) Verticillium wilt (b) Fusarium wilt
(c) Bacterial blight (d) Leaf curl

208. On cotton leaf, *Verticillium dahlia* causes
(a) Blight (b) Tiger strips
(c) Mosaic (d) Leaf curling

209. Which of the disease(s) induce floral abnormalities?
(a) Downy mildew of mustard
(b) Green ear of bajra
(c) White rust of crucifers
(d) Both (b) and (c)

210. Cotton wilt which is regarded as "the cancer of cotton"
(a) Fusarium wilt (b) Verticillium wilt
(c) Sudden wilt (d) None of these

211. Downy mildew of pea is caused by
(a) *Peronospora* (b) *Sclerospora*
(c) *Bremia* (d) *Plasmopara*

212. Downey mildew of pea is
(a) Airborne (b) Seed and soilborne
(c) Seedborne only (d) Soilborne only

213. Downey mildew pathogen of pea survives in the form of

(a) Acervuli (b) Oospores
(c) Zoospores (d) Sclerotia

214. Downey mildew of pea (*Peronospora viciae*) survives in the soil for

(a) 1–5 years (b) 5–10 years
(c) 10–15 years (d) 20–25 years

215. Seedborne inoculum of downy mildew of pea can be eradicated by hot water treatment of the seed at

(a) 50°C for 25 minutes (b) 60°C for 30 minutes
(c) 54°C for 25 minutes (d) 65°C for 25 minutes

216. Powdery mildew of pea is caused by

(a) An obligate pathogen (b) A necrotrophic pathogen
(c) A hemibiotrophic pathogen (d) A facultative parasite

217. Sexual fruiting body cleistothecium is produced by

(a) Downy mildew pathogen (b) Powdery mildew fungi
(c) Rust fungi (d) Wilt fungi

218. Which of the following is NOT true about cleistothecium?

(a) It is a perennating structure (b) It is a closed ascocarp
(c) It is a sexual fruiting body (d) It contains asexual spores

219. Powdery mildew of pea develops quickly in

(a) Warm and dry conditions (b) Warm and wet conditions
(c) Cool and dry conditions (d) Cool and wet conditions

220. Highly effective fungicide for powdery mildew disease

(a) Propiconazole (b) Carbendazim
(c) Tricyclazole (d) Sulphur

221. Pea rust fungus (*Uromyces viciae-fabae*) is

(a) Macrocyclic, autoecious rust
(b) Microcyclic, autoecious rust
(c) Macrocyclic, heteroecious rust
(d) Microcyclic, heteroecious rust

222. Pea rust fungus (*Uromyces pisi*) is

(a) Macrocyclic, autoecious rust
(b) Microcyclic, autoecious rust
(c) Macrocyclic, heteroecious rust
(d) Microcyclic, heteroecious rust

223. "Alligator skin" effect on mango fruit is caused by

(a) Anthracnose (b) Black tip
(c) Bacterial black spot (d) Mango malformation

224. Mango malformation disease is caused by

(a) *Colletotrichum* (b) *Oidium*
(c) *Xanthomonas* (d) *Fusarium*

225. Powdery mildew of mango is caused by

(a) *Colletotrichum* (b) *Oidium*

(c) *Xanthomonas* (d) *Fusarium*

226. Bacterial blight of mango is caused by

(a) *Dickeya* (b) *Ralstonia*

(c) *Xanthomonas* (d) *Pseudomonas*

227. Bacterial blight of mango pathogen enters through

(a) Lenticels (b) Stomata

(c) Wounds (d) All of these

228. "Koeli" disease of mango is caused by

(a) *Fusarium subglutinans* (b) *Cephaleuros virescens*

(c) *Oidium mangiferae* (d) SO_2 emitted from brick kilns

229. Black tip is a serious problem of

(a) Banana (b) Potato

(c) Mango (d) Rice

230. The incidence of black tip of mango can be minimized by spraying

(a) Borax (1%) (b) Caustic soda (0.8%)

(c) Washing soda (0.5%) (d) All of these

231. Powdery mildew of mango is caused by

(a) *Acrosporium* (b) *Podospheria*

(c) *Blumeria* (d) *Uncinula*

232. Bacterial blight and canker of mango is caused by

(a) *Xanthomanas* (b) *Clavibacter*

(c) *Ralstonia* (d) *Pseudomonas*

233. Citrus canker in India was first reported from

(a) Punjab (b) U.P.

(c) Nagpur (d) M.P.

234. Citrus canker was first reported In India by

(a) Luthra and Sattar (1942) (b) Buchanan and Gibbons (1974)

(c) Singh and Nene (1982) (d) Singh and Verma (1991)

235. Citrus canker causes

(a) Necrotic brown lesions

(b) Blister-like lesions

(c) Corky or spongy rough canker

(d) All of these

236. The main source of inoculum of citrus canker is

(a) Infected twigs and old lesions on standing tree

(b) Infected seed

(c) Infected plant parts present in the soil

(d) Alternate host

237. Citrus greening disease is caused by
(a) Phloem inhabiting fastidious bacteria
(b) Xylem inhabiting fastidious bacteria
(c) Fungus
(d) Virus

238. Blackleg of potato is caused by
(a) *Erwinia amylovora*
(b) *Erwinia carotovora*
(c) *Ralstonia solanacearum*
(d) *Agrobacterium tumefaciens*

239. *Candidatus Liberibacter* causes
(a) Yellow dragon of citrus
(b) Huanglongbing (HLB) of citrus
(c) Citrus greening
(d) All of these

240. Huanglongbing (HLB) disease of citrus is caused by
(a) Bacteria (b) Nematode
(c) Fungus (d) Virus

241. Huanglongbing (HLB) disease of citrus is transmitted by
(a) Aphid (b) Nematode
(c) Psyllid (d) Thrips

242. Citrus canker can be managed by spraying
(a) Copper oxychloride (b) Carbendazim
(c) Dithane M45 (d) Ridomil MZ72

243. Which of the following is resistant to citrus canker?
(a) Citron (*Citrus medica*)
(b) Mandarins (*C. reticulata*)
(c) Calmondin (*Citrofortunella* sp.)
(d) All of these

244. Citrus gummosis is caused by
(a) *Phytophthora* spp. (b) *Pythium* spp.
(c) *Rhizoctonia* spp. (d) *Fusarium* spp.

245. Gummosis in plants can be caused by
(a) Fungi, Oomycetes, and Bacteria
(b) Injuries and insect damage
(c) Nutrient deficiency
(d) All of these

246. *Phytophthora* spp. causing citrus gummosis survive as
(a) Chlamydospores in soil
(b) Oospores in soil
(c) Oospores in decaying roots
(d) All of these

247. Citrus gummosis can be managed by
(a) Painting the trunk with fungicidal paint of Mefenoxam MZ
(b) Soil drenching the root zone area of the infected tree with Mefenoxam MZ
(c) Spray of Fosetyl aluminium (Aliette)
(d) All of these

248. Citrus gummosis is identified by observing
(a) Oozing of gum (b) Chlorosis
(c) Root knot (d) Death of shoot tip

249. Huanglongbing disease of citrus is caused by
(a) *Spiroplasma citri* (b) *Botrytis cineria*
(c) *Citrus exocortis virus* (d) *Candidatus liberobactor*

250. Test used for diagnosis of bacterial infection from diseased plant parts?
(a) Starch test (b) Iodine test
(c) Ames test (d) Ooze test

251. Stubborn disease of citrus is caused by
(a) *Xyllela fastidiosa* (b) *Spiroplasma citri*
(c) *Xanthomonas citri* (d) *Uncinula necator*

252. In US, a citrus eradication programme was launched to eradicate
(a) Stubborn disease (b) Citrus canker
(c) Citrus greening (d) Citrus tristeza

253. Optimum temperature for infection and multiplication of *Citrus tristeza virus* is
(a) 10°C–15°C (b) 20°C–25°C
(c) 26°C–28°C (d) 30°C–35°C

254. Bordeaux mixture was developed in
(a) 1888 (b) 1885
(c) 1842 (d) 1941

255. Bordeaux mixture was developed by
(a) J.N. Simons (b) J. Katan
(c) P.M.A. Millardet (d) Y.L. Nene

256. Bordeaux mixture or Brodo mix is also known as
(a) Millardet-David (b) Millardet-Clark
(c) Millardet-Steven (d) None of the above

257. Bordeaux mixture does not contain
(a) Copper sulphate (b) Calcium oxide
(c) Water (d) Sodium carbonate

258. Downy mildew of grapevine is caused by
(a) *Bremia* (b) *Uncinula*
(c) *Peronospora* (d) *Plasmopara*

259. The ascomycete fungus *Elsinoë ampelina* causes
(a) Anthracnose of grapes (b) Grapevine black spot
(c) Bird's-eye rot of grapes (d) All of these

260. The asexual fruiting bodies produced by fungus *Elsinoë ampelina*

(a) Acervuli (b) Perithecia
(c) Pycnidia (d) Sporodochia

261. Powdery mildew of grapes is caused by

(a) *Bremia* (b) *Uncinula*
(c) *Peronospora* (d) *Blumeria*

262. Bird's-eye rot of grapes is caused by

(a) *Elsinoë* (b) *Uncinula*
(c) *Peronospora* (d) *Plasmopara*

263. The fungus *Elsinoë ampelina* (anthracnose of grapes) overwinters in the form of

(a) Sclerotia and acervuli
(b) Acervuli and oospores
(c) Chlamydospores and oospores
(d) Sclerotia and apothecia

264. In India, the disease was first reported from Kashmir Valley in 1935 on a variety

(a) Red Delicious (b) McIntosh
(c) Ambri (d) Ambrosia

265. Apple scab disease caused an epidemic in Kashmir in

(a) 1978 (b) 1977
(c) 1976 (d) 1973

266. Apple scab disease caused an epidemic in Kashmir (1973) and severely damaged the commercial cultivar

(a) Red Delicious (b) McIntosh
(c) Ambri (d) Ambrosia

267. Apple scab is caused by

(a) Fungi (b) Bacteria
(c) Mite (d) Nematode

268. Apple scab is caused by

(a) *Venturia* (b) *Uncinula*
(c) *Elsinoë* (d) *Erwinia*

269. *Venturia inaequalis* (apple scab) overwinters as

(a) Sclerotia (b) Acervuli
(c) Chlamydospores (d) Pseudothecia

270. Bicelled footprint or flame-shaped ascospores are produced by

(a) *Venturia* (b) *Uncinula*
(c) *Elsinoë* (d) *Sclerotinia*

271. In which disease, the decomposition of infected fallen leaves are induced by spraying 5% urea for interrupting disease cycle of pathogen?

(a) Apple scab (b) Late blight of potato
(c) Anthracnose of grapes (d) Citrus canker

272. EPIVEN is a computer simulation program for
(a) Gram blight (b) Early blight of potato
(c) Apple scab (d) Powdery mildew

273. Leaf curl in plants induced by
(a) Virus (b) Fungi
(c) Insects (d) All of these

274. Naked asci are produced by
(a) *Protomyces macrosporus* (b) *Taphrina deformans*
(c) *Botrytis cineria* (d) *Rhizoctonia solani*

275. The agent that carries *Erwinia amylovora* from place to place from infected to healthy apple trees?
(a) Fruit flies (b) Bees
(c) Ants (d) All of these

276. Apple scab fungus perpetuates through
(a) Apotheica (b) Pycnidium
(c) Pseudothecia (d) Perithecia

277. Which of the following disease lead to the discovery of the Bordeaux mixture by P.M.A. Millardet (1885)?
(a) Late blight of potato (b) Mango malformation
(c) Downy mildew of grapes (d) Powdery mildew of apple

278. *Venturia inaequalis*, is a
(a) Nectrotroph (b) Biotroph
(c) Hemibiotrophs (d) None of these

279. For interrupting the disease cycle of apple scab, the decomposition of infected fallen leaves is induced by spraying
(a) 0.5% urea (b) 5% zinc sulphate
(c) 5% urea (d) 0.5% zinc sulphate

280. Cleistothecium of *Podosharea leucotricha* is characterized by
(a) One ascus and appendages dichotomously branched at tip
(b) One ascus and appendages mycelloid at tip
(c) Several ascus and appendages dichotomously branched at tip
(d) Several ascus and appendages bulbus at base

281. Discovery of "Bordeaux mixture" is associated with
(a) Powdery mildew of apple (b) Downy mildew of grapes
(c) Late blight of potato (d) Fire blight of apple and pear

282. In grapes, "Bird's Eye-Rot" appearance is the characteristic symptom of
(a) Grey mold (b) Pierce's disease
(c) Anthracnose (d) Black rot

283. The conidial cycle (conidia-infection-conidia) of *Venturia inaequalis* completes within
(a) 8–10 days (b) 15–20 days
(c) 20–25 days (d) 25–30 days

284. Which of the following is a promising bioagent against *Venturia inaeqalis*?
(a) *Athelia bombicina* (b) *Fusarium sambucinum*
(c) *Pythium oligosporum* (d) *Aphanomyces raphani*

285. First bacterial plant disease was discovered by
(a) E.F. Smith (b) A. Fisher
(c) T.J. Burrill (d) D.W. Dye

286. Fire blight of apples and pears is caused by
(a) *Erwinia* (b) *Ralstonia*
(c) *Pectobacterium* (d) *Pseudomonas*

287. *Erwinia amylovora* penetrates apple plants through
(a) Hydothods (b) Nectaries
(c) Stomata (d) All of these

288. MARYBLYT™ and CougarBlight are forecasting models effectively used for
(a) Fire blight of apple (b) Late blight of potato
(c) Early blight of tomato (d) Bacterial blight of rice

289. *Erwinia amylovora* (fire blight of apple) can spread by wind and wind-driven rain within and between trees as
(a) Ooze (b) Strands (polysaccharide threads)
(c) Aerosols (d) All of these

290. Hairy roots of apples are induced by *Agrobacterium rhizogenes* because it has
(a) Ti-plasmid (b) Ri-plasmid
(c) Ru-plasmid (d) Tu-plasmid

291. Fire blight of apple and pear was first reported by
(a) Anton deBerry (b) Adolf Mayer
(c) E.F. Smith (d) T.J. Burril

292. T.J. Burril first studied bacteria as plant disease agents in
(a) 1878 (b) 1880
(c) 1978 (d) 1982

293. Which of the following is a natural genetic engineer of plants?
(a) Yeast (b) *Agrobacterium tumefaciens*
(c) *E. coli* (d) Mycoplasma

294. The primary infection in apple scab take place by
(a) Conidia (b) Ascospores
(c) Basidiospores (d) Zygospores

295. Apple scab develops in
(a) Cool wet weather (b) Warm dry weather
(c) Cool dry weather (d) Warm humid weather

296. The fungus *Venturia inaequalis* included in the subclass
(a) Hemiascomycetidae (b) Pleosporomycetidae
(c) Plectomycetidae (d) Euascomycetidae

297. Crown gall disease in the plant is caused by
(a) Bacterium (b) Fungus
(c) Virus (d) Nematode

298. Crown gall disease in plant is caused by
(a) *Rhizobium tumefaciens* (b) *Rhizoctonia solani*
(c) *Erwinia amylovora* (d) *Meliodogyne incognita*

299. Which of the following bacterium has been used extensively in genetic engineering?
(a) *Rhizobium tumefaciens* (b) *Rhizoctonia solani*
(c) *Erwinia amylovora* (d) *Meliodogyne incognita*

300. The bacterium which infects some plants and integrates part of its Ti plasmid into the plant genome and has been used to transfer foreign genes into a number of plant species
(a) *Rhizobium alamii* (b) *Rhizobium tumefaciens*
(c) *Erwinia amylovora* (d) *Bradyrhizobium arachides*

301. Part of the Ti plasmid which integrates into chromosomal DNA of the plants
(a) T-DNA (b) Opines
(c) ori T (d) Leu 2

302. Bacterial blight of cotton is
(a) Soilborne (b) Seedborne
(c) Airborne (d) Insectborne

303. Rapid death and collapse of seedlings in the seedbed is
(a) Soft rot (b) Damping off
(c) Anthracnose (d) Wilt

304. A localized wound, necrotic lesion, raised and corky is
(a) Dry rot (b) Wet rot
(c) Canker (d) Scab

305. Soft rot of fruits and vegetables is caused by the production of
(a) Cell wall degrading enzymes
(b) Auxins
(c) Cytokinins
(d) Ethylene

306. The symptoms of anthracnose appear in the form of
(a) Sudden death of plants
(b) Rapid necrosis in a downward direction
(c) Irregular sunken spot with grayish black center
(d) Rottening of plant tissues

307. *Fusarium oxysporum* belongs to the family
(a) Moniliaceae (b) Melanconiaceae
(c) Dematiaceace (d) Nectriaceae

308. Powdery mildew of cucurbits is caused by
(a) *Sclerospora graminicola*
(b) *Erysiphe cichoracearum*

(c) *Erysiphe polygoni*
(d) *Pseudoperonospora cubensis*

309. Blackarm is the most important disease of

(a) Rice (b) Citrus
(c) Cotton (d) Sugarcane

310. Hot-water treatment is useful for

(a) Loose smut of oat
(b) Loose smut of wheat and barley
(c) Karnal bunt of wheat
(d) None of these

311. Water-soaked areas are the characteristic symptoms of

(a) Fungal diseases (b) Bacterial diseases
(c) Viral diseases (d) Nematode diseases

312. "Citrus stubborn disease" (CSD) is caused by

(a) Spiroplasma (b) Phytoplasma
(c) Mycoplasma (d) Protoplasma

313. "Citrus stubborn disease" (CSD) is transmitted by

(a) Leaf hopper (b) Thrips
(c) Psyllid (d) Aphid

314. Most fungal diseases spread out in

(a) Dry and cool weather
(b) Wet and cool weather
(c) Dry and warm weather
(d) Wet and warm weather

315. Gram blight is a serious disease in

(a) Dry regions (b) Humid regions
(c) Hot regions (d) Cooler regions

316. Bordeaux mixture is mixed with a ratio of......... for the management of grape downy mildew.

(a) 4:4:50 (b) 2:3:50
(c) 2:2:50 (d) 2:2:100

317. Bordeaux mixture was discovered by Millardet in

(a) 1887 (b) 1885
(c) 1883 (d) 1882

318. Most plant viruses contain

(a) RNA (b) DNA
(c) Both RNA + DNA (d) None of these

319. Nucleic acids of most of the plant virus consist of

(a) ssRNA (b) ssDNA
(c) dsDNA (d) dsRNA

320. Tomato leaf curl disease is managed by using
(a) Trap crop (b) Yellow sticky traps
(c) Insecticides (d) All of these

321. *Mungbean yellow mosaic virus* belongs to
(a) Potex virus (b) Potyvirus group
(c) Geminivirus (d) Caulimovirus

322. In case of green ear disease of bajra, optimum temperature for conidia formation is
(a) 5°C–10°C (b) 10°C–15°C
(c) 8°C–20°C (d) 15°C–25°C

323. Early blight of potato and tomato is caused by
(a) *Fusarium solani* (b) *Alternaria solani*
(c) *Rhizoctonia solani* (d) *Fusarium oxysporum*

324. Ear cockle disease of wheat attacks on
(a) Roots (b) Stems
(c) Spike (d) Leaves

325. Tirak or bad opening of cotton bolls is mainly due to the deficiency of
(a) P (b) N
(c) K (d) Zn

326. Select the characteristic features of *Erwinia amylovora*
(a) Facultative anaerobe, gram-negative, rod-shaped, peritrichous flagella
(b) Facultative anaerobe, gram-positive, rod-shaped, lophotrichous flagella
(c) Anaerobic, gram-negative, rod-shaped, amphitrichous flagella
(d) Anaerobic, gram-positive, rod-shaped, polar flagella

327. Blight-Ban A506, a bioformulation used for fire blight of apple contains
(a) *Pseudomonas fluorescens* (b) *Metarhizium anisopliae*
(c) *Trichoderma harzianum* (d) *Erwinia amylovora*

328. Peach leaf curl pathogen (*Taphrina deformans*) survives in the form of
(a) Blastospores (b) Sclerotia
(c) Apothecia (d) Teliospores

329. Pathogen that produces ascospores on naked asci
(a) *Taphrina deformans* (b) *Neurospora crassa*
(c) *Erysiphe cichoracearum* (d) *Sclerospora graminicola*

330. Peach leaf curl can be managed by spraying
(a) Fytolon (b) Blitox-50
(c) Lime sulphur (d) All of these

331. Angular leaf spot of strawberry is caused by
(a) *Colletotrichum fragariae* (b) *Alternaria solani*
(c) *Mycosphaerella fragariae* (d) *Xanthomonas fragariae*

332. Select the characteristic features of *Xanthomonas fragariae* (angular leaf spot of strawberry)
(a) Aerobic, gram-positive, spore former, rod-shaped, single polar flagellum
(b) Aerobic, gram-negative, non-spore former, rod-shaped, single polar flagellum

(c) Aerobic, gram-positive, non-spore former, rod-shaped, peritrichous flagellum
(d) Aerobic, gram-negative, spore former, rod-shaped, peritrichous flagellum

333. *Xanthomonas fragariae* (angular leaf spot of strawberry) invades plants through the
(a) Stomata (b) Hydathodes
(c) Wounds (d) All of the above

334. Leaf spot of strawberry is caused by
(a) *Colletotrichum fragariae* (b) *Alternaria solani*
(c) *Mycosphaerella fragariae* (d) *Xanthomonas fragariae*

335. *Mycosphaerella fragariae* (leaf spot of strawberry) overwinters as
(a) Microsclerotia (b) Macrosclerotia
(c) Oospores (d) Acervuli

336. Black spot of strawberry is caused by
(a) *Colletotrichum* (b) *Alternaria*
(c) *Mycosphaerella* (d) *Xanthomonas*

337. The mycelium of *Alternaria solani* remains viable for a year or more in
(a) Dry infected leaves (b) Fresh infected leaves
(c) Dry infected stems (d) Dry infected roots

338. Black scurf of potato is caused by
(a) *Spongospora subterranea* (b) *Synchytrium endobioticum*
(c) *Rhizoctonia solani* (d) *Phytophthora infestans*

339. Wart disease of potato is caused by
(a) *Spongospora subterranea* (b) *Synchytrium endobioticum*
(c) *Rhizoctonia solani* (d) *Phytophthora infestans*

340. The pathogen which causes dark-brown to black cauliflower-like outgrowths on potato tuber eyes, stolon buds and the base of the stems
(a) *Spongospora subterranea* (b) *Synchytrium endobioticum*
(c) *Rhizoctonia solani* (d) *Phytophthora infestans*

341. "False wart" or "pseudo-wart" of potato is caused by
(a) *Spongospora subterranea* (b) *Synchytrium endobioticum*
(c) *Rhizoctonia solani* (d) Proliferation of eyes or chemical contaminant

342. Scab disease of potato can be managed best at pH
(a) 5.5–6.5 (b) 3.5–4.5
(c) 7.2–7.8 (d) 8.2–9.2

343. Ridomil MZ 72 can be used for the management of
(a) Late blight of potato (b) Powdery mildew of mango
(c) Anthracnose of bean (d) Early blight of tomato

344. Dehaluming is an effective cultural practice to manage
(a) Late blight of potato (b) Leaf roll of potato
(c) Black scurf of potato (d) Early blight of potato

345. In which of the following potato disease, dark-brown to black, cauliflower-like outgrowths appears on the tuber?

(a) Black scurf (b) Common scab
(c) Powdery scab (d) Wart disease

346. Blackleg and soft rot disease of potato is caused by

(a) *Phoma lingum* (b) *Pectobacterium carotovorum*
(c) *Fusarium solani* (d) *Rhizoctonia solani*

347. Which of the following serve as primary inoculum for late blight of potato?

(a) Infected leaves (b) Crop residue
(c) Infected tubers (d) None of these

348. Common scab disease of potato is caused by

(a) *Alternaria solani* (b) *Sclerotium rolfsii*
(c) *Spongospora subterranean* (d) *Streptomyces scabiei*

349. Stem canker and black scurf of potato is caused by

(a) *Rhizoctonia solani* (b) *Helmithosporium solani*
(c) *Spongospora subterranean* (d) *Streptomyces scabiei*

350. Black heart disease of potato is caused by

(a) Low oxygen and low temperature
(b) High oxygen and high temperature
(c) Low oxygen and high temperature
(d) High oxygen and low temperature

351. Gangrene disease of potato is caused by

(a) *Phoma solanicola* (b) *Spongospora subterranean*
(c) *Potato Virus* Y (d) *Rhizoctonia solani*

352. Late blight of potato is caused by

(a) *Alternaria solani* (b) *Phytophthora infestans*
(c) *Synchytrium endobioticum* (d) *Septoria lycopersici*

353. *Phytophthora infestans* produces zoospores at

(a) 16°C (b) 12°C
(c) 22°C (d) 18°C

354. The Irish potato famine (1840s) was caused by

(a) Early blight of potato (b) Late blight of potato
(c) Brown spot of rice (d) Wart disease of potato

355. The pathogen of late blight of potato was first described by

(a) Anton de Bary (b) M.J. Berkeley
(c) E.J. Butler (d) Louis Pasteur

356. The causal organism of the late blight of potato was named *Phytophthora infestans* by

(a) Anton de Bary (b) M.J. Berkeley
(c) E.J. Butler (d) H.M. Ward

357. The name "*Phytophthora*" is derived from the Greek word which means

(a) Plant destroyer (b) Plant protector
(c) Plant pathogen (d) Plant growth retardant

358. Infections of *Phytophthora infestans* generally begin in tuber from
(a) Tuber cracks (b) Tuber eyes
(c) Tuber lenticels (d) All of these

359. In *Phytophthora infestans* sexual reproduction may occur if
(a) Both mating types (A1 and B1) are present
(b) Both mating types (A1 and A2) are present
(c) A1 mating type is present and A2 mating type is absent
(d) A2 mating type is present and A1 mating type is absent

360. Which of the following potato variety is resistant to late blight?
(a) Kufri Naveen (b) Kufri Jeevan
(c) Kufri Alenkar (d) All of these

361. In which of the following disease "aerial tubers" develop in the axis of branches and petioles of potato?
(a) Early blight (b) Leaf roll
(c) Black scurf (d) Bacterial wilt

362. Reddening of leaf margins and tips of leaves, leaf rolling, usually leaflets rolled upwards and start from lower and oldest leaves are the major symptoms of the disease on leaves of potato caused by
(a) Phytoplasma (b) *Potato leaf roll virus*
(c) *Rhizoctonia solani* (d) *Potato virus Y*

363. *Globodera rostochiensis* is commonly known as the
(a) Golden nematode (b) Golden eelworm
(c) Yellow potato cyst nematode (d) All of these

364. Golden nematode is a pest of
(a) Soybean (b) Potato
(c) Kiwi (d) Maize

365. Cysts of the golden nematode can live in soil for
(a) 5 years (b) 10 years
(c) 20 years (d) 30 years

366. Potato cyst nematode is
(a) *Heterodera* spp. (b) *Meliodogyne* spp.
(c) *Ditylenchus* spp. (d) *Globodera* spp.

367. Potato disease that induces the development of small green aerial tubers
(a) Late blight (b) Black scurf
(c) Wart (d) Leaf roll

368. Which of the following is a *Rhizoctonia*-specific biocontrol agent?
(a) *Trichoderma harzianum* (b) *Gliocladium catenulatum*
(c) *Verticillium biguttatum* (d) *Pythium oligoandrum*

369. Wart disease was first reported in India from
(a) Darjeeling (b) Nanital
(c) Nilgiri (d) Mysore

370. Wart disease of potato is also known as
(a) Potato canker (b) Black scab
(c) Potato tumor (d) All of these

371. Select the correct statement about *Synchytrium endobioticum*
(a) It is an obligate parasite (b) It is a holocarpic parasite
(c) It is an endobiotic parasite (d) All of these

372. Resting sporangia of *Synchytrium endobioticum* may remain viable for
(a) 5–10 years (b) 10–15 years
(c) 15–20 years (d) 30–50 years

373. Potato wart disease is favored by
(a) Cool, wet soils (b) Cool, dry soils
(c) Warm, wet soils (d) Warm, dry soils

374. In India, wart disease of potato is confined to
(a) Kumaun hills (b) Nilgiri hills
(c) Kullu hills (d) Darjeeling hills

375. Plant disease that can be managed through domestic quarantine regulation
(a) Late blight of potato (b) Wheat rust
(c) Wart disease of potato (d) Rice blast

376. Potato leaf roll disease is also known as
(a) Potato leaf curl (b) Potato phloem necrosis
(c) Net necrosis (d) All of these

377. Net necrosis disease of potato is caused by
(a) Fungi (b) Aphid feeding
(c) Nematode (d) Virus

378. Net necrosis disease of potato is caused by
(a) *Potato virus X* (b) *Potato leaf roll virus*
(c) Aphid feeding (d) *Potato virus Y*

379. Potato disease that occurs during storage
(a) Late blight (b) Black heart
(c) Wart disease (d) Black scurf

380. Hard black crusty spots on potato tubers are produced by
(a) Late blight (b) Powdery scab
(c) Wart disease (d) Black scurf

381. Which of the following statement is NOT true about *Phytophthora infestans*?
(a) It produces finger-like haustoria
(b) It produces zig-zag sporangiophores
(c) It produces lemon-shaped papillate sporangia
(d) Low temperature favors germ tube production from zoospores

382. Optimum temperature for the growth of *Phytophthora infestans* is
(a) 10°C–14°C (b) 16°C–18°C
(c) 20°C–24°C (d) 25°C–28°C

383. The black wart disease of potato is caused by

(a) *Synchytrium endobioticum* (b) *Plasmopara viticola*
(c) *Rhizoctonia solani* (d) *Plasmodiophora brassicae*

384. The black wart disease of potato is specially confined to

(a) Nilgiri hills (b) Mussoori hills
(c) Darjeeling hills (d) Shimla hills

385. Black and hollow heart in potato occurs due to

(a) *Phytophthora infestans* (b) Overuse of fertilizers
(c) Zinc deficiency (d) Oxygen deficiency

386. *Potato leaf roll virus* is transmitted by

(a) Infected tubers and aphids
(b) Plant sap and aphids
(c) Infected tubers and nematode
(d) Pollen and plant sap

387. *Potato leaf roll virus* is transmitted by several species of aphid vectors in a

(a) Circulative, persistent, non-propagative manner
(b) Non-circulative, persistent, non-propagative manner
(c) Non-circulative, non-persistent, non-propagative manner
(d) Circulative, semi-persistent, propagative manner

388. Which of the following is the only virus infecting potato that is persistent in vector?

(a) PVY (b) PVX
(c) PLRV (d) PVM

389. Potato species that contain very high levels of resistance to *Potato leaf roll virus*

(a) *Solanum brevidens* (b) *Sonanum chacoense*
(c) *Solanum raphanifolium* (d) All of these

390. Zebra chip (ZC) disease of potato is caused by

(a) Fastidious bacterium (b) Virus
(c) Fungi (d) Oxygen deficiency

391. Zebra chip (ZC) disease of potato is caused by

(a) Zinc deficiency (b) Oxygen deficiency
(c) Fungi (d) *Candidatus Liberibacter solanacearum*

392. Internally seedborne bacterial disease

(a) Black rot of cabbage (b) Soft rot of carrot
(c) Pierce disease of grapes (d) Citrus canker

393. Zebra chip (ZC) disease of potato is transferred by

(a) Thrips (b) Aphid
(c) Fungi (d) Potato-tomato psyllid

394. Greening of potato tuber is caused by

(a) Exposer to sunlight (b) Oxygen deficiency
(c) *Fusarium sambucinum* (d) *Potato virus Y*

395. Mild mosaic is the general symptom that is clearly visible in the case of

(a) *Potato leaf roll virus* (b) *Potato Poty virus A*
(c) *Potato Poty virus Y* (d) *Potato Potexvirus X*

396. Phytoplasmas and spiroplasmas have

(a) Rigid cell wall (b) Thin cell wall
(c) Thick cell wall (d) No rigid cell wall

397. Powdery mildew of cucurbits is caused by

(a) *Sphaerotheca fuligena* and *Erysiphe cichoracearum*
(b) *Erysiphe pisi* and *Uncinula necator*
(c) *Erysiphe lycopersici* and *Leveillula taurica*
(d) *Erysiphe polygoni* and *Sphaerotheca fuliginea*

398. Downy mildew of cucurbits is caused by

(a) *Plasmopara viticola* (b) *Peronospora tabacina*
(c) *Peronospora effuse* (d) *Pseudoperonospora cubensis*

399. Anthracnose of cucurbits is caused by

(a) *Colletotrichum orbiculare* (b) *Didymella bryoniae*
(c) *Colletotrichum capsici* (d) *Corynespora cassiicola*

400. Target spot of cucurbits is caused by

(a) *Colletotrichum orbiculare* (b) *Didymella bryoniae*
(c) *Colletotrichum capsici* (d) *Corynespora cassiicola*

401. Gummy stem blight of cucurbits is caused by

(a) *Colletotrichum orbiculare* (b) *Didymella bryoniae*
(c) *Colletotrichum capsici* (d) *Corynespora cassiicola*

402. The conidia of cucurbit powdery mildew fungi (*Sphaerotheca fuligena* and *Erysiphe cichoracearum*) can be distinguished from each other on the basis of

(a) Shape of conidia (b) Color of conidia
(c) Size of conidia (d) Presence of fibrosin bodies

403. A feature that enables a firm identification of the cucurbit powdery mildew pathogens?

(a) Shape of conidia (b) Color of conidia
(c) Size of conidia (d) Presence of fibrosin bodies

404. Arka Manik is a highly resistant variety of watermelon under Bangalore conditions against

(a) Powdery mildew (b) Anthracnose
(c) Gummy stem blight (d) Downy mildew

405. Muskmelon variety resistant to powdery mildew disease

(a) Arka Rajhans (b) Diguria
(c) Haragola (d) All of these

406. Bacterial wilt (*Erwinia tracheiphila*) impacts all cucurbit crops EXCEPT

(a) Watermelon (b) Pumpkin
(c) Muskmelon (d) Bitter guard

407. Watermelon is generally resistant to

(a) Powdery mildew
(b) Anthracnose
(c) Gummy stem blight
(d) Bacterial wilt

408. Bacterial wilt of cucurbits was first studied by

(a) E.F. Smith
(b) E.J. Butler
(c) J.P. Verma
(d) R.S. Singh

409. Bacterial wilt of cucurbits is caused by

(a) *Pseudomonas syringae*
(b) *Erwinia tracheiphila*
(c) *Fusarium oxysporum*
(d) *Ralstonia solanacearum*

410. Which of the following statement about *Erwinia tracheiphila* is/are correct?

1. It is Gram-positive.
2. It is phloem-limited.
3. It is peritrichously flagellated.
4. It is transmitted by adults of striped and spotted cucumber beetles.
5. It is a facultative anaerobic bacterium.
6. It causes bacterial wilt of maize.

Select the correct answer from the code given below:

Code:

(a) Only 1, 2, 6
(b) Only 1, 2, 3, 4
(c) Only 3, 4, 5
(d) Only 2, 4, 5, 6

411. Cucurbit wilt bacterium (*Erwinia tracheiphila*) does not survive in

(a) Cucurbit seeds
(b) Weeds
(c) Plant debris
(d) Digestive tract of cucumber beetles

412. Cucurbit wilt bacterium (*Erwinia tracheiphila*) survives in the digestive tract (gut) of

(a) Striped cucumber beetle
(b) Spotted cucumber beetle
(c) Both (a) and (b)
(d) None of these

413. Which of the following is/are very effective method for the management of bacterial wilt of cucurbits?

1. Use of trap crop (buttercup squash).
2. Following crop rotation for 3-year.
3. Rouging out infected plants.
4. Spraying insecticidal soap containing neem oil.
5. Dusting the crop with rotenone 1%.
6. Using fast-killing insecticides like Methoxychlor or Carbryl

Select the correct answer from the code given below:

Code:

(a) Only 1, 2, 6
(b) Only 1, 2, 4, 6
(c) Only 3, 4, 5
(d) All 1, 2, 3, 4, 5, 6

414. Fusarium wilt of cucurbits is

(a) Only seedborne
(b) Only soilborne
(c) Both seed and soilborne
(d) Soil, seed, and airborne

415. Fusarium wilt of cucurbits can be quite severe in the fields with
(a) Light and sandy soils
(b) pH 5.5–6.5
(c) High nitrogen content (especially NH_4—nitrogen)
(d) All of these

416. Symptom produced by gummy stem blight of cucurbits is
(a) Leaf spot and stem canker (b) Vine wilt
(c) Black fruit rot (d) All of these

417. Fruiting bodies produced by *Didymella bryoniae*
(a) Pycnidia (b) Acervuli
(c) Apothecia (d) Sclerotia

418. Which of the following disease of onion is known as "Ebola of onion"?
(a) Onion yellow dwarf (b) Stemphyllium blight
(c) Irish yellow spot (d) Purple blotch

419. Powdery mildew of chilli and pepper is caused by
(a) *Leveillula taurica* (b) *Blumeria graminis* f. sp. *tritici*
(c) *Erysiphe polygoni* (d) *Podosphaera pannosa*

420. Which one of the following does not causes vascular disease in chilli?
(a) *Verticillium dahliae* (b) *Colletotrichum truncatum*
(c) *Ralstonia solanacearum* (d) *Fusarium oxysporum*

421. Bacterial leaf spot of chilli is caused by
(a) *Cercospora capsici* (b) *Xanthomonas campestris* pv. *vesicatoria*
(c) *Phytophthora capsici* (d) *Erwinia carotovora* subsp. *carotovora*

422. Anthracnose and fruit rot of chilli is caused by
(a) *Colletotrichum capsici* (b) *Colletotrichum truncatum*
(c) *Colletotrichum acutatum* (d) All of these

423. *Colletotrichum* species have
(a) Intramural necrotrophic lifestyles
(b) Hemibiotrophic lifestyles
(c) Quiescent and endophytic lifestyles
(d) All of these

424. Anthracnose of chilli is
(a) Seed and soilborne (b) Waterborne
(c) Airborne (d) All of these

425. Chilli leaf curl is a
(a) Whitefly-borne–begomovirus disease
(b) Aphid-borne–polerovirus disease
(c) Thrips-borne–caulimovirus disease
(d) Leaf hopper-borne–begomovirus disease

426. Begomoviruses (Geminviridae) have
(a) Circular, ssDNA genomes (b) Circular, dsDNA genomes
(c) Linear, ssRNA genomes (d) Linear, dsRNA genomes

427. Seedborne inoculum of *Chilli leaf curl virus* can be destroyed by soaking seeds in
(a) Trisodium orthriphosphate solution
(b) Imidacloprid solution
(c) Neem oil solution
(d) Streptocyclin solution

428. In onion and garlic, *Alternaria porri* causes
(a) Early blight
(b) Airborne disease
(c) Purple blotch
(d) Externally seedborne disease

429. Purple blotch disease of onion and garlic is caused by
(a) *Alternaria porri*
(b) *Stemphyllium vesicarium*
(c) *Aspergillus niger*
(d) *Peronospora destructor*

430. Purple blotch is the most destructive disease of
(a) Onions
(b) Garlic
(c) Leeks and chives
(d) All of these

431. *Alternaria porri* (purple blotch of onion) overwinters in
(a) Crop residue
(b) Soil surface
(c) Seed
(d) All of these

432. Purple blotch of onion and garlic is favored by
(a) Free moisture
(b) High humidity (80%–90%)
(c) Moderate temperature (25°C–30°C)
(d) All of these

433. Hot water soaking of onion seed proved to be the best treatment for reducing purple blotch pathogens at
(a) 50°C for 20 minutes
(b) 54°C for 15 minutes
(c) 60°C for 20 minutes
(d) 60°C for 30 minutes

434. Tolerant cultivar of garlic to purple blotch disease is
(a) Red Comet F1
(b) Bombay Red
(c) Red Creole
(d) All of these

435. Onion variety moderately resistant to purple blotch disease
(a) Agri found Light Red
(b) Red Globe
(c) VL Piyaz 3
(d) All of these

436. Leaf spot of turmeric is caused by
(a) *Rhizoctonia bataticola*
(b) *Colletotrichum capsici*
(c) *Taphrina maculans*
(d) *Pythium graminicolum*

437. Dry rot of turmeric is caused by
(a) *Rhizoctonia bataticola*
(b) *Colletotrichum capsici*
(c) *Taphrina maculans*
(d) *Pythium graminicolum*

438. Leaf blotch of turmeric is caused by
(a) *Rhizoctonia bataticola*
(b) *Colletotrichum capsici*
(c) *Taphrina maculans*
(d) *Pythium graminicolum*

439. Resistant/tolerant variety of turmeric to leaf spot is
(a) Sudarshan and Suroma (b) Rashmi and Pratibha
(c) Kedaram and Alleppey (d) All of these

440. Stem gall of coriander is caused by
(a) Bacteria (b) Nematode
(c) Aphid (d) Fungi

441. Stem gall of coriander is caused by
(a) *Rhizobium tumefaciens* (b) *Plasmopara viticola*
(c) *Protomyces macrosporus* (d) *Meliodogyne incognita*

442. Stem gall disease of coriander was first time reported from India by
(a) Sydow and Butler (1911) (b) Nene and Thapliyal (1979)
(c) Nagrajan and Joshi (1971) (d) Unger (1833)

443. Stem gall of coriander is
(a) Seed and soilborne (b) Air and soilborne
(c) Only seedborne (d) Only airborne

444. *Protomyces macrosporus* (stem gall of coriander) survives in the form of
(a) Sclerotia (b) Apothecia
(c) Chlamydospores (d) Acervuli

445. Resistant variety to stem gall of coriander
(a) Hisar Bhoomi (b) Pant Haritma
(c) Rajndra Swati (d) All of these

446. Powdery mildew of coriander is caused by
(a) *Erysiphe pisi* (b) *Erysiphe polygoni*
(c) *Leveillula taurica* (d) *Erysiphe lycopersici*

447. The stem gall of coriander is caused by
(a) *Taphrina deformans* (b) *Protomyces macrosporus*
(c) *Xyllela fastidiosa* (d) *Rhizobium tumefaciens*

448. The pathogen of stem gall of coriander survive in soil as
(a) Sclerotia (b) Oospores
(c) Chlamydospores (d) None of these

449. Inflorescence blight of marigold is caused by
(a) *Alternaria zinniae* (b) *Alternaria alternata*
(c) *Colletotrichum capsici* (d) *Alternaria tagetica*

450. Rose species resistant to powdery mildew disease is
(a) *Rosa moschata* (b) *Rosa bourboniana*
(c) *Rosa rugosa* (d) *Rosa indica*

451. Black spot of rose is caused by
(a) *Diplodia rosarum* (b) *Diplocarpon rosae*
(c) *Mycospharella musicola* (d) *Sphaerotheca pannosa*

452. Lesions of black spot of rose can be distinguished from others by
(a) Fringed margin (b) Even margin
(c) Angular margin (d) Raised margin

453. Dieback of rose is caused by

(a) *Diplodia rosarum*
(b) *Diplocarpon rosae*
(c) *Rhizoctonia bataticola*
(d) *Colletotrichum capsici*

454. Which of the following is resistant to black spot disease?

(a) *Rosa laevigata*
(b) *Rosa multiflora*
(c) *Rosa roxburghii*
(d) All of these

455. Powdery mildew of rose is caused by

(a) *Podosphaera leucotricha*
(b) *Erysiphe polygoni*
(c) *Leveillula taurica*
(d) *Podosphaera pannosa*

456. Fruiting bodies produced by *Marssonina rosae* (black spot of rose)

(a) Asexual–sporodochia; sexual–perithecia
(b) Asexual–acervuli; sexual–apothecia
(c) Asexual–pycnidia; sexual–cleistothecia
(d) Asexual–synnemata; sexual–apothecia

457. The complete infection cycle of powdery mildew of rose occurs at the temperature

(a) 5°C–15°C
(b) 15°C–25°C
(c) 20°C–30°C
(d) 25°C–35°C

458. Which one of the following is NOT required for the development of powdery mildew disease of rose?

(a) Relative humidity 75%–79%
(b) Temperature 15°C–25°C
(c) Warm, dry weather
(d) Free water on the plant surface

459. Powdery mildew of cucurbits is caused by

(a) *Erysiphe cichoracearum*
(b) *Sphaerotheca fuliginea*
(c) *Erysiphe cichoracearum* and *Sphaerotheca fuliginea*
(d) None of these

460. Purple blotch disease of onion and garlic is caused by

(a) *Alternaria pori*
(b) *Alternaria solani*
(c) *Botrytis cineria*
(d) *Rhizoctonia solani*

461. Chilli leaf curl is transmitted by

(a) Chilli mite (*Polyphagotarsonemus latus*)
(b) Whitefly (*Bemisia tabaci*)
(c) Thrips (*Scirtothrips dorsalis*)
(d) All of these

462. Pathogen of stem gall of coriander survives as

(a) Oospores
(b) Teliospores
(c) Conidia
(d) Chlamydospores

463. The major source of primary inoculum for citrus canker is

(a) Infected twigs and old lesions on tree
(b) Alternate host

(c) Infected plants on soil
(d) Seed

464. Irish famine occurred during 1842-45 which was caused by
(a) Rice blast (b) Late blight of potato
(c) Wheat rust disease (d) Brown spot of rice

465. Tomato leaf curl can be managed by spraying
(a) Mancozeb (b) Triazophos
(c) Ridomil (d) Tricyclazole

466. Blackheart of potato is a
(a) Postharvest disorder (b) Bacterial disease
(c) Fungal disease (d) Viral disease

467. Green wilt in many solanaceous crops is caused by
(a) *Verticillium alboatrum* (b) *Fusarium solani*
(c) *Ralstonia solanacearum* (d) *Rhizoctonia solani*

468. Potassium application in plants increases
(a) Disease resistance (b) Drought resistance
(c) Resistance to frost injury (d) Resistance to flood

469. Which one of the following pairs is NOT correctly matched?
(a) *Bract mosaic virus*–Banana
(b) *Leaf curl virus*–Tomato
(c) *Yellow vein mosaic virus*–Potato
(d) *Ring spot virus*–Papaya

470. Yellow spots in citrus are caused by the deficiency of
(a) Zinc (b) Copper
(c) Phosphorus (d) Molybdenum

471. Deficiency of which one of the following elements causes dieback and decline in citrus?
(a) Cobalt (b) Copper
(c) Manganese (d) Magnesium

472. Select the wrong pair
(a) Excessive cell enlargement–Hypertrophy
(b) Excessive cell division–Hyperplasia
(c) Excessive elongation–Etiolation
(d) Plant death–Necrosis

473. Which of one the following variety of barley is nematode resistant?
(a) Jyoti (b) Ratna
(c) Karan-19 (d) Raj Kiran (RD-387)

474. *Hemelia vastatrix* causes
(a) Bean rust (b) Coffee leaf rust
(c) Yellow rust of wheat (d) Sunflower rust

475. Nonmotile spores are characteristics of
(a) Zygomycetes (b) Chytridiomycetes
(c) Myxomycetes (d) Oomycetes

476. Parasexuality occurs
(a) Within heterokaryotic hyphae (b) Between dikaryotic hyphae
(c) Between homokaryotic hyphae (d) Between heterokaryotic hyphae

477. Single polar flagellum is characteristics features of
(a) *Xanthomonas* (b) *Erwinia*
(c) *Clavibacter* (d) *Agrobacterium*

478. Oomycetes are distinguished from fungi by all of the characteristics below EXCEPT
(a) Source of nutrition (b) Spore motility
(c) Cellulose walls (d) Pattern of mitosis

479. Fungal hyphae with two genetically distinct nuclei are said to be
(a) Heterokaryotic (b) Monokaryotic
(c) Dikaryotic (d) Karyotic

480. Nutritionally, fungi may be best characterized as
(a) Photosynthetic autotrophs (b) Chemosynthetic autotrophs
(c) Absorptive heterotrophs (d) Ingestive heterotrophs

481. Pseudomycelium is formed in
(a) *Synchytrium* (b) *Rhizopus*
(c) Yeast (d) *Pythium*

482. Cleistothecium is a sexual fruiting body produced by
(a) *Albugo candida* (b) *Pythium butleri*
(c) *Rhizopus nigricans* (d) *Erysiphe polygoni*

483. The fruiting body of *Colletotrichum capsici* is
(a) Acervulus (b) Perithecia
(c) Pycnidia (d) Sporodochium

484. *Phytophthora, Venturia inaequalis*, and *Fusarium* spp. are
(a) Perthotroph (b) Facultative saprophyte
(c) Obligate parasite (d) Facultative parasite

485. Rust-causing fungi are
(a) *Ustilago* spp. (b) *Colletotrichum* spp.
(c) *Fusarium* spp. (d) *Puccinia* spp.

486. Which of the following fungus does NOT cause rust disease in plants?
(a) *Uromyces* (b) *Ustilago*
(c) *Puccinia* (d) *Hemileia*

487. Red rust of mango is caused by
(a) *Puccinia* (b) *Cephaleuros*
(c) *Albugo* (d) *Xanthomonas*

488. Excessive cell division leading to the formation of tumors, galls, knots, and cysts is known as
(a) Hypertrophy (b) Hyperplasia
(c) Hypoplasia (d) Atrophy

489. Canker, curl, witches broom, and hair roots are formed as a result of
(a) Hypertrophy (b) Hyperplasia
(c) Hypoplasia (d) Atrophy

490. Typical example of monocyclic disease is
(a) Late blight of potato (b) Apple scab
(c) Loose smut of wheat (d) Yellow rust of wheat

491. Loose smut of wheat is
(a) Soilborne disease (b) Airborne disease
(c) Internally seedborne disease (d) Externally seedborne disease

492. Brown rot of potato is caused by
(a) *Rhizoctonia solani* (b) *Ralstonia solanacearum*
(c) *Phytophthora infestans* (d) *Fusarium solani*

493. Downward bending of the leaf from its normal position is called
(a) Leaf curl (b) Epinasty
(c) Etiolation (d) Puckering

494. Wilt of arhar is caused by
(a) *Colletotrichum* (b) *Fusarium*
(c) *Rhizoctonia* (d) *Macrophomina*

495. Movement and transport of potato growing in Darjeeling district of West Bengal to other parts of India is prohibited due to domestic regulations for
(a) Potato wart (b) Golden cyst nematode
(c) Late blight of potato (d) Brown rot of potato

496. "Pokkh boeing" disease is associated with
(a) Rice (b) Wheat
(c) Potato (d) Sugarcane

497. Death of plant tissue caused by the pathogen is called
(a) Mottling (b) Mosaic
(c) Necrosis (d) Chlorosis

498. Which one is NOT a sign of a fungal pathogen?
(a) Spores (b) Mycelia
(c) Sclerotia (d) Endospores

499. Fungi associated with damping off disease are
(a) *Pythium* spp. (b) *Fusarium* spp.
(c) *Rhizoctonia* spp. (d) All of these

500. Repeating spore for the infection of rust fungi is
(a) Aeciospores (b) Teliospores
(c) Urediospores (d) Pycniospores

501. Smut diseases infect plants at
(a) Seedling stage (b) Ear emergence stage
(c) Tillering stage (d) Root formation stage

502. Which of the following strategy will NOT be effective against airborne plant diseases?
(a) Foliar spray of fungicides (b) Host resistance
(c) Biological control (d) Plant quarantine

503. The strategy that will NOT be effective against airborne plant diseases
(a) Seed treatment with fungicides (b) Crop rotation
(c) Plant quarantine (d) All of these

504. Sign of a fungal pathogen is
(a) Spores (b) Mycelium
(c) Sclerotia (d) All of these

505. Ratoon stunting disease (RSD) of sugarcane is caused by
(a) Phytoplasma (b) Virus
(c) Bacteria (d) Fungi

506. "Egg-shaped" conidia are produced by
(a) *Erysiphe cichoracearum* (b) *Podosphaera leucotricha*
(c) *Uncinula necator* (d) *Phylloctinia corylea*

507. V-shaped yellow spots progressing from leaf margins towards the center, caused by
(a) Brown spot of crucifers (b) White rust of crucifers
(c) Black rot of crucifers (d) Club root of crucifers

508. Red rot disease is associated with
(a) Sorghum (b) Sugarcane
(c) Maize (d) Pearl millet

509. Sporangia are produced in chains by
(a) *Pythium* (b) *Albugo*
(c) *Peronospora* (d) *Saprolegnia*

510. Disease that is managed by adding lime to the soil
(a) Powdery scab of potato (b) Downy mildew of grape
(c) Club root of crucifers (d) Common scab of potato

511. Acervuli with seate arising through epidermis, is the characteristic of
(a) *Cercospora* (b) *Rhizoctonia*
(c) *Diplocarpon* (d) *Colletotrichum*

512. Hyperplasia refers to
(a) Increased cell division (b) Cell death
(c) Cell enlargement (d) Decreased cell division

513. An internally seedborne fungal disease is
(a) Smut of ragi (b) Red rot of sugarcane
(c) Loose smut of wheat (d) Grain smut of jowar

514. Effective fungicide for the management of blast disease of rice is
(a) Mancozeb (b) Tricyclazole
(c) Fosteyl-Al (d) Plantvax

515. Diseases that interferes with the conduction of water in plants?
(a) Wilts (b) Canker
(c) Leaf spots (d) Root knot

516. The sexual spore produced by downy mildew pathogen is

(a) Oospore (b) Zygospore
(c) Ascospore (d) Basidiospore

517. *Oidium mangiferae* causes

(a) Downy mildew of grapes (b) Powdery mildew of mango
(c) Anthracnose of mango (d) Powdery mildew of pea

518. Early blight of tomato is caused by

(a) *Alternaria triticina* (b) *Alternaria brassicae*
(c) *Alternaria solani* (d) *Alternaria porri*

519. Red rust of mango is caused by

(a) Fungi (b) Bacteria
(c) Algae (d) Red spider mite

520. Which of the following pathogen is NOT a necrotroph?

(a) *Fusarium* (b) *Septoria*
(c) *Albugo* (d) *Alternaria*

521. Acervuli with setae (hair-like or spine-like) that erupts through the epidermis of host plants, is characteristic feature of

(a) *Fusarium* (b) *Septoria*
(c) *Colletotrichum* (d) *Blumeria*

522. Acervuli contains numerous

(a) Chlamydospores (b) Ascospores
(c) Oospores (d) Conidia

523. *Capnodium* causes

(a) Sooty mold (b) Red rust
(c) Loose smut (d) Powdery mildew

524. *Colletotrichum capsici* produces

(a) Microsclerotia (b) Microconidia and macroconidia
(c) Acervuli (d) All of these

525. The fungal structures produced by *Rhizoctonia solani*

(a) Conidia, sclerotia, apothecia, haustoria
(b) Conidia, acervuli, apothecia, appressorium
(c) Mycelium, sclerotia, basidiospores, infection cushion
(d) Zoospores, pycnidia, basidiospores, infection cushion

526. The structures produced by the fungus *Botrytis cinerea*

(a) Conidia, sclerotia, apothecia, haustoria
(b) Conidia, acervuli, apothecia, appressorium
(c) Mycelium, sclerotia, basidiospores, infection cushion
(d) Zoospores, pycnidia, basidiospores, infection cushion

527. Which of the following is known to cause sooty mold in plants?

(a) *Ustilago* (b) *Sclerotinia*
(c) *Capnodium* (d) *Cercospora*

528. Net blotch of barley is caused by
(a) *Bipolaris oryzae* (b) *Drechslera teres*
(c) *Stemphylium solani* (d) *Exserohilum turcicum*

529. Geniculate conidiophores are NOT produced by
(a) *Bipolaris* (b) *Helminthosporium*
(c) *Pyricularia* (d) *Curvularia*

530. Fungal genera having cylindrical spores
(a) *Bipolaris* (b) *Helminthosporium*
(c) *Exserohilum* (d) All of these

531. Chemical weapon produced by *Pythium* species used in tissue disintegration and maceration
(a) Pectolytic enzymes (b) Cellulolytic enzymes
(c) Toxins (d) Both (a) and (b)

532. Symptoms of pre-emergence damping-off appear as
(a) Seed and seedling rot (b) Collar rot
(c) Radicle rot (d) Stem rot and leaf spot

533. White creamy pustules appear in
(a) White rust of mustard (b) Downy mildew of grapes
(c) Powdery mildew of wheat (d) Puccinia rust of wheat

534. Sporangia of *Albugo candida* in chains, are separated from each other by a gelatinus disc called
(a) Disjunctor (b) Isthmus
(c) Separation disc (d) All of these

535. Pathogen belonging to the order.............causes downy mildew disease in plants
(a) Erysiphales (b) Pythiales
(c) Peronosporales (d) Hypocreales

536. Which of the following statement is TRUE about *Synchytrium endobioticum*?
(a) It is an obligate parasite (b) It causes galls and warty outgrowth on tubers
(c) It is a quarantine pest (d) All of these

537. Symptoms induced by *Taphrina* in plants
(a) Galls and deformation (b) Necrosis
(c) Witch's broom (d) All of these

538. Which of the following pair is correct?
(a) Peach leaf curl–*Taphrina deformans*
(b) Witch's broom of cherries–*Taphrina epiphylla*
(c) Turmeric leaf blotch–*Taphrina maculans*
(d) All of these

539. Storage fungus which infects grains with high moisture content
(a) *Aspergillus glaucus* (b) *Aspergillus restrictus*
(c) *Aspergillus halophilicus* (d) All of these

540. Which of the following disease is NOT caused by "pseudofungi"?
(a) Downy mildew (b) White rust
(c) Late blight of potato (d) Powdery mildew

541. Endophytic mycelium is found in
(a) *Erysiphe* (b) *Phyllactinia*
(c) *Uncinula* (d) *Leveillula*

542. Conidiophores and conidia resemble a "grape-like cluster" of
(a) *Monilia* (b) *Botrytis*
(c) *Alternaria* (d) *Penicillium*

543. Fungus *Claviceps* causes
(a) Sugary disease (b) Ergot disease
(c) Honeydew disease (d) All of these

544. The well-known hallucinating drug of the century, LSD (d-lysergic acid diethylamide) is obtained from
(a) *Aspergillus glaucus* (b) *Penicillium notatum*
(c) *Sclerotinia sclerotiorum* (d) *Claviceps paspalis*

545. Teleomorphs are known only for some species of *Fusarium* EXCEPT
(a) *Fusarium oxysporum* (b) *Fusarium udum*
(c) *Fusarium solani* (d) *Fusarium graminearum*

546. Symptoms produced by both viral and fungal plant pathogens
(a) Chlorosis (b) Leaf curl
(c) Necrosis (d) All of these

547. Which of the following symptom is NOT produced by bacterial and viral plant pathogens?
(a) Blight (b) Spots
(c) Downy mildew (d) Yellowing

548. Symptoms produced by both fungal and bacterial plant pathogens
(a) Blight (b) Spots
(c) Galls (d) All of these

549. Gall symptoms in plants are produced by
(a) Fungi (b) Bacteria
(c) Nematode (d) All of these

550. Drying of twigs from tip to downward is called
(a) Chlorosis (b) Blight
(c) Wilting (d) Dieback

551. *Cephaleuros virescens* have
(a) Chlorophyll a, b, and carotenoid
(b) Cellulose, xylan, and mannan in cell wall
(c) Rhizoids arising from the thallus
(d) All of these

552. Nutritional deficiencies/disorders are also known as

(a) Non-parasitic diseases (b) Abiotic diseases
(c) Physiological disorders (d) All of these

553. Cauliflower plant showing mottled and necrotic leaves with thinner and dry leaf lamina and distortion and death of growing tissues. These symptoms may be caused by

(a) *Alternaria brassicae* (b) *Plasmodiophora brassicae*
(c) Molybdenum deficiency (d) *Cauliflower mosaic virus*

554. The disease causing sudden and complete wilting of plants in patches. Infected plants are showing rotting of the entire roots, except taproots and few laterals. Stem near the soil level showing the large number of black pycnidial bodies. Identify the pathogen

(a) *Ralstonia solanacearum* (b) *Macrophomina phaseolina*
(c) *Ascochyta rabiei* (d) *Fusarium oxysporum*

555. Which of the following is NOT an entomopathogenic fungus?

(a) *Beauveria bassiana* (b) *Metarhizium anisopliae*
(c) *Paecilomyces fumosoroseus* (d) *Trichoderma asperellum*

556. Rotting of emerged seedlings at the ground level resulting in topping down of seedlings. The disease is

(a) Root rot (b) Post-emergence damping off
(c) Wilting (d) Pre-emergence damping off

557. Loss of turgidity and dropping of leaves and shoots is called

(a) Leaf roll (b) Curling
(c) Wilting (d) Epinasty

558. Loss of chlorophyll from leaves is called

(a) Necrosis (b) Chlorosis
(c) Discoloration (d) Pigmentation

559. Infected grain filled with sooty black powdery masses of spores is called

(a) Smut (b) Bunt
(c) Sooty mold (d) Rust

560. Which of the following is a sign of bacterial disease?

(a) Ooze
(b) Water-soaked lesions
(c) Bacterial streaming in water from cut stem
(d) All of these

561. Sign cannot be seen with

(a) Viral diseases (b) Bacterial diseases
(c) Fungal diseases (d) Nematode diseases

562. Biotic stresses of the plant do NOT include

(a) Weeds (b) Disorders
(c) Pests (d) Diseases

563. Compare to canker, scab is also a corky outgrowth but confirmed to

(a) Epidermis and not deep-seated
(b) Epidermis and deep-seated

(c) Parenchymatous tissues and deep-seated
(d) Cortex tissues and deep-seated

564. In bacterial canker disease, the plant tissues involved in producing cankerous outgrowths
(a) Epidermis
(b) Parenchymatous tissues
(c) Cambium layer
(d) All of these

565. A characteristic symptom "yellow halo" develop in
(a) Leaf spot diseases (b) Blight diseases
(c) Wilt diseases (d) Root rot diseases

566. Dissolution of middle lamella by enzymatic degradation and disintegration causes
(a) Leaf spot (b) Soft rot
(c) Wilt (d) Blight

567. The soft rot causing plant pathogens are
(a) *Alternaria* and *Colletotrichum* species
(b) *Pectobacterium* and *Dickeya* species
(c) *Fusarium* and *Rhizoctonia* species
(d) *Uromyces* and *Albugo* species

568. The bacterium that cause citrus canker enters leaves through
(a) Stomata (b) Wounds
(c) Cuticle (d) Both (a) and (b)

569. Formation of corky outgrowth on the surface of leaves, twigs, and fruits, necrosis of host tissues is called
(a) Soft rot (b) Scab
(c) Crown gall (d) Canker

570. T-DNA genes catalyze the formation of
(a) Auxin (b) Cytokinin
(c) Opines (d) All of these

571. The production of tyloses in the xylem vessels is a typical response to infection by
(a) Vascular wilt pathogens (b) Leaf spot pathogens
(c) Root rot pathogens (d) Crown gall pathogens

572. Which one of the following pathogens causes sooty mold in plants?
(a) *Ustilago* (b) *Sclerotinia*
(c) *Cercospora* (d) *Capnodium*

573. *Cephaleuros* is a
(a) Green algae (b) Red algae
(c) Brown algae (d) Blue green algae

574. *Cephaleuros* belong to the phylum
(a) Chlorophyta (b) Glaucophyta
(c) Rhodophyta (d) Embryophyta

575. In red rust disease, the rusty-red color of pustules is due to the presence of
(a) Carotenes (b) Haematochrome
(c) Phycobilins (d) Xanthophylls

576. The powdery mildew diseases are favored by
(a) Cool and dry weather (b) Cool and wet weather
(c) Warm and low humid (d) Warm and high humid

577. First bacterial plant pathogen whose genome has been completely sequenced is
(a) *Xyllela fastidiosa* (b) *Xanthomonas campestris*
(c) *Ralstonia solanacearum* (d) *Erwinia amylovora*

578. Which of the following disease is caused by flagellated protozoa?
(a) Phloem necrosis of coffee (b) Hart rot of coconut
(c) Sudden wilt of oil palm (d) All of these

579. White rust pathogen "*Albugo candida*" belongs to
(a) Basidiomycetes (b) Ascomycetes
(c) Oomycetes (d) Deuteromycetes

580. "Exanthema" in *Citrus* spp. occurs due to the deficiency of
(a) Copper (b) Iron
(c) Zinc (d) Potash

581. Fungal structure that is NOT the part of the infection process
(a) Appressorium (b) Crozier
(c) Haustorium (d) Penetration peg

582. Systemic acquired resistance stimulating fungicide is
(a) Acibenzolar S-methyl (b) Mancozeb
(c) Carbendazim (d) Calixin

583. Anti-oomycetes fungicide is
(a) Zoxamide (b) Fosetyl-Al
(c) Phenylamide (d) All of these

584. Pahala blight of sugarcane is due to the deficiency of
(a) Boron (b) Zinc
(c) Manganese (d) Magnesium

585. Black rot disease is associated with
(a) Sugarcane (b) Sweet potato
(c) Cauliflower (d) Carrot

586. Loose smut of cereal fungus (*Ustilago nuda*) survives as
(a) Teliospores (b) Dormant mycelium
(c) Sclerotia (d) None of these

587. *Exobasidium vexans* completes its life cycle in
(a) 11–28 days (b) 10 days
(c) 60 days (d) 6–8 days

588. Feliocellosis or mottling or trenching in citrus is caused by the deficiency of
(a) Zinc (b) Copper
(c) Iron (d) Magnesium

589. Copper deficiency in citrus is called as
(a) Exanthema (b) Red rust
(c) Foliocellosis (d) All of these

590. Web blight of mungbean caused by
(a) *Alternaria solani* (b) *Fusarium oxysporum*
(c) *Rhizoctonia bataticola* (d) *Rhizoctonia solani*

591. Little leaf of brinjal is transmitted through
(a) *Cestius phycitis* (b) *Amrasca biguttula biguttula*
(c) *Aphis cracivora* (d) Both (a) and (b)

592. Closteroviruses are unique in being transmitted by
(a) Aphids (b) Whiteflies
(c) Mealybugs (d) All of these

593. *Citrus tristeza virus* (CTV) mainly causes a distinct syndrome known as
(a) Tristeza (b) Stem pitting
(c) Seedling yellows (d) All of these

594. Nursery leaf small cardamom is caused by
(a) *Phyllactinia elletariae* (b) *Colletotrichum gloeosporide*
(c) *Pythium aphanidermatum* (d) *Fusarium moniliforme*

595. The disease which does not produce a foul smell
(a) White rust of mustard (b) Karnal bunt of wheat
(c) Red rot of sugarcane (d) Stalk rot of maize

596. The nature of chickpea blight (*Mycosphaerella rabiei*) is
(a) Airborne (b) Seedborne
(c) Soilborne (d) Both (a) and (b)

597. Downy mildew pathogen of onion perpetuates through
(a) Bulbs (b) Soil
(c) Saprophytic mycelium (d) All of these

598. Physical evidence of the pathogen is called
(a) Symptom (b) Infection
(c) Disease (d) Sign

599. Infection of the powdery mildew to host plant results in an increase in
(a) Photosynthesis (b) Respiration
(c) Traspiration (d) Guttation

600. The inherited ability of a plant species to resist attack by pests or diseases is called
(a) Disease escape (b) Host-plant resistance
(c) Disease severity (d) Disease incidence

601. Green ear disease perpetuates in the form of
(a) Oospore (b) Cleistothecium
(c) Sclerotium (d) Apothecium

602. Which of the following is NOT a host of *Striga*?
(a) Maize (b) Mustard
(c) Sorghum (d) Millets

603. Black tip of mango is due to
(a) Boron deficiency (b) High rainfall
(c) High temperature (d) Low oxygen level

604. The infection source of the stem rust of wheat is
(a) Foothills of Himalayas (b) Nilgiri and Pulney hills
(c) Darjeeling (d) Central plain

605. The temperature at which *Phytophthora* germinates or penetrate through zoospores is
(a) 12°C (b) 24°C
(c) 24°C (d) 28°C

606. Powdery mildew of chilli and pepper is caused by
(a) *Leveillula taurica* (b) *Blumeria graminis* f.sp. *tritici*
(c) *Erysiphe polygoni* (d) *Podosphaera pannosa*

607. Which one of the following does not causes vascular disease in chilli?
(a) *Verticillium dahliae* (b) *Colletotrichum truncatum*
(c) *Ralstonia solanacearum* (d) *Fusarium oxysporum*

608. Bacterial leaf spot of chilli is caused by
(a) *Cercospora capsici* (b) *Xanthomonas campestris* pv. *vesicatoria*
(c) *Phytophthora capsici* (d) *Pectobacterium carotovora* subsp. *carotovora*

609. *Gliocladium catenulatum* is a potential biological control agent of
(a) Damping-off in bedding plants
(b) Bacterial blight of rice in nursery
(c) Southern blight of apple in nursery
(d) Bacterial wilt of tomato in polyhouse

610. Green wilt in many solanaceous crops is caused by
(a) *Verticillium albo-atrum* (b) *Fusarium solani*
(c) *Ralstonia solanacearum* (d) *Rhizoctonia solani*

611. Which one of the following is a facultative biotrophic pathogen?
(a) *Puccinia graminis tritici* (b) *Blumeria graminis tritici*
(c) *Phytophthora infestans* (d) All of these

612. Conversion of inflorescence to a green leafy structure is called
(a) Virescence (b) Atrophy
(c) Phyllody (d) Dwarfing

613. Most widely used fungicide for rust fungi is
(a) Vitavax (b) Plantvax
(c) Carbendazim (d) Dithane M-45

614. Formation of cork layers prevents the invasion of pathogen in
(a) Black scurf of potato (b) Common scab of potato
(c) Wart disease of potato (d) Blackleg of potato

615. Cirrhus, a horn-like structure present in
(a) *Colletotrichum falcatum* (b) *Ascochyta rabei*
(c) *Pyricularia oryzae* (d) *Rhizoctonia solani*

616. ***Citrus tristeza virus* is transmitted by**
(a) Thrips (b) Beetles
(c) Aphids (d) Leaf hopper

617. Rust of broad bean is caused by
(a) *Uromyces fabae* (b) *Hemileia vastatrix*
(c) *Puccinia tritici* (d) All of these

618. Which is non-parasitic disease?
(a) Black tip of mango (b) Black heart of rice
(c) Tip burn of rice (d) All of these

619. Period between inoculation and appearance of the first visible disease symptom is
(a) Penetration period (b) Incubation period
(c) Latent period (d) Infection period

620. The main source of inoculum for *Xanthomonas citri* pv. *citri* causing citrus canker
(a) Bacteria present in infected plant parts in the soil
(b) Bacteria carried through seeds
(c) Infected twigs and old lesions on the standing tree
(d) Alternate host

621. The annual reoccurrence of black rust of wheat in the plains is through
(a) Uredospores (b) Teleutospores
(c) Aeciospores (d) Basidiospores

622. V-shaped yellowish spots progressing from the leaf margin towards the center, is the characteristic symptoms of
(a) Brown spot of crucifers (b) Black rot of crucifers
(c) Club root of crucifers (d) White rust of crucifers

623. Damping-off caused by *Pythium* is often associated with
(a) Poor drainage (b) Well-aerated soil
(c) High temperature (d) Resistant host

624. Most powdery mildew fungi attacking trees belong to the genus
(a) *Leveillula* (b) *Microsphaera*
(c) *Blumeria* (d) *Erysiphe*

625. An area of diseased tissue is called
(a) Lesion (b) Necrosis
(c) Spot (d) Chlorosis

626. Host-alternating rust is termed as
(a) Autoecious (b) Heterotroph
(c) Heteroecious (d) Auototroph

627. Rusts that complete their life cycle on a single host are called
(a) Autoecious (b) Heterotroph
(c) Heteroecious (d) Auototroph

628. Powdery mildew disease is caused by
(a) *Erysiphe* (b) *Podosphaera*
(c) *Oïdium* (d) All of these

629. As a rule, symptoms of non-infectious diseases in the field appear
(a) Uniform (b) Patchy
(c) Random (d) None of these

630. A well-defined dead area of a leaf, stem, or root is called
(a) Necrotic lesion (b) Dieback
(c) Necrosis (d) Rot

631. Phomopsis blight is associated with
(a) Potato (b) Tomato
(c) Capsicum (d) Eggplant

632. The optimum temperature for powdery mildew development is
(a) 10°C–15°C (b) 20°C–25°C
(c) 15°C–20°C (d) 28°C–35°C

633. The most common and widespread plant-parasitic nematode is
(a) *Meloidogyne* spp. (b) *Radopholus* spp.
(c) *Tylenchulus* spp. (d) *Pratylenchus* spp.

634. Zoosporic plant pathogen that does not belong to Oomycota
(a) *Labyrinthula terrestris* (b) *Synchytrium macrosporum*
(c) *Plasmodiophora brassicae* (d) All of these

635. White blisters of crucifer is caused by
(a) *Pythium debarynam* (b) *Albugo candida*
(c) *Alternaria brassicae* (d) *Plasmodiophora brassicae*

636. Select the wrong pair
(a) Ergot of bajra—*Claviceps fusiformis*
(b) False smut of rice—*Villosiclava virens*
(c) Ergot of rye—*Claviceps purpurea*
(d) Red rot of sugarcane—*Colletotrichum gloeosporiodes*

637. Club root of crucifers (*Plasmodiophora brasicae*) can be managed by
(a) Raising pH of soil (b) Decreasing soil pH
(c) Both (a) and (b) (d) None of these

638. Damping-off disease in nursery is caused by
(a) *Pythium ultimum* (b) *Pythium aphanidermatum*
(c) *Pythium debaryanum* (d) All of these

639. The optimum temperature for the attack of foot rot of papaya is
(a) 22°C (b) 28°C
(c) 32°C (d) 36°C

640. The effective management of late blight of potato is possible by the use of
(a) Sanitation measure (b) Foliar spray of Metalaxyl
(c) Foliar spray of Mancozeb (d) All of these

641. Citrus gummosis disease can be identified by
(a) Death of shoot (b) Oozing of gum
(c) Root galls (d) Chlorosis

642. ***Albugo candida*** **causes white blisters of crucifers is**
(a) Obligate parasites (b) Obligate saprophytes
(c) Facultative saprophytes (d) Facultative parasites

643. The most characteristic symptom of Fusarium wilt of cotton is
(a) Yellowing of tissue (b) Discoloration of vascular tissues
(c) Necrosis (d) All of these

644. The incidence of black scurf of potato is more in
(a) Sandy soil (b) Clay soil
(c) Alluvial soil (d) Loam soil

645. Black heart is a physiological disorder of
(a) Tomato (b) Capsicum
(c) Potato (d) Cabbage

646. Select the wrong pair
(a) Mustard—White rust (b) Paddy—Brown spot
(c) Wheat—Red rust (d) Groundnut—Tikka disease

647. When the death of terminal shoots led to the death of whole plant, it is called
(a) Anthracnose (b) Blight
(c) Wilt (d) Dieback

648. ***Ralstonia solanacearum*** **causes**
(a) Wilting of tomato (b) Leaf curl of tomato
(c) Bunchy top of banana (d) Soft rot of potato

649. ***Barbery vulgaris*** **is the alternative host for**
(a) *Hemeilia vastatrix* (b) *Puccinia recondita*
(c) *Puccinia graminis tritici* (d) *Melampsora lini*

650. Effective fungicide for the control of late blight of potato
(a) Ridomil MZ (b) Iprodione
(c) Bavistin (d) Hinosan

651. ***Colletotrichum falcatum*** **causes**
(a) Anthracnose of chilli (b) Red rot of sugarcane
(c) Sett rot of sugarcane (d) Anthracnose of grapes

652. ***Tomato leaf curl virus*** **is transmitted by**
(a) *Myzus persicae* (b) *Bemisia tabaci*
(c) *Aphis gossypii* (d) *Thrips palmi*

653. The most important soilborne disease in the nursery is
(a) Leaf spot (b) Damping off
(c) Blight (d) Anthracnose

654. The vector of ***Cotton leaf curl virus***
(a) Jassid (b) Aphid
(c) Pink bollworm (d) Whitefly

655. Capsid of the virus is composed of
(a) Carbohydrates (b) Amino acids
(c) Nucleic acids (d) Lipids

656. Scab disease of potato can be controlled best at pH
(a) 5.5–6.5 (b) 3.5–4.5
(c) 7.2–7.8 (d) 8.2–9.2

657. Mosaic symptoms are associated with
(a) Bacteria (b) Phytoplasma
(c) Virus (d) Fungi

658. Pathogen associated with "Green ear disease" of bajra perpetuates in the soil as
(a) Oospores (b) Sclerotia
(c) Conidia (d) Chlamydospores

659. Which of the following is an ascocarp?
(a) Pycnidium (b) Perithecium
(c) Sporodochium (d) Conidiomata

660. Which of the following bacterium has the distinction of being first bacterium causing a plant disease?
(a) *Agrobacterium tumefaciens* (b) *Pseudomonas syringae*
(c) *Xanthomonas citri* (d) *Erwinia amylovora*

661. Peach leaf curl is caused by
(a) *Leaf curl virus* (b) *Mycosphaerella fragariae*
(c) *Agrobacterium tumefaciens* (d) *Taphrina deformans*

662. FAO declared, 2020, as an international year of
(a) Soil health (b) Pulses
(c) Water conservation (d) Plant health

663. Citrus gummosis is caused by
(a) *Phytophthora citropthora*
(b) *Phytophthora nicotianae* var. *parasitica*
(c) *Phytophthora palmivora*
(d) All of the above

664. Loose smut of wheat can be controlled by hot water treatment of seed at
(a) 48°C for 4 minutes (b) 54°C for 10 minutes
(c) 50°C for 3 minutes (d) 60°C for 10 minutes

665. A common ooze test is performed for diagnosing
(a) Viral diseases (b) Nematode diseases
(c) Fungal diseases (d) Bacterial diseases

666. Disease cycle of the most plant pathogen can be disrupted by
(a) Crop rotation (b) Destruction of collateral hosts
(c) Fallow conditions (d) All of these

667. Which of the following disease caused the epidemic in past?
(a) Late blight of potato (b) Coffee rust
(c) Brown spot of rice (d) All of these

668. Fungicide Thiram is used for
(a) Foliar spray (b) Soil drenching
(c) Seed treatment (d) Blossom spray

669. The abnormal increase in the size of a plant organ is due to
(a) Hyperplasia (b) Hypertrophy
(c) Both (a) and (b) (d) None of these

670. Term necrosis indicates
(a) Death of tissue or cells
(b) Destruction of chlorophyll of lamina
(c) Hypertrophy and hyperplasia
(d) Charcoal-like powder

671. Which of the following chemical is highly effective against powdery mildew pathogens?
(a) Copper fungicide (b) Sulphur fungicide
(c) Carbendazim (d) Carboxin

672. Hot water treatment is widely used for the control of
(a) Airborne pathogens (b) Seedborne pathogens
(c) Soilborne pathogens (d) All of these

673. Symptom of a disease characterized by the progressive death of shoots, branches, and roots, generally starting at the tip, known as
(a) Canker (b) Dieback
(c) Sooty mold (d) Blight

674. Grassy-shoot disease of sugarcane is caused by
(a) Spiroplasma (b) Bacteria
(c) Virus (d) Phytoplasma

675. Which of the following pair of diseases and pathogens is NOT correct?
(a) Flag smut of wheat–*Urocystis agropyri*
(b) Smut of sugarcane–*Ustilago maydis*
(c) Black arm of cotton–*Xanthomonas campestris* pv. *malvacearum*
(d) Ratoon stunting of sugarcane–*Clavibacter xyli* subsp. *xyli*

676. Loose smut of wheat is
(a) Internally seedborne
(b) Externally seedborne
(c) Internally and externally seedborne
(d) None of these

677. In grape plants, under favorable weather conditions, large numbers of oily spots may develop and coalesce to cover most of the leaf surface. After suitably warm, humid nights, a white downy fungal growth (sporangia) appear on the underside of the leaves and other infected plant parts. The disease is
(a) Anthracnose (b) Downy mildew
(c) Powdery mildew (d) White rust

678. Streptomycin sulphate will be effective against
(a) Nematode (b) Fungi
(c) Virus (d) Bacteria

679. Which of the following is a polycyclic disease?
(a) Rusts (b) Powdery mildew
(c) Late blight of potato (d) All of these

680. Yellowing of normally green tissue due to chlorophyll destruction or failure of chlorophyll formation, known as

(a) Enation (b) Mosaic
(c) Chlorosis (d) Epinasty

681. The book "Plant Diseases," the bible of Plant Pathology, is written by

(a) Dr. Y.L. Nene (b) Dr. R.S. Singh
(c) Dr. U.S. Singh (d) Dr. G.N. Agrios

682. Mycoplasmas are NOT sensitive to

(a) Tetracycline (b) Chloramphenicol
(c) Penicillin (d) Erythromycin

683. First plant pathogenic bacteria whose genome was sequenced?

(a) *Xanthomonas oryzae* pv. *oryzae* (b) *Agrobacterium tumefaciens*
(c) *Xylella fastidiosa* (d) *Pseudomonas syringae* pv. *syringae*

684. Which of the following diseases is known as polyetic disease?

(a) Panama wilt of banana (b) Late blight of potato
(c) Wilt of guava (d) Wheat rust

685. In India, domestic plant quarantine is operated against

(a) Wart disease of potato (b) Cyst nematode of potato
(c) Wheat rust (d) Both (a) and (b)

686. The disease cycle of the most plant pathogen can be disrupted by

(a) Crop rotation (b) Destruction of collateral hosts
(c) Fallow conditions (d) All of these

687. Little leaf of brinjal is transmitted by

(a) *Jesus indicus* (b) *Bemisia tabaci*
(c) *Pentalonia nigronervosa* (d) *Hishimonus phycitis*

688. Huanglongbing disease of citrus is transmitted by

(a) Aphid (b) Psyllid
(c) Leaf hopper (d) Whitefly

689. Muriform conidia are produced by

(a) *Fusarium* (b) *Helminthosporium*
(c) *Alternaria* (d) *Pyrenophora*

690. Flageller arrangement of *Erwinia amylovora* is

(a) Monotrichous (b) Amphitrichous
(c) Lophotrichous (d) Peritrichous

691. Bunchy top of papaya is caused by

(a) Virus (b) Phytoplasma
(c) Viroid (d) Fungus

692. "Shoe-string" symptoms in papaya are produced in

(a) Mosaic disease (b) Ring spot diseases
(c) Leaf curl disease (d) Anthracnose

693. "Sun up" and "Rainbow" are resistant transgenic against
(a) Mango malformation (b) Downy mildew of sunflower
(c) Papaya ring spot (d) Sigatoka leaf spot of banana

694. Dolak Sind and Lucknow-49 are resistant cultivars against the
(a) Guava wilt (b) Panama wilt of banana
(c) Anthracnose of guava (d) Apple scab

695. Bunchy top disease of banana was introduced in India from
(a) 1946 (b) 1956
(c) 1840 (d) 1940

696. Which of the following pathogen is known as fungi-like organisms?
(a) *Pythium* (b) *Phytophthora*
(c) *Peronospora* (d) All of these

697. Rust of pea is caused by
(a) *Uromyces fabae* (b) *Puccinia coronate*
(c) *Uromyces appendiculatus* (d) *Uromyces ciceris-arietini*

698. Which of the following pathogen is a facultative parasite?
(a) *Phytophthora* (b) *Pythium*
(c) *Fusarium* (d) *Sclerotinia*

699. Which of the following disease caused the epidemic in past?
(a) Late blight of potato (b) Coffee rust
(c) Brown spot of rice (d) All of these

700. Which of the following disease is caused by a non-viral agent?
(a) Chilli leaf curl (b) Papaya leaf curl
(c) Peach leaf curl (d) Cotton leaf curl

701. Which of the following diseases is known as polyetic disease?
(a) Panama wilt of banana (b) Late blight of potato
(c) Wilt of guava (d) Wheat rust

702. In cross protection process, pre-inoculation of the host plant is done with
(a) Mild strain of the same virus
(b) Virulent strain of the same virus
(c) Mild strain of the different virus
(d) Virulent strain of the different virus

703. Muriform conidia are produced by
(a) *Fusarium* (b) *Helminthosporium*
(c) *Alternaria* (d) *Pyrenophora*

704. Flageller arrangement of *Erwinia amylovora* is
(a) Monotrichous (b) Amphitrichous
(c) Lophotrichous (d) Peritrichous

705. Which of the following pathogen does NOT contains nucleic acid?
(a) Prion (b) Viroid
(c) Phytoplasma (d) Virion

706. Which of the following is NOT a ture fungi?
(a) *Puccinia* (b) *Rhizoctonia*
(c) *Fusarium* (d) *Peronospora*

707. Which of the following pathogen is a facultative parasite?
(a) *Phytophthora* (b) *Pythium*
(c) *Rhizoctonia* (d) *Sclerotinia*

708. In grapes, "Bird's Eye-Rot" appearance is the characteristic symptom of
(a) Grey mold (b) Pierce's disease
(c) Anthracnose (d) Black rot

709. In India, domestic plant quarantine is operated against
(a) Late blight of potato (b) Root knot of okra
(c) Wheat rust (d) Cyst nematode of potato

710. Abnormally small plant growth
(a) Epinasty (b) Witches' broom
(c) Dieback (d) Stunting

711. Little leaf of brinjal is caused by
(a) Nematode (b) Virus
(c) Phytoplasma (d) Fungi

712. Huanglongbing disease of citrus is transmitted by
(a) Aphid (b) Psyllid
(c) Leaf hopper (d) Whitefly

713. Muriform conidia are produced by
(a) *Fusarium* (b) *Phytophthora*
(c) *Alternaria* (d) *Rhizoctonia*

714. The characteristic symptoms of *Plum pox virus* on *Prunus* sp. are
(a) Distortion and ring spots on leaves
(b) Severely deformed and bumpy fruit
(c) Discoloration of seeds
(d) All of these

715. Bunchy top disease of banana was introduced in India in
(a) 1946 (b) 1956
(c) 1840 (d) 1940

716. Fruiting body produced by *Phomopsis vexans* is
(a) Acervuli (b) Pycnidia
(c) Apothecia (d) Perithecia

717. Father of Indian Plant Pathology
(a) Antone de Barry (b) B.B. Mundkar
(c) E.J. Butler (d) K.C. Mehta

718. Bacterium *Erwinia tracheiphila* survives in
(a) Soil
(b) Seed

(c) Striped and spotted cucumber beetles
(d) All of these

719. Blossom end rot of tomato is caused by
(a) Fungus (b) Boron deficiency
(c) Calcium deficiency (d) Iron deficiency

720. Purple blotch of onion and garlic is caused by
(a) *Urocystis cepulae* (b) *Stemphylium vesicarium*
(c) *Alternaria porri* (d) *Alternaria mali*

721. Which one of the following diseases is NOT transmitted by white fly?
(a) Chilli leaf curl (b) Papaya leaf curl
(c) Peach leaf curl (d) Cotton leaf curl

722. *Tomato spotted wilt virus* is transmitted by
(a) Aphid (b) Thrips
(c) White fly (d) Plant hopper

723. Which one of the following chemicals is highly effective against powdery mildew pathogens?
(a) Copper fungicide (b) Sulphur fungicides
(c) Carbendazim (d) Carboxin

724. The bacterial wilt of cucurbits is caused by
(a) *Erwinia tracheiphila* (b) *Erwinia amylovora*
(c) *Erwinia herbicola* (d) *Erwinia chrysanthemi*

725. Onion blast is caused by
(a) *Pyricularia grisia* (b) *Cercopspora alli*
(c) *Botrytis allii* (d) *Alternaria porri*

726. In which of the following disease, the internal tissues of the bark appear dry and give honeycomb appearance?
(a) Ring spot of papaya (b) Anthracnose of papaya
(c) Root and foot rot of papaya (d) Bunchy top of papaya

727. Pathogen of stem gall of coriander survives in the form of
(a) Oospores (b) Blastospores
(c) Chlamydospores (d) Sclerotia

728. Huanglongbing (HLB) is a devastating disease of citrus caused by
(a) Fungi (b) Bacteria
(c) Virus (d) Zinc deficiency

729. Which of the following nematode causes root knot disease in plants?
(a) *Meloidogyne* (b) *Trichodorus*
(c) *Pratylenchus* (d) *Xiphinema*

730. Term necrosis indicates
(a) Death of tissue or cells (b) Destruction of chlorophyll of lamina
(c) Hypertrophy and hyperplasia (d) Charcoal-like powder

731. Generally .... ascospores are present per ascus.

(a) 4 (b) 6
(c) 8 (d) 12

732. The abnormal increase in the size of a plant organ is due to

(a) Hyperplasia (b) Hypertrophy
(c) Both (a) and (b) (d) None of these

733. Bean rust is caused by

(a) *Uromyces ciceris-arietini* (b) *Uromyces appendiculatus*
(c) *Puccinia polysora* (d) *Puccinia graminis*

734. Fungicide Vitavax is used for

(a) Foliar spray (b) Soil drenching
(c) Seed treatment (d) Blossom spray

735. Which of the following disease is caused by *Plasmopara*?

(a) Green ear disease of bajra (b) Late blight of potato
(c) Downy mildew of grapes (d) Damping off in seedlings

736. Chickpea wilt can be managed by

(a) Spraying fungicides
(b) Cultivation of a tolerant variety
(c) Using disease-free seed for sowing
(d) Application of biocontrol agents

737. *Colletotrichum falcatum* is

(a) Facultative parasite (b) Saprophyte
(c) Parasite and saprophyte (d) All of these

738. Ear cockle of wheat can be controlled by

(a) Seed treatment with Captan (b) Proper cultural practices
(c) Using sterilized soil (d) All of these

739. *Ascochyta rabiei* survives in the soil in the form of

(a) Dormant mycelium (b) Resting sporangia
(c) Conidia (d) Pycnidia

740. The suitable temperature for the development of citus canker is

(a) 10°C–15°C (b) 12°C–16°C
(c) 25°C–30°C (d) 30°C–35°C

741. Stripe rust of wheat is favored by

(a) Cool moist weather (b) Warm humid weather
(c) Warm dry weather (d) Hot humid weather

742. Early blight of potato occurs

(a) Early in the season when the temperature is low
(b) Early in the season when the temperature is moderate
(c) Early in the season when the temperature is high
(d) Early in the season when the temperature is cool

743. ***Puccinia graminis tritici*** **survives adverse environmental conditions in the form of**
(a) Dormant mycelium (b) Uredospores
(c) Teliospores (d) Basidiospores and pycniospore

744. ***Anguina tritici*** **gets contaminated with** ***Corynebacterium tritici*** **in the**
(a) Spikelets (b) Soil
(c) Seed (d) All of these

745. ***Taphrina deformans*** **survives adverse environmental conditions in the form of**
(a) Ascospore (b) Dormant mycelium
(c) Resting sporangia (d) All of these

746. Plant galls are produced by
(a) Viruses (b) Nematodes
(c) Fungi and bacteria (d) All of these

747. Most of the plant viruses are transmitted by
(a) Whiteflies (b) Plant hopper
(c) Leaf hopper (d) Aphids

748. Symptom-based identification of virus diseases are not reliable because
(a) Symptoms are not influenced by the environment
(b) Symptoms may be induced by mixed infection
(c) Viral symptoms appear similar
(d) Symptoms are general host reaction

749. Downy mildews are named so because they
(a) Develop on the lower side of leaves
(b) Appear only on the lower surface of leaves
(c) Develop a downy mass on the leaf
(d) Appear down on the roots

750. Powdery mildew of mango overwinters in
(a) Dead twigs on the affected tree (b) Diseased debris
(c) Alternate hosts (d) Living twigs

751. Hypertrophy is best described as
(a) Abnormal plant growth due to increased cell division
(b) Abnormal plant growth due to abnormal cell enlargement
(c) Normal plant growth due to increased cell division
(d) None of these

752. Damping-off is best described as
(a) Destruction of seedlings near soil line by pathogen and falling on the ground
(b) Destruction of seedlings by pathogen at the middle of shoot and falling on the ground
(c) Destruction seedlings by wind and falling on the ground
(d) Destruction of seedlings by hailstorms and falling on the ground

753. Forma specialis (f. sp.) is a category of the pathogen that infects
(a) Wide range of host plant
(b) Host plants within a genus or species
(c) Few host plants
(d) Non-host plants

754. Yellowing of normal green tissue due to chlorophyll degradation is known as
(a) Necrosis (b) Mosaic
(c) Chlorosis (d) Scorching

755. The sexual, thick-walled, and resting spores of the rust fungi are known as
(a) Basidiospores (b) Urediospores
(c) Teliospores (d) Aeciospores

756. The term witches' broom means
(a) Stunting of the whole plant
(b) Reduction in leaf size
(c) Dense clustering of branches
(d) Shredding leaves

757. *Banana bunchy top virus* (BBTV) is transmitted through
(a) Black aphids (b) Banana whitefly
(c) Black jassids (d) None of these

758. Sugarcane mosaic disease was first recorded in 1892 from
(a) Java (b) Japan
(c) Brazil (d) USA

759. *Citrus tristeza virus* belongs to the group
(a) Begomovirus (b) Nepovirus
(c) Closterovirus (d) None of these

760. The spore produced during II stage of the life cycle of wheat rust fungi
(a) Aeciospores (b) Teleutospores
(c) Uredospores (d) Basidiospores

761. Wheat rust fungi are an obligate parasite that belongs to the family
(a) Uredinaceae (b) Pucciniaceae
(c) Melampsoraceae (d) Ustilaginaceae

762. The heteroecious wheat rust species typically develop fruiting bodies on another host
(a) Pycnia and Aecia (b) Telia and Uredia
(c) Telia and Basidia (d) Both (b) and (c)

763. Plant disease symptoms expressing an overdevelopment in size (hypertrophy) or number of cells (hyperplasia) is
(a) Galls and Witches' Broom
(b) Curling and rolling
(c) Epinasty and Scab
(d) All of the above

764. Wilting is favored by
(a) Hot, moist weather (b) Cool, dry weather
(c) Cool, moist weather (d) Hot, dry weather

765. *Potato leaf roll virus* is
(a) Tospovirus (b) Sap transmissible
(c) Luteovirus (d) All of these

766. The insect vectors of viruses can be managed by spraying
(a) Dithane M-45 (b) Metasystox
(c) Captan (d) All of these

767. Potato phyllody is caused by
(a) Phytoplasma (b) Fungi
(c) Nematodes (d) Viruses

768. Which one of the following statements is NOT correct about *Potato leaf roll virus*?
(a) It causes the rolling of leaves
(b) It is a sap-transmissible virus
(c) It is a luteovirus
(d) It is transmitted by aphids

769. Which of the following are biotrophs?
(a) *Synchytrium* and *Albugo*
(b) Downy mildew pathogen, *Puccinia*, and *Melampsora*
(c) Powdery mildew fungi
(d) All of these

770. Golden nematode is a pest of
(a) Potato (b) Okra
(c) Rice (d) Soybean

771. Potato varieties namely, Kufri Jyoti and Kufri Jeevan evolved at the Central Potato Research Institute (CPRI), have shown a high degree of field resistance to
(a) Late blight (b) Early blight
(c) Golden cyst nematode (d) Black scurf

772. *Globodera rostochiensis* is commonly known as the
(a) Golden nematode
(b) Golden eelworm
(c) Yellow potato cyst nematode
(d) All of these

773. Symptoms caused by potato golden cyst nematode is
(a) Surface cracking (b) Reduced root system
(c) Cysts on the root surface (d) All of these

774. Ooze test is performed for the diagnosis of
(a) Bacterium (b) Virus
(c) Fungi (d) Nematodes

775. The first link in the infection chain is
(a) Survival (b) Infection
(c) Penetration (d) Dispersal

776. Most serious disease in rose after pruning is
(a) Canker (b) Dieback
(c) Stem rot (d) Rose wilt

777. Seeds also play an important role in
(a) Survival of pathogens
(b) Nutrition of pathogens

(c) Transmission of pathogens
(d) All of these

778. ***Bipolaris maydis*** **(Southern corn leaf blight), the infection can lead to severe leaf chlorosis by producing some toxins such as T-toxin that attack on.............and affect the photosynthesis.**

(a) Mitochondria (b) Nucleus
(c) Ribosome (d) Vacuole

779. ***Xylella fastidiosa*** **causes**

(a) Phony peach (b) Pierce's disease of grape
(c) Citrus variegated chlorosis (d) All of these

780. The loss of leaves from a plant is called

(a) Canker (b) Defoliation
(c) Epinasty (d) Clipping

781. The ring rot pathogen (*Clavibacter michiganensis* subsp. *sepedonicus*) of potato is

(a) Short, motile, Gram-positive rod
(b) Short, nonmotile, Gram-positive rod
(c) Long, motile, Gram-negative rod
(d) Long, nonmotile, Gram-positive rod

782. The bacterial ring rot of potato is characterized by

(a) Interveinal chlorosis
(b) Curling of leaves and wilting of foliage
(c) Decay in tubers
(d) All of these

783. Rust fungi usually enter their host through stomata. Which of the following rust pathogen infect their hosts through the cuticle?

(a) *Phakopsora pachyrhizi*
(b) *Puccinia psidii*
(c) *Ravenelia humphreyana*
(d) All of these

784. Marigold has nematicidal activities but it can serve as a host for

(a) Stubby-root nematode (b) Reniform nematode
(c) Sting and awl nematode (d) All of these

785. Marigold is highly nematicidal for

(a) Stubby-root nematode (b) Reniform nematode
(c) Sting and awl nematode (d) Southern root-knot nematode

786. Cotton root rot disease is caused by

(a) *Pythium aphanidermatum* (b) *Macrophomina phasiolina*
(c) *Fusarium oxysporum* (d) *Rhizoctonia bataticola*

787. The symptom appears as fluffy white growth with minute black dots on the surface of wheat leaves, the disease is

(a) Powdery mildew (b) Septoria leaf spot
(c) Downy mildew (d) Black rust

788. Speckled leaf blotch in wheat is caused by

(a) *Puccinia striiformis* (b) *Urocystis agropyri*
(c) *Alternaria triticina* (d) *Septoria tritici*

789. In virus infected-plants the following symptoms can be observed

(a) Mosaic formation and stunted growth
(b) Leaf rolling and curling
(c) Yellowing and vein clearing
(d) All of these

790. The first sign of stripe rust on wheat is the appearance of

(a) Brown pustules (b) Yellow spots
(c) Black spots (d) Yellow streaks

791. Consider the following:

(i) Black rust of wheat
(ii) Bean common mosaic
(iii) Fusarium wilt of tomato

Select the correct sequence of the above as seedborne, soilborne, and airborne disease, respectively.

(a) (i)—(ii)—(iii) (b) (i)—(iii)—(ii)
(c) (ii)—(i)—(iii) (d) (ii)—(iii)—(i)

792. After initiation of crown gall disease (*Agrobacterium tumefaciens*), its spread and development depend on

(a) Growth and reproduction of bacteria
(b) Auxin production by bacteria
(c) Non-plasmid genome of bacteria
(d) Motility of bacteria

793. Which of the following pair is responsible for "tundu disease" of wheat?

(a) Nematode and bacterium (b) Bacterium and fungus
(c) Nematode and fungus (d) Bacterium and virus

794. Brown rot of peach caused by

(a) *Monilinia fructicola* (b) *Taphrina deformans*
(c) *Nectria galligena* (d) *Podosphaera leucotricha*

795. False loose smut is a fungal disease of barley caused by

(a) *Ustilago hordei* (b) *Ustilago nuda*
(c) *Ustilago tritici* (d) *Ustilago nigra*

796. Powdery mildew on apple blossom cluster is caused by

(a) *Uncinula necator* (b) *Venturia inaequalis*
(c) *Sphaerotheca fuliginea* (d) *Podosphaera leucotricha*

797. The "physiological races" of the rust fungi are separable by the patterns of pathogenicity and virulence on

(a) Cultivated hosts (b) Collateral hosts
(c) Weed hosts (d) Differential hosts

798. Wheat blast disease is caused by
(a) *Botrytis alli* (b) *Pyricularia oryzae*
(c) *Magnaporthe oryzae Triticum* (d) *Bipolaris sorokiniana*

799. In 1931, karnal bunt of wheat was first time reported in Karnal (Haryana) by
(a) R.S. Singh (b) D.F. Dustur
(c) B.B. Mundkur (d) K.K. Mitra

800. The ring spot disease in sugarcane is caused by
(a) *Sclerospora sacchari* (b) *Leptosphaeria sacchari*
(c) *Puccinia melanocephala* (d) *Rhizoctonia solani*

801. Citrus greening disease is mainly confined to
(a) East India (b) West India
(c) North India (d) South India

802. Which of the following spores are formed by *Puccinia* on wheat host?
(a) Urediniospores, pycniospores, basidiospores
(b) Urediniospores, teliospores, basidiospores
(c) Pycniospores, aeciospores, basidiospores
(d) Urediniospores, aeciospores, teliospores

803. Which of the following spore is formed by *Puccinia* on barberry?
(a) Urediniospores (b) Basidiospores
(c) Teliospores (d) Aeciospores

804. Which of the following pair is mismatched?
(a) Coffee rust—*Hemileia vastatrix*
(b) Wheat stem rust— *Puccinia graminis*
(c) White pine blister rust—*Cronartium ribicola*
(d) Oat crown rust—*Puccinia graminis* f. sp. *avenae*

805. The teliospores of *Puccinia graminis* divides by meiosis and the resulting haploid nuclei are packaged in
(a) Two basidiospores (b) Four basidiospores
(c) Six basidiospores (d) Eight basidiospores

806. Apple replant problem is caused by
(a) *Phytophthora* sp. and *Pythium* sp.
(b) *Fusarium* sp. and *Rhizoctonia*
(c) *Pratylenchus penetrans*
(d) All of these

807. Urediospore samples of wheat rust fungi can be stored for long-term in
(a) Refrigerator (b) Dried samples
(c) Sterilized soil (d) Liquid nitrogen

808. Stripe rust of wheat develops successfully within the range of
(a) 10°C–15°C (b) 15°C–25°C
(c) 18°C–30°C (d) 25°C–35°C

809. Which of the following pair of the pathogen and its lifestyle is INCORRECT?

(a) *Zymoseptoria tritici*—Necrotroph
(b) *Blumeria graminis* f.sp. *tritici*—Biotroph
(c) *Albugo candida*—Biotroph
(d) *Fusarium oxysporum*—Hemibiotroph

810. A canker pathogen can cause symptoms by

(a) Killing the vascular cambium and phloem tissue
(b) Growing through the wood and killing the entire stem
(c) Killing the heartwood of a tree
(d) Killing only the vascular cambium

811. Which plant pathogen serves as a tool for transgenic plant development?

(a) *Angunia tritici* (b) *Podosphaera xanthii*
(c) *Agrobacterium tumefacians* (d) *Venturia inaequalis*

812. Which one of the following is a bacterial disease?

(a) Black rot of crucifers (b) Blackleg of crucifers
(c) Club root of crucifers (d) None of these

813. The Molya disease of wheat is caused by

(a) *Heterodera avenae* (b) *Anguina tritici*
(c) *Rotylenchus robustus* (d) *Pratylenchus penetrans*

814. An example of internally seedborne disease is

(a) Loose smut of wheat (b) Rust of wheat
(c) Rust of groundnut (d) Covered smut of wheat

815. Blackleg of the potato is caused by

(a) Fungus (b) Bacteria
(c) Virus (d) Nematode

24.2 MATCH THE FOLLOWING

1. (i) Flag smut of wheat — (A) *Puccinia striiformis* f. sp. *tritici*
(ii) Stripe rust of wheat — (B) *Neovossia indica*
(iii) Powdery mildew of wheat — (C) *Blumeria graminis* f. sp. *tritici*
(iv) Ear cockle of wheat — (D) *Urocystis agropyri*
(v) Karnal bunt of wheat — (E) *Anguina tritici*

***Ans*:** (i)—( ), (ii)—( ), (iii)—( ), (iv)—( ), (v)—( )

2. (i) Molya disease of wheat — (A) *Rathayibacter tritici*
(ii) Black chaff of wheat — (B) *Septoria nodorum*
(iii) Yellow year rot of wheat — (C) *Anguina tritici*
(iv) Glume blotch of wheat — (D) *Xanthomonas translucens* f. sp. *undulosa*
(v) Ear cockle of wheat — (E) *Heterodera avenae*

***Ans*:** (i)—( ), (ii)—( ), (iii)—( ), (iv)—( ), (v)—( )

3. (i) Stripe rust of wheat (A) *Puccinia triticina*
(ii) Brown leaf rusts of rye (B) *Puccinia coronata* var. *avenae*
(iii) Leaf rust of barley (C) *Puccinia recondita* f. sp. *secalis*
(iv) Leaf rust of wheat (D) *Puccinia hordei*
(v) Crown rust of oats (E) *Puccinia striiformis* f. sp. *tritici*

Ans: (i)—( ), (ii)—( ), (iii)—( ), (iv)—( ), (v)—( )

4. (i) Red rot of sugarcane (A) *Ustilago scitaminea*
(ii) Eye spot of sugarcane (B) *Fusarium* spp.
(iii) Red sheath of sugarcane (C) *Drecheslera sacchari*
(iv) Pokkah boeng of sugarcane (D) *Sclerotium rolfsii*
(v) Smut of sugarcane (E) *Colletotrichum falcatum*

Ans: (i)—( ), (ii)—( ), (iii)—( ), (iv)—( ), (v)—( )

5. (i) Ratoon stunting of sugarcane (A) *Xanthomonas rubrilineans*
(ii) Gummosis of sugarcane (B) *Ceratocystis paradoxa*
(iii) Downy mildew of sugarcane (C) *Xanthomonas vasculorum*
(iv) Red stripe of sugarcane (D) *Sclerospora sacchari*
(v) Pineapple disease of sugarcane (E) *Leifsonia xyli* subsp. *xyli*

Ans: **(i)—( ), (ii)—( ), (iii)—( ), (iv)—( ), (v)—( )**

6. (i) Grassy shoot of sugarcane (A) *Colletotrichum falcatum*
(ii) Red rot of sugarcane (B) Phytoplasma
(iii) Red stripe of sugarcane (C) *Leifsonia xyli* subsp. *xyli*
(iv) Ratoon stunting disease of sugarcane (D) *Ustilago scitaminea*
(v) Smut of sugarcane (E) *Xanthomonas rubrilineans*

Ans: (i)—( ), (ii)—( ), (iii)—( ), (iv)—( ), (v)—( )

7. (i) Downy mildew of sunflower (A) *Sclerotinia sclerotiorum*
(ii) Powdery mildew of sunflower (B) *Puccinia helianthi*
(iii) Stem necroses of sunflower (C) *Plasmopara halstedii*
(iv) Stem rot of sunflower (D) *Golovinomyces cichoracearum*
(v) Rust of sunflower (E) *Tobacco streak virus*

Ans: (i)—( ), (ii)—( ), (iii)—( ), (iv)—( ), (v)—( )

8. (i) Club root of mustard (A) *Leptosphaerella maculans*
(ii) Black leg of mustard (B) *Peronospora brassicae*
(iii) Downy mildew of mustard (C) *Xanthomonas campestris* pv. *campestris*
(iv) White rust of mustard (D) *Plasmodiophora brassicae*
(v) Black rot of mustard (E) *Albugo candida*

Ans: (i)—( ), (ii)—( ), (iii)—( ), (iv)—( ), (v)—( )

9. (i) Chickpea wilt (A) *Botrytis cinerea*
(ii) Grey mold of chickpea (B) *Uromyces ciceris arientini*
(iii) Ascochyta blight of chickpea (C) *Opercullela padwikii*
(iv) Rust of chickpea (D) *Ascochyta rabiei*
(v) Root rot of chickpea (E) *Fusarium oxysporum* f. sp. *ciceri*

Ans: (i)—( ), (ii)—( ), (iii)—( ), (iv)—( ), (v)—( )

10. (i) Pink seed of chickpea — (A) *Macrophomina phaseolina*
(ii) Grey mold of chickpea — (B) *Thielaviopsis basicola*
(iii) Dry root rot of chickpea — (C) *Pseudomonas radiciperda*
(iv) Black streak root rot of chickpea — (D) *Botrytis cinerea*
(v) Bacterial root rot of chickpea — (E) *Erwinia rhapontici*

Ans: (i)—( ), (ii)—( ), (iii)—( ), (iv)—( ), (v)—( )

11. (i) Chickpea blight — (A) *Ramularia areola*
(ii) Sigatoka leaf spot of banana — (B) *Capnodium citri*
(iii) Grey mildew of cotton — (C) *Ascochyta rabiei*
(iv) Southern corn leaf blight — (D) *Bipolaris maydis*
(v) Sooty mold of citrus — (E) *Cercospora musae*

Ans: (i)—( ), (ii)—( ), (iii)—( ), (iv)—( ), (v)—( )

12. (i) Ascochyta blight of chickpea — (A) *Tobacco streak virus*
(ii) Club root of mustard — (B) Phytoplasma
(iii) Stem necroses of sunflower — (C) *Plasmodiophora brassicae*
(iv) Grassy shoot of sugarcane — (D) *Anguina tritici*
(v) Ear cockle of wheat — (E) *Ascochyta rabiei*

Ans: (i)—( ), (ii)—( ), (iii)—( ), (iv)—( ), (v)—( )

13. (i) Flag smut of wheat — (A) *Leifsonia xyli* subsp. *xyli*
(ii) Karnal bunt of wheat — (B) *Neovossia indica*
(iii) Ratoon stunting disease of sugarcane — (C) *Urocystis agropyri*
(iv) Molya disease of wheat — (D) *Colletotrichum falcatum*
(v) Red rot of sugarcane — (E) *Heterodera avenae*

Ans: (i)—( ), (ii)—( ), (iii)—( ), (iv)—( ), (v)—( )

14. (i) Lentil wilt — (A) *Rhizoctonia solani*
(ii) Dry root rot of lentil — (B) *Uromyces fabae*
(iii) Wet root rot of lentil — (C) *Erysiphe pisi* and *Oidiopsis taurica*
(iv) Powdery mildew of lentil — (D) *Fusarium oxysporum* f. sp. *lentis*
(v) Rust of lentil — (E) *Rhizoctonia bataticola*

Ans: (i)—( ), (ii)—( ), (iii)—( ), (iv)—( ), (v)—( )

15. (i) Anthracnose of cotton — (A) *Puccinia cacabata*
(ii) Areolate mildew of cotton — (B) *Ramularia gossypii*
(iii) Blight of cotton — (C) *Puccinia schedonnardi*
(iv) Cotton rust — (D) *Ascochyta gossypii*
(v) Southwestern cotton rust — (E) *Glomerella gossypii*

Ans: (i)—( ), (ii)—( ), (iii)—( ), (iv)—( ), (v)—( )

16. (i) Black arm of cotton — (A) *Cotton leaf curl virus*
(ii) Leaf curl of cotton — (B) *Agrobacterium tumefaciens*
(iii) Crown gall of cotton — (C) *Xanthomonas campestris* pv. *malvacearum*
(iv) Lint degradation of cotton — (D) *Phakopsora gossypii*
(v) Tropical cotton rust — (E) *Pantoea agglomerans*

Ans: (i)—( ), (ii)—( ), (iii)—( ), (iv)—( ), (v)—( )

17. (i) Downy mildew of pea — (A) *Uromyces pisi* and *U. viciae-fabae*
(ii) Powdery mildew of pea — (B) *Peronospora viciae*
(iii) Leaf spot of pea — (C) *Sclerotinia sclerotiorum*
(iv) White mold of pea — (D) *Ascochyta pisi*
(v) Pea rust — (E) *Erysiphe pisi*

Ans**:** (i)—( ), (ii)—( ), (iii)—( ), (iv)—( ), (v)—( )

18. (i) Black tip of mango — (A) *Glomerella cingulata*
(ii) Bacterial blight of mango — (B) *Acrosporium mangiferae*
(iii) Powdery mildew of mango — (C) *Fusarium mangiferae* var. *subglutinans*
(iv) Malformation of mango — (D) *Xanthomonas axonopodis* pv. *mangiferaeindicae*
(v) Anthracnose of mango — (E) Boron deficiency

Ans**:** (i)—( ), (ii)—( ), (iii)—( ), (iv)—( ), (v)—( )

19. (i) Citrus gummosis — (A) *Diaporthe citri*
(ii) Flyspeck of citrus — (B) *Colletotrichum gloeosporioides*
(iii) Melanose of citrus — (C) *Capnodium citri*
(iv) Anthracnose of citrus — (D) *Schizothyrium pomi*
(v) Sooty mold of citrus — (E) *Phytophthora* spp.

Ans**:** (i)—( ), (ii)—( ), (iii)—( ), (iv)—( ), (v)—( )

20. (i) Citrus gummosis — (A) *Candidatus Liberibacter asiaticus*
(ii) Citrus variegated chlorosis — (B) *Spiroplasma citri*
(iii) Yellow dragon of citrus — (C) *Xanthomonas citri* subsp. *citri*
(iv) Citrus stubborn — (D) *Xylella fastidiosa*
(v) Citrus canker — (E) *Phytophthora* spp.

Ans**:** (i)—( ), (ii)—( ), (iii)—( ), (iv)—( ), (v)—( )

21. (i) Citrus mal secco — (A) *Candidatus Liberibacter asiaticus*
(ii) Citrus bark cracking — (B) *Spiroplasma citri*
(iii) Citrus yellow dragon — (C) *Xanthomonas citri* subsp. *citri*
(iv) Citrus stubborn — (D) *Citrus exocortis viroid*
(v) Citrus canker — (E) *Plenodomus tracheiphilus*

Ans**:** (i)—( ), (ii)—( ), (iii)—( ), (iv)—( ), (v)—( )

22. (i) Downy mildew of pea — (A) *Rhizoctonia bataticola*
(ii) Black arm of cotton — (B) *Peronospora viciae*
(iii) Dry root rot of lentil — (C) *Thielaviopsis basicola*
(iv) Black streak root rot of gram — (D) *Phoma lingam*
(v) Black leg of crucifers — (E) *Xanthomonas campestris* pv. *malvacearum*

Ans**:** (i)—( ), (ii)—( ), (iii)—( ), (iv)—( ), (v)—( )

23. (i) Angular leaf spot of grapes — (A) *Uncinula necator*
(ii) Downy mildew of grapes — (B) *Botrytis cinerea*
(iii) Powdery mildew of grapes — (C) *Plasmopara viticola*

(iv) Gray mold of grapes	(D) *Sphaceloma ampelinum*
(v) Anthracnose of grapes	(E) *Mycosphaerella angulata*

***Ans*:** (i)—( ), (ii)—( ), (iii)—( ), (iv)—( ), (v)—( )

24. (i) Pierce's disease of grapes	(A) *Uncinula necator*
(ii) Downy mildew of grapes	(B) *Agrobacterium vitis*
(iii) Powdery mildew of grapes	(C) *Plasmopara viticola*
(iv) Crown gall of grapes	(D) *Sphaceloma ampelinum*
(v) Anthracnose of grapes	(E) *Xylella fastidiosa*

***Ans*:** (i)—( ), (ii)—( ), (iii)—( ), (iv)—( ), (v)—( )

25. (i) Apple scab	(A) *Gymnosporangium juniperi-virginianae*
(ii) Apple powdery mildew	(B) *Venturia inaequalis*
(iii) Cedar apple rust	(C) *Podosphaera leucotricha*
(iv) Japanese apple rust	(D) *Erwinia amylovora*
(v) Fire blight of apple and pear	(E) *Gymnosporangium yamadae*

***Ans*:** (i)—( ), (ii)—( ), (iii)—( ), (iv)—( ), (v)—( )

26. (i) Blister spot of apple	(A) *Agrobacterium tumefaciens*
(ii) Crown gall of apple	(B) *Apple stem pitting virus*
(iii) Fire blight of apple	(C) *Agrobacterium rhizogenes*
(iv) Hairy root of apple	(D) *Erwinia amylovora*
(v) Apple stem pitting	(E) *Pseudomonas syringae* pv. *papulans*

***Ans*:** (i)—( ), (ii)—( ), (iii)—( ), (iv)—( ), (v)—( )

27. (i) Hairy root of apple	(A) *Erwinia amylovora*
(ii) Apple scab	(B) *Podosphaera leucotricha*
(iii) Powdery mildew	(C) *Agrobacterium rhizogenes*
(iv) Fire blight of apple	(D) *Marssonina coronaria*
(v) Apple blotch	(E) *Venturia inaequalis*

***Ans*:** (i)—( ), (ii)—( ), (iii)—( ), (iv)—( ), (v)—( )

28. (i) Downy mildew of grapes	(A) *Peronospora viciae*
(ii) Fire blight of apple	(B) *Candidatus Liberibacter asiaticus*
(iii) Downy mildew of pea	(C) *Erwinia amylovora*
(iv) Yellow dragon	(D) Boron deficiency
(v) Black tip of mango	(E) *Plasmopara viticola*

***Ans*:** (i)—( ), (ii)—( ), (iii)—( ), (iv)—( ), (v)—( )

29. (i) Brown rot of peach	(A) *Taphrina deformans*
(ii) Powdery mildew of peach	(B) *Wilsonomyces carpophilus*
(iii) Peach shot hole	(C) *Podosphaera leucotricha*
(iv) Peach rust	(D) *Tranzschelia discolor* f. sp. *persica*
(v) Peach leaf curl	(E) *Monilinia fructicola*

***Ans*:** (i)—( ), (ii)—( ), (iii)—( ), (iv)—( ), (v)—( )

30. (i) Bacterial canker of peach — (A) *Xanthomonas arboricola* pv. *pruni*
(ii) Phony peach — (B) *Agrobacterium tumefaciens*
(iii) Bacterial spot of peach — (C) *Taphrina deformans*
(iv) Crown gall of peach — (D) *Pseudomonas syringae* pv. *syringae*
(v) Peach leaf curl — (E) *Xylella fastidiosa*

Ans: (i)—( ), (ii)—( ), (iii)—( ), (iv)—( ), (v)—( )

31. (i) Black leaf spot of strawberry — (A) *Botrytis cinerea*
(ii) Leaf spot of strawberry — (B) *Rhizoctonia fragariae*
(iii) Gray mold of strawberry — (C) *Colletotrichum gloeosporioides*
(iv) Anther and pistil blight of strawberry — (D) *Sphaerotheca macularis*
(v) Powdery mildew of strawberry — (E) *Cercospora fragariae*

Ans: (i)—( ), (ii)—( ), (iii)—( ), (iv)—( ), (v)—( )

32. (i) Leaf rust of strawberry — (A) *Xanthomonas fragariae*
(ii) Root gall of strawberry — (B) *Phragmidium potentillae*
(iii) Angular leaf spot of strawberry — (C) *Ralstonia solanacearum*
(iv) Cauliflower disease of strawberry — (D) *Synchytrium fragariae*
(v) Bacterial wilt of strawberry — (E) *Rhodococcus fascians* and *Aphelanchoides fragariae*

Ans: (i)—( ), (ii)—( ), (iii)—( ), (iv)—( ), (v)—( )

33. (i) Early blight of potato — (A) *Rhizoctonia solani*
(ii) Late blight of potato — (B) *Synchytrium endobioticum*
(iii) Wart of potato — (C) *Phytophthora infestans*
(iv) Canker and black scurf of potato — (D) *Phoma solanicola* f. sp. *foveata*
(v) Gangrene of potato — (E) *Alternaria solani*

Ans: (i)—( ), (ii)—( ), (iii)—( ), (iv)—( ), (v)—( )

34. (i) Ring rot of potato — (A) *Streptomyces scabiei*
(ii) Common scab of potato — (B) *Clavibacter michiganensis* subsp. *sepedonicus*
(iii) Wart of potato — (C) *Spongospora subterranea* f. sp. *subterranea*
(iv) Powdery scab of potato — (D) *Phoma solanicola* f. sp. *foveata*
(v) Gangrene of potato — (E) *Synchytrium endobioticum*

Ans: (i)—( ), (ii)—( ), (iii)—( ), (iv)—( ), (v)—( )

35. (i) Pink eye of potato — (A) *Globodera rostochiensis*
(ii) Potato cyst nematode — (B) *Potato leaf roll virus*
(iii) Potato leaf roll — (C) Oxygen deficiency
(iv) Potato black heart — (D) *Potato virus Y*
(v) Potato rugose mosaic — (E) *Pseudomonas fluorescens*

Ans: (i)—( ), (ii)—( ), (iii)—( ), (iv)—( ), (v)—( )

36. (i) Downy mildew of cucurbits — (A) *Stagonosporopsis caricae*
(ii) Gummy stem blight of cucurbits — (B) *Podosphaera xanthii*
(iii) Powdery mildew of cucurbits — (C) *Pseudoperonospora cubensis*

(iv) Angular leaf spot of cucurbits — (D) *Erwinia tracheiphila*
(v) Bacterial wilt of cucurbits — (E) *Pseudomonas syringae* pv. *lachrymans*

Ans: (i)—( ), (ii)—( ), (iii)—( ), (iv)—( ), (v)—( )

37. (i) Soft rot of ginger — (A) *Macrophomina phaseolina*
(ii) Ginger wilt — (B) *Fusarium oxysporum* f. sp. *zingiberi*
(iii) Leaf spot of ginger — (C) *Pythium* spp.
(iv) Dry rot of ginger rhizome — (D) *Pellicularia filamentosa*
(v) Thread blight of ginger — (E) *Phyllosticta zingiberi*

Ans: (i)—( ), (ii)—( ), (iii)—( ), (iv)—( ), (v)—( )

38. (i) Leaf spot of ginger — (A) *Erwinia carotovora*
(ii) Thread blight of ginger — (B) *Pellicularia filamentosa*
(iii) Basal rot/white rot of ginger — (C) *Zinger chlorotic fleck virus*
(iv) Chlorotic fleck of ginger — (D) *Phyllosticta zingiberi*
(v) Soft rot of ginger — (E) *Sclerotium rolfsii*

Ans: (i)—( ), (ii)—( ), (iii)—( ), (iv)—( ), (v)—( )

39. (i) Purple blotch of onion — (A) *Puccinia allii*
(ii) Rust of onion — (B) *Colletotrichum circinans*
(iii) Onion smudge — (C) *Alternaria porri*
(iv) Southern blight of onion — (D) *Urocystis colchici*
(v) Onion smut — (E) *Athelia rolfsii*

Ans: (i)—( ), (ii)—( ), (iii)—( ), (iv)—( ), (v)—( )

40. (i) Yellow dwarf of onion — (A) *Colletotrichum gloeosporioides*
(ii) Center rot of onion — (B) *Pantoea allii*
(iii) Onion smudge — (C) *Peronospora destructor*
(iv) Twister disease of onion — (D) *Onion yellow dwarf virus*
(v) Downy mildew of onion — (E) *Colletotrichum circinans*

Ans: (i)—( ), (ii)—( ), (iii)—( ), (iv)—( ), (v)—( )

41. (i) Purple blotch of onion — (A) *Erwinia carotovora*
(ii) Soft rot of ginger — (B) *Pseudoperonospora cubensis*
(iii) Downy mildew of cucurbits — (C) *Alternaria porri*
(iv) Wart of potato — (D) *Taphrina deformans*
(v) Peach leaf curl — (E) *Synchytrium endobioticum*

Ans: (i)—( ), (ii)—( ), (iii)—( ), (iv)—( ), (v)—( )

42. (i) Bacterial wilt of chilli — (A) *Cercospora capsici*
(ii) Anthracnose and fruit rot of chilli — (B) *Ralstonia solanacearum*
(iii) Frogeye leaf spot of chilli — (C) *Leveillula taurica*
(iv) Powdery mildew of chilli — (D) *Clavibacter michiganensis* subsp. *michiganensis*
(v) Bacterial canker of chilli — (E) *Colletotrichum* spp.

Ans: (i)—( ), (ii)—( ), (iii)—( ), (iv)—( ), (v)—( )

43. (i) Bacterial wilt of chilli — (A) *Cercospora capsici*
(ii) Leaf curl of chilli — (B) *Chilli leaf curl virus*
(iii) Frogeye leaf spot of chilli — (C) *Leveillula taurica*
(iv) Powdery mildew of chilli — (D) Calcium deficiency
(v) Blossom-end rot — (E) *Ralstonia solanacearum*

Ans: (i)—( ), (ii)—( ), (iii)—( ), (iv)—( ), (v)—( )

44. (i) Leaf spot of turmeric — (A) *Erysiphe polygoni*
(ii) Stem gall of coriander — (B) *Colletotrichum capsici*
(iii) Powdery mildew of coriander — (C) *Protomyces macrosporus*
(iv) Aster yellows of marigold — (D) *Taphrina maculans*
(v) Leaf blotch of turmeric — (E) Phytoplasma

Ans: (i)—( ), (ii)—( ), (iii)—( ), (iv)—( ), (v)—( )

45. (i) Powdery mildew of rose — (A) *Sphaceloma rosarum*
(ii) Black leaf spot of rose — (B) *Podosphaera pannosa*
(iii) Downy mildew of rose — (C) *Diplocarpon rosae*
(iv) Rust of rose — (D) *Peronospora sparsa*
(v) Anthracnose of rose — (E) *Phragmidium mucronatum*

Ans: (i)—( ), (ii)—( ), (iii)—( ), (iv)—( ), (v)—( )

46. (i) Aflatoxin — (A) *Penicillium chrysogenum*
(ii) Penicillin — (B) *Aspergillus oryzae*
(iii) Citric acid — (C) *Aspergillus niger*
(iv) Amylase — (D) *Monoascus purpurescens*
(v) Red-pigmented rice — (E) *Aspergillus flavus*

Ans: (i)—( ), (ii)—( ), (iii)—( ), (iv)—( ), (v)—( )

47. (i) Localized necrotic lesions on leaves — (A) Gummosis
(ii) Black sunken circular to angular lesions with black dots — (B) Sooty mold
(iii) Slightly raised or sunken lesions with cracks — (C) Scab
(iv) Superficial dark black sooty growth — (D) Anthracnose
(v) Exudation of gummy substances from infected parts — (E) Spots

Ans: (i)—( ), (ii)—( ), (iii)—( ), (iv)—( ), (v)—( )

48. (i) Rapid browning and death of leaves — (A) Downy mildew
(ii) White mildew growth on the lower surface of the leaves — (B) Dieback
(iii) Drying of twigs from tip to downward — (C) Blight
(iv) White or creamy, shiny, irregular pustules on the lower surface of the leaves — (D) Powdery mildew

(v) White, small, irregular powdery patches on the upper surface of leaves	(E) White blisters

Ans: (i)—( ), (ii)—( ), (iii)—( ), (iv)—( ), (v)—( )

49.

(i) Maize common smut	(A) *Etyloma oryzae*
(ii) Rice leaf smut	(B) *Uromyces appendiculatus*
(iii) Flax rust	(C) *Puccinia helianthi*
(iv) Sunflower rust	(D) *Melampsora lini*
(v) Bean rust	(E) *Ustilago maydis*

Ans: (i)—( ), (ii)—( ), (iii)—( ), (iv)—( ), (v)—( )

50.

(i) Grey mold	(A) *Puccinia striiformis*
(ii) Stripe rust of wheat	(B) *Thanatephorus cucumeris*
(iii) Rice blast	(C) *Phytophthora infestans*
(iv) Sheath blight of rice	(D) *Magnaporthe grisea*
(v) Late blight of potato	(E) *Botrytis cinerea*

Ans: (i)—( ), (ii)—( ), (iii)—( ), (iv)—( ), (v)—( )

51.

(i) Apple scab	(A) *Podosphaera leucotricha*
(ii) Powdery mildew of apple	(B) *Venturia inaequalis*
(iii) Powdery mildew of grape	(C) *Ustilago nuda*
(iv) Leaf rust of coffee	(D) *Uncinula necator*
(v) Loose smut of barley	(E) *Hemileia vastatrix*

Ans: (i)—( ), (ii)—( ), (iii)—( ), (iv)—( ), (v)—( )

52.

(i) *Plasmodiophora brassicae*	(A) Basidiomycetes
(ii) *Albugo candida*	(B) Myxomycetes
(iii) *Synchytrium endoboiticum*	(C) Ascomycetes
(iv) *Taphrina deformans*	(D) Chytridiomycetes
(v) *Ustilago maydis*	(E) Oomycetes

Ans: (i)—( ), (ii)—( ), (iii)—( ), (iv)—( ), (v)—( )

53.

(i) Clubroot of crucifers	(A) *Ustilago maydis*
(ii) White rust of mustard	(B) *Taphrina deformans*
(iii) Potato wart	(C) *Albugo candida*
(iv) Peach leaf curl	(D) *Plasmodiophora brassicae*
(v) Corn smut	(E) *Synchytrium endoboiticum*

Ans: (i)—( ), (ii)—( ), (iii)—( ), (iv)—( ), (v)—( )

54.

(i) Propiconazole	(A) Blitox-50
(ii) Carbendazim	(B) Beam
(iii) Tricyclazole	(C) Bavistin
(iv) Copper oxychloride	(D) Vitavax
(v) Carboxin	(E) Tilt

Ans: (i)—( ), (ii)—( ), (iii)—( ), (iv)—( ), (v)—( )

55.	(A)
(i) Closterovirus	(A) Half-smooth
(ii) Gemini	(B) Plant destroyer
(iii) *Phytophthora*	(C) A spindle or thread
(iv) Tungro	(D) Twins
(v) *Hemileia*	(E) Degenerated growth

Ans: (i)—( ), (ii)—( ), (iii)—( ), (iv)—( ), (v)—( )

56.	
(i) White mold	(A) *Botrytis cineria*
(ii) Blue mold	(B) *Aspergillus niger*
(iii) Black mold	(C) *Sclerotinia sclerotiorum*
(iv) Grey mold	(D) *Penicillium digitatum*
(v) Green mold	(E) *Penicillium expansum*

Ans: (i)—( ), (ii)—( ), (iii)—( ), (iv)—( ), (v)—( )

57.	
(i) Leaf smut of rice	(A) *Urocystis agropyri*
(ii) Flag smut of wheat	(B) *Tilletia tritici*
(iii) Stinking smut of wheat	(C) *Entyloma oryzae*
(iv) Smut of pearl millet	(D) *Sporisorium reilianum*
(v) Head smut of sorghum	(E) *Moesziomyces bullatus*

Ans: (i)—( ), (ii)—( ), (iii)—( ), (iv)—( ), (v)—( )

58.	
(i) Tissue maceration	(A) Blight
(ii) Rapid and extensive necrosis of leaves	(B) Wilt
(iii) Corky outgrowth on the plant surface	(C) Rust
(iv) Plugging of vascular tissues	(D) Canker
(v) Eruption of the epidermis	(E) Soft rot

Ans: (i)—( ), (ii)—( ), (iii)—( ), (iv)—( ), (v)—( )

59.	
(i) Wilt disease	(A) *Colletotrichum*
(ii) Citrus canker	(B) *Fusarium*
(iii) Red rot of sugarcane	(C) *Xanthomonas*
(iv) Powdery mildew	(D) *Pyricularia*
(v) Rice blast	(E) *Erysiphe*

Ans: (i)—( ), (ii)—( ), (iii)—( ), (iv)—( ), (v)—( )

60.	
(i) *Phytophthora*	(A) Bacteria
(ii) *Meloidogyne*	(B) Fungi like organism
(iii) *Puccinia*	(C) Algae
(iv) *Cephalurus*	(D) True fungi
(v) *Ralstonia*	(E) Nematode

Ans: (i)—( ), (ii)—( ), (iii)—( ), (iv)—( ), (v)—( )

61.	
(i) Citrus canker	(A) Algae
(ii) Late blight of potato	(B) Bacteria
(iii) White mold of pea	(C) Fungi like organism
(iv) Red rust of tea	(D) Nematode
(v) Root knot of tomato	(E) True fungi

Ans: (i)—( ), (ii)—( ), (iii)—( ), (iv)—( ), (v)—( )

62. (i) Red stripe of sugarcane — (A) *Acidovorax avenae* subsp. *avenae*
(ii) Sugarcane grassy shoot — (B) *Xanthomonas axonopodis* pv. *vasculorum*
(iii) Bacterial leaf blight of sugarcane — (C) *Xanthomonas albilineans*
(iv) Sugarcane ratoon stunting disease — (D) Phytoplasma
(v) Gummosis of sugarcane — (E) *Clavibacter xyli* subsp. *xyli*

Ans: (i)—( ), (ii)—( ), (iii)—( ), (iv)—( ), (v)—( )

63. (i) Red stripe of sugarcane — (A) Typical white (pencil-line) stripes on leaves along the veins
(ii) Sugarcane grassy shoot — (B) Yellow to reddish with brown dots, commas, or short lines at the nodal region
(iii) Leaf scald of sugarcane — (C) Parallel red streaks on leaves
(iv) Sugarcane ratoon stunting disease — (D) Whip-like structure comprising of central host tissue surrounded by a thin layer of black teliospores covered by a thin silvery white membrane
(v) Sugarcane smut — (E) Numerous, thin, slender chlorotic tillers

Ans: (i)—( ), (ii)—( ), (iii)—( ), (iv)—( ), (v)—( )

64. (i) Powdery mildew of grapes — (A) *Phyllactinia corylea*
(ii) Powdery mildew of rose — (B) *Acrosporium mangiferae*
(iii) Powdery mildew of wheat — (C) *Uncinula necator*
(iv) Powdery mildew of shisham — (D) *Sphaerotheca pannosa*
(v) Powdery mildew of mango — (E) *Blumeria graminis tritici*

Ans: (i)—( ), (ii)—( ), (iii)—( ), (iv)—( ), (v)—( )

65. (i) Late blight of potato — (A) *Xanthomonas axonopodis* pv. *malvacesrum*
(ii) Colocacia blight — (B) *Albugo candida*
(iii) White rust of crucifers — (C) *Phytophthora colocasiae*
(iv) Angular leaf spot of cotton — (D) *Erwinia amylovora*
(v) Fire blight of apple — (E) *Phytophthora infestans*

Ans: (i)—( ), (ii)—( ), (iii)—( ), (iv)—( ), (v)—( )

66. (i) Ring rot of potato — (A) *Burkholderia gladicola*
(ii) Onion rot — (B) *Xanthomonas axonopodis* pv. *citri*
(iii) Soft rot of carrot — (C) *Clavibacter michiganensis* subsp. *sepedonicus*
(iv) Citrus canker — (D) *Rhizobium tumefaciens*
(v) Crown gall of dicot plants — (E) *Pectobacterium carotovora* pv. *carotovora*

Ans: (i)—( ), (ii)—( ), (iii)—( ), (iv)—( ), (v)—( )

67. (i) Pierce's disease of grapevine — (A) *Spiroplasma citri*
(ii) Citrus greening — (B) *Rathayibacter tritici*
(iii) Tundu disease of wheat — (C) *Candidatus Liberibacter*
(iv) Scab of potato — (D) *Xylella fastidiosa*
(v) Citrus stubborn — (E) *Streptomyces scabies*

Ans: (i)—( ), (ii)—( ), (iii)—( ), (iv)—( ), (v)—( )

68. (i) Blackleg of potato — (A) *Clavibacter michiganensis* subsp. *sepedonicus*
(ii) Zebra chip of potato — (B) *Ralstonia solanacearum*
(iii) Bacterial ring rot of potato — (C) *Dickeya dadantii* and *Pectobacterium carotovorum*
(iv) Bacterial wilt or brown rot of potato — (D) *Candidatus Liberibacter solanacearum*
(v) Potato tuber soft rot diseases — (E) *Pectobacterium atrosepticum*

Ans: (i)—( ), (ii)—( ), (iii)—( ), (iv)—( ), (v)—( )

69. (i) Potato — (A) Cadang-cadang
(ii) Groundnut — (B) Moko disease
(iii) Banana — (C) Blister blight
(iv) Coffee — (D) Tikka disease
(v) Coconut — (E) Late blight

Ans: (i)—( ), (ii)—( ), (iii)—( ), (iv)—( ), (v)—( )

70. (i) Vascular disease — (A) Loose smut
(ii) Seedborne disease — (B) Root knot
(iii) Foliar disease — (C) Eyespot of cereals
(iv) Stem-based disease — (D) Wilt
(v) Root disease — (E) Mildew and rust

Ans: (i)—( ), (ii)—( ), (iii)—( ), (iv)—( ), (v)—( )

71. (i) Abnormal increase in size of organ — (A) Necrosis
(ii) Abnormal increase in number of cells of the organ — (B) Atrophy
(iii) Inhibition of growth or dwarfing — (C) Etiolation
(iv) Death of plant tissue — (D) Hypertrophy
(v) Excessive shoot elongation and chlorosis, induced in poor light or by growth hormones — (E) Hyperplasia

Ans: (i)—( ), (ii)—( ), (iii)—( ), (iv)—( ), (v)—( )

72. (i) Dead areas of tissue on the foliage, large and irregular in shape — (A) Spot
(ii) An obvious, defined lesion or area of diseased tissue, on a leaf — (B) Epinasty
(iii) The yellowing or whitening of green plant parts as a result of chlorophyll breakdown or production failure — (C) Gall
(iv) An overgrowth or swelling of some plant part — (D) Chlorosis
(v) An abnormal, downward-curving growth or movement of a leaf resulting from more rapid growth of cells on the upper side than on the lower side of the leaf stalk — (E) Blotch

Ans: (i)—( ), (ii)—( ), (iii)—( ), (iv)—( ), (v)—( )

Answers: Multiple Choice Questions (MCQs)

1.	d	2.	c	3.	b	4.	a	5.	b	6.	d	7.	d	8.	a	9.	d	10.	a
11.	c	12.	a	13.	d	14.	b	15.	c	16.	a	17.	c	18.	b	19.	c	20.	d
21.	c	22.	d	23.	c	24.	b	25.	c	26.	d	27.	c	28.	d	29.	c	30.	b
31.	b	32.	b	33.	b	34.	d	35.	c	36.	b	37.	d	38.	b	39	c	40.	a
41.	c	42.	d	43.	d	44.	b	45.	c	46.	d	47.	b	48.	b	49.	d	50.	d
51.	d	52.	a	53.	c	54.	d	55.	d	56.	b	57.	b	58.	a	59.	d	60.	c
61.	a	62.	a	63.	d	64.	d	65.	d	66.	d	67.	d	68.	d	69.	d	70.	a
71.	d	72.	b	73.	c	74.	b	75.	b	76.	c	77.	d	78.	a	79.	a	80.	b
81.	d	82.	b	83.	d	84.	b	85.	b	86.	b	87.	c	88.	d	89.	d	90.	d
91.	d	92.	b	93.	d	94.	b	95.	b	96.	d	97.	b	98.	c	99.	a	100.	a
101.	a	102.	b	103.	b	104.	a	105.	b	106.	d	107.	d	108.	a	109.	a	110.	d
111.	d	112.	b	113.	c	114.	b	115.	d	116.	d	117.	a	118.	c	119.	c	120.	b
121.	a	122.	a	123.	d	124.	c	125.	c	126.	b	127.	b	128.	b	129.	d	130.	a
131.	d	132.	d	133.	d	134.	d	135.	c	136.	d	137.	d	138.	d	139.	d	140.	b
141.	a	142.	b	143.	a	144.	a	145.	a	146.	a	147.	b	148.	d	149.	c	150.	d
151.	b	152.	b	153.	d	154.	b	155.	a	156.	a	157.	d	158.	a	159.	b	160.	d
161.	a	162.	a	163.	d	164.	d	165.	a	166.	c	167.	a	168.	c	169.	c	170.	b
171.	d	172.	d	173.	d	174.	a	175.	d	176.	b	177.	b	178.	a	179.	d	180.	b
181.	d	182.	d	183.	d	184.	a	185.	d	186.	d	187.	d	188.	a	189.	c	190.	a
191.	d	192.	d	193.	d	194.	d	195.	a	196.	b	197.	a	198.	d	199.	a	200.	d
201.	d	202.	b	203.	b	204.	d	205.	c	206.	a	207.	a	208.	b	209.	d	210.	b
211.	a	212.	b	213.	b	214.	c	215.	a	216.	a	217.	b	218.	d	219.	a	220.	c
221.	a	222.	c	223.	a	224.	d	225.	b	226.	c	227.	d	228.	d	229.	c	230.	d
231.	a	232.	a	233.	a	234.	a	235.	d	236.	a	237.	a	238.	b	239.	d	240.	a
241.	c	242.	a	243.	d	244.	a	245.	d	246.	d	247.	d	248.	a	249.	d	250.	d
251.	b	252.	b	253.	d	254.	b	255.	c	256.	a	257.	d	258.	d	259.	d	260.	a
261.	b	262.	a	263.	a	264.	c	265.	d	266.	a	267.	a	268.	a	269.	d	270.	a
271.	a	272.	c	273.	d	274.	b	275.	d	276.	c	277.	c	278.	c	279.	c	280.	a
281.	b	282.	c	283.	a	284.	a	285.	c	286.	a	287.	d	288.	a	289.	d	290.	b
291.	d	292.	a	293.	b	294.	b	295.	a	296.	b	297.	a	298.	a	299.	a	300.	b
301.	a	302.	b	303.	b	304.	c	305.	a	306.	c	307.	d	308.	d	309.	c	310.	b
311.	b	312.	a	313.	a	314.	d	315.	b	316.	a	317.	b	318.	a	319.	a	320.	d
321.	c	322.	c	323.	b	324.	c	325.	b	326.	a	327.	a	328.	a	329.	a	330.	d
331.	d	332.	b	333.	d	334.	c	335.	b	336.	a	337.	a	338.	c	339.	b	340.	b
341.	d	342.	a	343.	a	344.	a	345.	d	346.	b	347.	c	348.	d	349.	a	350.	c
351.	a	352.	b	353.	b	354.	b	355.	b	356.	b	357.	a	358.	d	359.	b	360.	d

361. c **362.** b **363.** d **364.** b **365.** d **366.** d **367.** b **368.** a **369.** a **370.** b
371. d **372.** d **373.** a **374.** d **375.** c **376.** d **377.** d **378.** b **379.** b **380.** d
381. d **382.** c **383.** a **384.** c **385.** d **386.** a **387.** a **388.** c **389.** d **390.** a
391. d **392.** a **393.** d **394.** a **395.** b **396.** d **397.** a **398.** d **399.** a **400.** d
401. b **402.** d **403.** d **404.** a **405.** d **406.** a **407.** d **408.** a **409.** b **410.** c
411. a **412.** c **413.** d **414.** c **415.** d **416.** d **417.** a **418.** b **419.** a **420.** b
421. b **422.** d **423.** d **424.** d **425.** a **426.** a **427.** a **428.** c **429.** a **430.** d
431. d **432.** d **433.** a **434.** d **435.** d **436.** b **437.** a **438.** c **439.** d **440.** d
441. c **442.** a **443.** a **444.** c **445.** d **446.** b **447.** b **448.** c **449.** d **450.** c
451. a **452.** a **453.** a **454.** d **455.** d **456.** b **457.** b **458.** d **459.** c **460.** a
461. d **462.** d **463.** a **464.** b **465.** b **466.** a **467.** c **468.** a **469.** c **470.** d
471. b **472.** d **473.** d **474.** b **475.** a **476.** a **477.** a **478.** c **479.** a **480.** c
481. c **482.** d **483.** a **484.** b **485.** d **486.** b **487.** b **488.** b **489.** b **490.** c
491. c **492.** b **493.** b **494.** b **495.** a **496.** d **497.** c **498.** d **499.** d **500.** c
501. b **502.** d **503.** d **504.** d **505.** a **506.** b **507.** c **508.** b **509.** b **510.** d
511. d **512.** a **513.** c **514.** b **515.** a **516.** a **517.** b **518.** c **519.** c **520.** c
521. c **522.** d **523.** a **524.** d **525.** c **526.** a **527.** c **528.** b **529.** a **530.** d
531. d **532.** a **533.** a **534.** d **535.** c **536.** d **537.** d **538.** d **539.** d **540.** d
541. d **542.** b **543.** d **544.** d **545.** a **546.** d **547.** c **548.** d **549.** d **550.** d
551. d **552.** d **553.** c **554.** b **555.** d **556.** b **557.** c **558.** b **559.** b **560.** d
561. a **562.** b **563.** a **564.** d **565.** a **566.** b **567.** b **568.** d **569.** d **570.** d
571. a **572.** d **573.** a **574.** a **575.** b **576.** a **577.** a **578.** d **579.** c **580.** a
581. b **582.** a **583.** d **584.** c **585.** c **586.** c **587.** b **588.** b **589.** d **590.** d
591. d **592.** d **593.** d **594.** a **595.** a **596.** d **597.** a **598.** d **599.** a **600.** b
601. a **602.** b **603.** a **604.** b **605.** a **606.** a **607.** b **608.** b **609.** a **610.** c
611. c **612.** c **613.** b **614.** a **615.** b **616.** c **617.** a **618.** d **619.** b **620.** b
621. a **622.** b **623.** a **624.** b **625.** a **626.** c **627.** a **628.** d **629.** a **630.** a
631. d **632.** b **633.** a **634.** d **635.** b **636.** d **637.** a **638.** d **639.** d **640.** d
641. b **642.** b **643.** d **644.** a **645.** c **646.** c **647.** d **648.** a **649.** c **650.** a
651. b **652.** b **653.** b **654.** d **655.** b **656.** a **657.** c **658.** a **659.** b **660.** d
661. d **662.** d **663.** d **664.** b **665.** d **666.** d **667.** d **668.** c **669.** c **670.** a
671. b **672.** b **673.** b **674.** d **675.** b **676.** c **677.** b **678.** d **679.** d **680.** c
681. b **682.** c **683.** c **684.** c **685.** d **686.** d **687.** d **688.** b **689.** c **690.** d
691. b **692.** b **693.** c **694.** a **695.** d **696.** d **697.** a **698.** b **699.** d **700.** c
701. c **702.** a **703.** c **704.** d **705.** a **706.** d **707.** b **708.** c **709.** d **710.** d
711. c **712.** b **713.** c **714.** d **715.** d **716.** b **717.** c **718.** c **719.** c **720.** c
721. c **722.** b **723.** b **724.** a **725.** c **726.** c **727.** c **728.** b **729.** a **730.** a
731. c **732.** c **733.** b **734.** b **735.** c **736.** b **737.** d **738.** d **739.** d **740.** c

741. a	**742.** b	**743.** c	**744.** a	**745.** a	**746.** d	**747.** d	**748.** b	**749.** a	**750.** d
751. b	**752.** a	**753.** b	**754.** c	**755.** c	**756.** c	**757.** a	**758.** a	**759.** c	**760.** a
761. b	**762.** a	**763.** d	**764.** d	**765.** c	**766.** b	**767.** a	**768.** b	**769.** d	**770.** a
771. a	**772.** d	**773.** d	**774.** a	**775.** a	**776.** b	**777.** d	**778.** a	**779.** d	**780.** b
781. a	**782.** d	**783.** d	**784.** d	**785.** d	**786.** d	**787.** a	**788.** d	**789.** d	**790.** d
791. d	**792.** b	**793.** a	**794.** a	**795.** d	**796.** d	**797.** d	**798.** c	**799.** d	**800.** b
801. c	**802.** b	**803.** a	**804.** d	**805.** b	**806.** d	**807.** d	**808.** a	**809.** d	**810.** a
811. c	**812.** a	**813.** a	**814.** a	**815.** b					

Answers: Match the Following

1. (i)—(D), (ii)—(A), (iii)—(C), (iv)—(E), (v)—(B)
2. (i)—(E), (ii)—(D), (iii)—(A), (iv)—(B), (v)—(C)
3. (i)—(E), (ii)—(C), (iii)—(D), (iv)—(A), (v)—(B)
4. (i)—(E), (ii)—(C), (iii)—(D), (iv)—(B), (v)—(A)
5. (i)—(E), (ii)—(C), (iii)—(D), (iv)—(A), (v)—(B)
6. (i)—(B), (ii)—(A), (iii)—(E), (iv)—(C), (v)—(D)
7. (i)—(C), (ii)—(D), (iii)—(E), (iv)—(A), (v)—(B)
8. (i)—(D), (ii)—(A), (iii)—(B), (iv)—(E), (v)—(C)
9. (i)—(E), (ii)—(A), (iii)—(D), (iv)—(B), (v)—(C)
10. (i)—(E), (ii)—(D), (iii)—(A), (iv)—(B), (v)—(C)
11. (i)—(C), (ii)—(E), (iii)—(A), (iv)—(D), (v)—(B)
12. (i)—(E), (ii)—(C), (iii)—(A), (iv)—(B), (v)—(D)
13. (i)—(C), (ii)—(B), (iii)—(A), (iv)—(E), (v)—(D)
14. (i)—(D), (ii)—(E), (iii)—(A), (iv)—(C), (v)—(B)
15. (i)—(E), (ii)—(B), (iii)—(D), (iv)—(C), (v)—(A)
16. (i)—(C), (ii)—(A), (iii)—(B), (iv)—(E), (v)—(D)
17. (i)—(B), (ii)—(E), (iii)—(D), (iv)—(C), (v)—(A)
18. (i)—(E), (ii)—(D), (iii)—(B), (iv)—(C), (v)—(A)
19. (i)—(E), (ii)—(D), (iii)—(A), (iv)—(B), (v)—(C)
20. (i)—(E), (ii)—(D), (iii)—(A), (iv)—(B), (v)—(C)
21. (i)—(E), (ii)—(D), (iii)—(A), (iv)—(B), (v)—(C)
22. (i)—(B), (ii)—(E), (iii)—(A), (iv)—(C), (v)—(D)
23. (i)—(E), (ii)—(C), (iii)—(A), (iv)—(B), (v)—(D)
24. (i)—(E), (ii)—(C), (iii)—(A), (iv)—(B), (v)—(D)
25. (i)—(B), (ii)—(C), (iii)—(A), (iv)—(E), (v)—(D)
26. (i)—(E), (ii)—(A), (iii)—(D), (iv)—(C), (v)—(B)
27. (i)—(C), (ii)—(E), (iii)—(B), (iv)—(A), (v)—(D)
28. (i)—(E), (ii)—(C), (iii)—(A), (iv)—(B), (v)—(D)
29. (i)—(E), (ii)—(C), (iii)—(B), (iv)—(D), (v)—(A)
30. (i)—(D), (ii)—(E), (iii)—(A), (iv)—(B), (v)—(C)
31. (i)—(C), (ii)—(E), (iii)—(A), (iv)—(B), (v)—(D)
32. (i)—(B), (ii)—(D), (iii)—(A), (iv)—(E), (v)—(C)

33. (i)—(E), (ii)—(C), (iii)—(B), (iv)—(A), (v)—(D)
34. (i)—(B), (ii)—(A), (iii)—(E), (iv)—(C), (v)—(D)
35. (i)—(E), (ii)—(A), (iii)—(B), (iv)—(C), (v)—(D)
36. (i)—(C), (ii)—(A), (iii)—(B), (iv)—(E), (v)—(D)
37. (i)—(C), (ii)—(B), (iii)—(E), (iv)—(A), (v)—(D)
38. (i)—(D), (ii)—(B), (iii)—(E), (iv)—(C), (v)—(A)
39. (i)—(C), (ii)—(A), (iii)—(B), (iv)—(E), (v)—(D)
40. (i)—(D), (ii)—(B), (iii)—(A), (iv)—(E), (v)—(C)
41. (i)—(C), (ii)—(A), (iii)—(B), (iv)—(E), (v)—(D)
42. (i)—(B), (ii)—(E), (iii)—(A), (iv)—(C), (v)—(D)
43. (i)—(E), (ii)—(B), (iii)—(A), (iv)—(C), (v)—(D)
44. (i)—(D), (ii)—(C), (iii)—(A), (iv)—(E), (v)—(B)
45. (i)—(B), (ii)—(C), (iii)—(D), (iv)—(E), (v)—(A)
46. (i)—(E), (ii)—(A), (iii)—(C), (iv)—(B), (v)—(D)
47. (i)—(E), (ii)—(D), (iii)—(C), (iv)—(B), (v)—(A)
48. (i)—(C), (ii)—(A), (iii)—(B), (iv)—(E), (v)—(D)
49. (i)—(E), (ii)—(A), (iii)—(D), (iv)—(C), (v)—(B)
50. (i)—(E), (ii)—(A), (iii)—(D), (iv)—(B), (v)—(C)
51. (i)—(B), (ii)—(A), (iii)—(D), (iv)—(E), (v)—(C)
52. (i)—(B), (ii)—(E), (iii)—(D), (iv)—(C), (v)—(A)
53. (i)—(D), (ii)—(C), (iii)—(E), (iv)—(B), (v)—(A)
54. (i)—(E), (ii)—(C), (iii)—(B), (iv)—(A), (v)—(D)
55. (i)—(C), (ii)—(D), (iii)—(B), (iv)—(E), (v)—(A)
56. (i)—(C), (ii)—(E), (iii)—(B), (iv)—(A), (v)—(D)
57. (i)—(C), (ii)—(A), (iii)—(B), (iv)—(E), (v)—(D)
58. (i)—(E), (ii)—(A), (iii)—(D), (iv)—(B), (v)—(C)
59. (i)—(B), (ii)—(C), (iii)—(A), (iv)—(E), (v)—(D)
60. (i)—(B), (ii)—(E), (iii)—(D), (iv)—(C), (v)—(A)
61. (i)—(B), (ii)—(C), (iii)—(E), (iv)—(A), (v)—(D)
62. (i)—(C), (ii)—(D), (iii)—(A), (iv)—(E), (v)—(B)
63. (i)—(C), (ii)—(E), (iii)—(A), (iv)—(B), (v)—(D)
64. (i)—(C), (ii)—(D), (iii)—(E), (iv)—(A), (v)—(B)
65. (i)—(E), (ii)—(C), (iii)—(B), (iv)—(A), (v)—(D)
66. (i)—(C), (ii)—(A), (iii)—(E), (iv)—(B), (v)—(D)
67. (i)—(D), (ii)—(C), (iii)—(B), (iv)—(E), (v)—(A)
68. (i)—(E), (ii)—(D), (iii)—(A), (iv)—(B), (v)—(C)
69. (i)—(E), (ii)—(D), (iii)—(B), (iv)—(C), (v)—(A)
70. (i)—(D), (ii)—(A), (iii)—(E), (iv)—(C), (v)—(B)
71. (i)—(D), (ii)—(E), (iii)—(B), (iv)—(A), (v)—(C)
72. (i)—(E), (ii)—(A), (iii)—(D), (iv)—(C), (v)—(B)

REFERENCES

Books

Aglave, B. 2019. *Handbook of Plant Disease Identification and Management*. CRC Press and Disease Management. Taylor & Francis Group, pp. 751.

Awasthi, L.P. 2015. *Recent Advances in the Diagnosis and Management of Plant Diseases*. Springer India, pp. 285.

Chakrabarti, D.K. 2011. *Mango Malformation*. Springer-Science + Business Media B.V. pp. 147.

Chand, G. and Kumar, S. 2016. *Crop Diseases and Their Management: Integrated Approaches*. Apple Academic Press, Inc., pp. 385.

Chaube, H.S. and Pundhir, V.S. 2012. *Crop Diseases and Their Management*. PHI Learning, pp. 703.

Cooke, T., Persley, D. and House, S. 2009. *Diseases of Fruit Crops in Australia*. CSIRO Publishing, pp. 269.

Elad, Y., Williamson, B., Tudzynski, P. and Delen, N. 2007. *Botrytis: Biology, Pathology and Control*. Springer, pp. 393.

Gatak, A. and Ansar, M. 2020. *The Vegetable Pathosystem: Ecology, Disease Mechanism, and Management*. Apple Academic Press Inc, pp. 527.

Horst, R.K. 2001. *Westcott's Plant Disease Handbook*. Sixth edition. Kluwer Academic Publishers, pp. 733.

Huang, Y-K., Li, W-F., Zhang, R-Y. and Wang, X-Y. 2018. *Color illustration of diagnosis and control for modern sugarcane diseases, pests, and weeds*. China Agriculture Press and Springer Nature Singapore Pte Ltd., pp. 420.

Husaini, A.M. and Ner, D. 2016. *Strawberry: Growth, Development and Diseases*. CABI. pp. 313.

Loebenstein, G., Cemer, V. and Thottappill, G. 2003. *Virus and Virus-like Diseases of Major Crops in Developing Countries*. Springer-Science + Business Media, B.V., pp. 791.

Mukerji, K.G. 2004. *Disease Management of Fruits and Vegetables*, Vol. 1, Kluwer Academic Publishers, pp. 553.

Murray, T.D., Parry, D.W. and Cattlin, N.D. 2009. *A Colour Handbook of Diseases of Small Grain Cereal Crops*. Manson Publishing Ltd, pp. 139.

Nair, K.P.P. 2013. *The Agronomy and Economy of Turmeric and Ginger: The Invaluable Medicinal Spice Crops*. Elsevier Inc., pp. 513.

Nyvall, R.F. 1999. *Field Crop Diseases*. Third Edition. Iowa State University Press, pp. 969.

Ploetz, R.C. 2003. *Diseases of Tropical Fruit Crops*. CABI Publishing, pp. 499.

Saharan, G.S. and Mehta, N. 2008. *Sclerotinia Diseases of Crop Plants: Biology*, Ecology. Springer-Science + Business Media B.V., pp. 481.

Saharan, G.S. Mehta, N., and Meena, P.D. 2016. *Alternaria Diseases of Crucifers: Biology, Ecology and Disease Management*. Springer-Science + Business Media Singapore, pp. 293.

Saharan, G.S., Mehta, N. and Meena, P.D. 2017. *Downy Mildew Disease of Crucifers: Biology, Ecology and Disease Management*. Springer Nature Singapore, pp. 349.

Saharan, G.S., Verma, P.R., Meena, P.D. and Kumar, A. 2014. *White Rust of Crucifers: Biology, Ecology and Management*. Springer India, pp. 239.

Sathe, T.V., Shinde, K.P., Shaikh, A.L. and Raut D.K. 2009. *Sugarcane Pests and Diseases*. Manglam Publications, Dehli, pp. 174.

Sharma, I. 2012. *Disease Resistance in Wheat*. CABI, pp. 313.

Singh, B.P., Singh, G., Kumar, K., Nayak, S.C. and Srinivasa, N. 2020. *Management of Fungal Pathogens in Pulses: Current Status and Future Challenges*. Springer Nature Switzerland AG, pp. 219.

Singh, D.P. 2018. *Management of Wheat and Barley Diseases*. Apple Academic Press Inc. pp. 645.

Singh, R.S. 2018. *Plant Diseases*. Tenth edition. Medtech Science Press, pp. 821.

Srivastava, J.N. and Singh, A.K. 2021. *Diseases of Field Crops: Diagnosis and Management: Cereals, Small Millets, and Fiber Crops*. Vol 1, Apple Academic Press, Inc. pp. 387.

Srivastava, J.N. and Singh, A.K. 2021. *Diseases of Field Crops: Diagnosis and Management: Pulses, Oil Seeds, Narcotics, and Sugar Crops*. Vol 1, Apple Academic Press, Inc., pp. 365.

Tennant, P. and Fermin, G. 2015. *Virus Diseases of Tropical and Subtropical Crops*. CABI Publishing, pp. 241.

Thind, B.S. 2020. *Phytopathogenic Bacteria and Plant Diseases*. CRC Press Taylor & Francis Group, pp. 367.

Waller, J.M., Bigger M. and Hillocks, R.J. 2007. *Coffee Pests, Diseases and Their Management*. CABI Publishing, pp. 423.

Research/Review Articles/Technical Bulletin

Acharya, B., Mackasmiel, L., Taheri, A., et al. 2021. Identification of bacterial wilt (*Erwinia tracheiphila*) resistances in USDA melon collection. *Plants*, 10, 1972. https://doi.org/10.3390/plants10091972.

Anuradha, Sanghera, G.S., Kashyap, L., et al. 2018. Incidence of pokkah boeng disease on sugarcane clones and varieties in Punjab. *Agric. Res. J.* 55(4): 754–756.

Arauz, L.F. 2000. Mango anthracnose: current options for integrated management. *Plant Disease*, 84 (6): 600–611.

Awasthi, L.P. and Verma, H.N. 2017. Current status of viral diseases of potato and their ecofriendly management-A critical review. *Virology: Research & Reviews*, 1(4):1-16.

Barnes, G., Saunders, D.G.O. and Williamson, T. 2020. Banishing barberry: The history of *Berberis vulgaris* prevalence and wheat stem rust incidence across Britain. *Plant Pathology*, 69: 1193–1202.

Bhardwaj, S.C., Singh, G.P., Gangwar, O.P., et al. 2019. Status of wheat rust research and progress in rust management-Indian context. *Agronomy*, 9, 892; doi:10.3390/agronomy9120892.

Biggs, A.R. and Turechek, W.W. 2010. Fire blight of Apples and pears: Epidemiological concepts comprising the Maryblyt forecasting program. *Plant Health Progress*, doi:10.1094/PHP-2010-0315-01-RS.

Bishnoi, S.K., He, X., Phuke, R.M., et al. 2020. Karnal Bunt: A Re-Emerging Old Foe of Wheat. *Front. Plant Sci.* 11:569057. doi: 10.3389/fpls.2020.569057.

Blecherta, O. and Debenerb, T. 2005. Morphological characterization of the interaction between *Diplocarpon rosae* and various rose species. *Plant Pathology*, 54, 82–90.

Bolton, M.D., Kolmer, J.A. and Garvin, D.F. 2008. Wheat leaf rust caused *by Puccinia triticina. Molecular Plant Pathology*, 9(5), 563–575.

Bourras, S., Rouxel, T., and Meyer, M. 2015. *Agrobacterium tumefaciens* gene transfer: How a plant pathogen hacks the nuclei of plant and nonplant organisms. *Phytopathology*, 105:1288–1301.

Carisse, O. and Morissette-Thomas, V. 2013. Epidemiology of grape anthracnose: Factors associated with defoliation of grape leaves infected by *Elsinoë ampelina. Plant Disease*, 97:222–30.

Carris, L.M., Castlebury, L.A. and Goates, B.J. 2006. Nonsystemic bunt fungi–*Tilletia indica* and *T. horrida*: A review of history, systematics, and biology. *Annu. Rev. Phytopathol.* 44:113–33.

Chang, K.F., Hwang, S.F., Ahmed, H.U., et al. 2013. Yield loss and management of downy mildew on field pea in Alberta, Canada. *Crop Protection*, 46, 23–28.

Chen, K., Wang, L., Chen, H., et al. 2021. Complete genome sequence analysis of the peanut pathogen *Ralstonia solanacearum* strain Rs-P.362200. *BMC Microbiology*, 21:118 https://doi.org/10.1186/s12866-021-02157-7.

Chen, W., Wellings, C., Chen, X., et al. 2014. Wheat stripe (yellow) rust caused by *Puccinia striiformis* f. sp. *tritici. Molecular Plant Pathology*, 15(5): 433–446.

Christie P.J. and Gordon, J.E. 2014. The *Agrobacterium* Ti Plasmids. *Microbiol Spectr.* 2(6). doi:10.1128/microbiolspec.PLAS-0010-2013.

Dala-Paula, B.M., Plotto, A., Bai, J., Manthey, J.A., et al. 2019. Effect of huanglongbing or greening disease on orange juice quality: a review. Front. *Plant Sci.* 9:1976. doi: 10.3389/fpls.2018.01976.

Davis, R.M., Colyer, P.D., Rothrock, C.S. et al. 2006. Fusarium wilt of cotton: population diversity and implications for management. *Plant Disease*, 90(6):692–703.

De Silva, D.D., Groenewald, J.Z., Crous, P.W., et al. 2019. Identification, prevalence, and pathogenicity of *Colletotrichum* species causing anthracnose of *Capsicum annuum* in Asia. *IMA Fungus*, 10:8 https://doi.org/10.1186/s43008-019-0001-y.

Domínguez-Serrano, D., García-Velasco, R., Mora-Herrera, M.E., et al. 2016. The rose powdery mildew (*Podosphaera pannosa*). *Publicado como ENSAYO en Agrociencia* 50: 901–917.

Duveiller, E., P.K. Singh, M. Mezzalama, R.P. Singh, et al. 2012. Wheat Diseases and Pests: A Guide for Field Identification (2nd Edition). *CIMMYT*, Mexico, D.F., Mexico, pp.138.

Emebiri, L., Singh, S., Tan, M-K., et al. 2019. Unravelling the complex genetics of karnal bunt (*Tilletia indica*) resistance in common wheat (*Triticum aestivum*) by genetic linkage and genome-wide association analyses. *Gene, Genomes, Genetics*, 9:1437–1447.

Ference, C.M., Gochez, A.M., Behlau, F., et al. 2018. Recent advances in the understanding of *Xanthomonas citri* ssp. *citri* pathogenesis and citrus canker disease management. *Molecular Plant Pathology*, 19(6): 1302–1318.

Figlan, S., Ntushelo, K., Mwadzingeni, L., et al. 2020. Breeding wheat for durable leaf rust resistance in Southern Africa: variability, distribution, current control strategies, challenges and future prospects. *Front. Plant Sci.* 11:549. doi: 10.3389/fpls.2020.00549.

Figueroa, M., Hammond-Kosack, K.E. and Solomon, P.S. 2018. A review of wheat diseases–a field perspective. *Molecular Plant Pathology*, 19(6):1523–1536.

Fondevilla, S. and Rubiales, D. 2012. Powdery mildew control in pea. *A review. Agron. Sustain. Dev.* 32: 401–409.

Freeman, S., Shtienberg, D., Maymon, M., et al. 2014. New insights into mango malformation disease epidemiology lead to a new integrated management strategy for subtropical environments. *Plant Disease*, 98(11): 1456–1466.

Gabrekiristos, E. and Demiyo, T. 2020. Hot Pepper Fusarium Wilt (*Fusarium oxysporum* f. sp. *capsici*): Epidemics, characteristic features, and management options. *Journal of Agricultural Science*, 12 (10).

Gagnevin, L. and Pruvost, O. 2001. Epidemiology and control of mango bacterial black spot. *Plant Disease*, 85(9): 928–935.

Gray, S., Boer, S.D., Lorenzen, J., et al. 2010. *Potato virus Y*: An evolving concern for potato crops in the United States and Canada. *Plant Disease*, 94(12): 1384–1397.

Haack, S.E., Walse, S.S., Nguyen, K. et al. 2019. Management of *Xanthomonas fragariae* with pre- and postharvest treatments to overcome trade barriers for California strawberries. *Plant Disease*, 103: 1256–1263.

Hami, A., Rasool, R.S., Khan, N.A., et al. 2021. Morpho-molecular identification and first report of *Fusarium equiseti* in causing chilli wilt from Kashmir (Northern Himalayas). *Scientific Reports*, 11:3610. https://doi.org/10.1038/s41598-021-82854-5.

Hammond-Kosack, K.E. and Jones, J.D.G. 2015. *Responses to plant pathogens*. pp. 984-1050. In book: Biochemistry & Molecular Biology of Plants; editors Buchanan, B.B., Gruissem, W. and Jones, R.L. pp. 1264, John Wiley & Sons, Ltd.

Hassan, M.Z., Rahim, M.A., Natarajan, S.K., et al. 2018. Gummy stem blight resistance in melon: Inheritance pattern and development of molecular markers. *Int. J. Mol. Sci.* 19, 2914. doi:10.3390/ijms19102914.

Hiebert, C.W., Moscou, M.J., Hewitt, T., et al. 2020. Stem rust resistance in wheat is suppressed by a subunit of the mediator complex. *Nature Communications*, 11:1123. https://doi.org/10.1038/s41467-020-14937-2.

Hossain, M.I., Ahmad, K., Siddiqui, Y., et al. 2020. Current and prospective strategies on detecting and managing *Colletotrichum falcatum* causing red rot of sugarcane. *Agronomy*, 10, 1253. doi:10.3390/agronomy10091253.

Jendoubi, W., Bouhadida, M., Boukteb, A., et al. 2017. Fusarium wilt affecting chickpea crop. *Agriculture*, 7, 23. doi:10.3390/agriculture7030023.

Jiang, G., Wei, Z., Xu, J., et al. 2017. Bacterial wilt in China: History, current status, and future perspectives. *Front. Plant Sci.* 8:1549. doi: 10.3389/fpls.2017.01549.

Jimenez-Díaz, R.M., Castillo, P., Jimenez-Gasco, M.D.M., et al. 2015. Fusarium wilt of chick-peas: Biology, ecology, and management. *Crop Protection.* http://dx.doi.org/10.1016/j.cro-pro.2015.02.023.

Keinath, A.P. 2013. Diagnostic guide for gummy stem blight and black rot on cucurbits. *Plant Health Progress.* doi:10.1094/PHP-2013-1024-01-DG.

Land, C.J., Lawrence, K.S., Burmester, C.H., et al. 2017. Cultivar, irrigation, and soil contribution to the enhancement of Verticillium wilt disease in cotton. *Crop Protection*, 96, 1–6.

Lebeda, A. and Cohen, Y. 2011. Cucurbit downy mildew (*Pseudoperonospora cubensis*)-biology, ecology, epidemiology, host-pathogen interaction, and control Eur. *J. Plant Pathol.* 129:157–192.

Li, X., Zhang, Y., Ding, C., et al. 2017. Temporal patterns of cotton Fusarium and Verticillium wilt in Jiangsu coastal areas of China. *Scientific Reports*, 7:12581. DOI:10.1038/s41598-017-12985-1.

Marques, J.P.R., Appezzato-da-G, B., Piepenbring, M., et al. 2017. Sugarcane smut: shedding light on the development of the whip-shaped sorus. *Annals of Botany*, 119:815–827.

Mathew, F., Harveson, R., Block, C., et al. 2020. *Sclerotinia sclerotiorum* diseases of sunflower (white mold). *Plant Health Instructor.* DOI: 10.1094/PHI-I-2020-1201-01.

Meyer, M., Bacha, N., Tesfaye, T., et al. 2021. Wheat rust epidemics damage Ethiopian wheat production: A decade of field disease surveillance reveals national-scale trends in past out-breaks. *PLoS ONE* 16(2): e0245697. https://doi.org/10.1371/journal. pone.0245697.

Mongkolporna, O. and Taylor, P.W.J. 2018. Chili anthracnose: *Colletotrichum* taxonomy and pathogenicity. *Plant Pathology*, 67, 1255–1263.

Mundy, D.C., Elmer, P., Wood, P. et al. 2022. A Review of Cultural practices for Botrytis bunch rot management in New Zealand vineyards. *Plants*, 11, 3004. https://doi.org/10.3390/plants11213004.

Negussie T.G. and Pretorius, Z.A. 2008. Yield loss of lentil caused by *Uromyces viciae-fabae*. South *African Journal of Plant and Soil*, 25:1, 32–41.

Negussie, T. and Pretorius, Z.A. 2012. Lentil rust: present status and future prospects. *Crop Protection*, 32, 119–128.

Negussie, T., Pretorius, Z.A., McTaggart, A.R. et al. 2015. National Diagnostic Protocol for *Uromyces viciae-fabae* Subcommittee on Plant Health Diagnostics. *NDP*32 V1.

Obidiegwu, E.J., Flath, K. and Gebhardt, C. 2014. Managing potato wart: a review of present research status and future perspective. *Theor. Appl. Genet.* 127:763–780.

Pande, S., Galloway, J., Gaur, P.M., et al. 2006. Botrytis grey mould of chickpea: a review of biology, epidemiology, and disease management. *Australian Journal of Agricultural Research*, 57, 1137–1150.

Petrasch, S., Knapp, S.J., Van Kan, J.A.L. et al. 2019. Grey mould of strawberry, a devastating disease caused by the ubiquitous necrotrophic fungal pathogen *Botrytis cinerea*. *Molecular Plant Pathology*, 20(6): 877–892.

Pietrusi-ska, A. and Tratwal, A. 2020. Characteristics of powdery mildew and its importance for wheat grown in Poland. *Plant Protection Science*, 56 (3): 141–153.

Pscheidt, J.W. and Rodriguez, T.G. 2018. Diseases of Rose. *In*: McGovern R., Elmer, W. (eds): Handbook of Florists' Crops Diseases. *Handbook of Plant Disease Management*. Springer, Cham. https://doi.org/10.1007/978-3-319-39670-5_23.

Ram, B. and Singh, K.P. 2004. Smuts of wheat: A review. *Indian Phytopath*. 57(2): 125–134.

Ranga Rao, G.V., Rameshwar Rao, V. and Ghaffar, M.A. 2013. Handbook on Chickpea and Pigeonpea Insect Pests Identification and Management. *Information Bulletin* No. 57, Patancheru, Andhra Pradesh 502 324, India: *International Crops Research Institute for the Semi-Arid Tropics*, pp. 96.

Rivera-Burgos, L.A., Silverman, E., Sari, N. and Wehner, T.C. 2021. Evaluation of resistance to gummy stem blight in a population of recombinant inbred lines of Watermelon × Citron. *HortScience*, 56(3): 380–388.

Rojas, S.E., Batzer, J.C., Bearitle, G.A., et al. 2015. Bacterial wilt of cucurbits: Resurrecting a classical pathosystem. *Plant Disease*, 99 (5): 564–574.

Roy, B., Dubey, S., Ghosh, A., et al. 2021. Simulation of leaf curl disease dynamics in chili for strategic management options. *Scientifc Reports*, 11:1010. https://doi.org/10.1038/s41598-020-79937-0.

Rush, C.M., Stein, J.M., Bowden, R.L., et al. 2005. Status of karnal bunt of wheat in United States 1996-2004. *Plant Disease*, 89 (3): 212–223.

Salotti, I. and Rossi, V.A. 2021. Mechanistic weather-driven model for *Ascochyta rabiei* infection, and disease development in chickpea. *Plants*, 10, 464. https://doi.org/10.3390/plants10030464.

Savory, E.A., Granke, L.L., Quesada-Ocampo, L.M., et al. 2011. The cucurbit downy mildew pathogen *Pseudoperonospora cubensis*. *Molecular Plant Pathology*, 12(3): 217–226.

Saxena, A., Raghuwanshi, R., Gupta, V.K. et al. 2016. Chilli anthracnose: the epidemiology and management. *Front. Microbiol.* 7:1527. doi: 10.3389/fmicb.2016.01527.

Shaheen, N., Khan, U.M., Azhar, M.T., et al. 2021. Genetics and genomics of Fusarium wilt of chilies: A Review. *Agronomy*, 11, 2162.

Shahid, M.S., Shafiq, M., Raza, A., et al. 2019. Molecular and biological characterization of Chilli leaf curl virus and associated tomato leaf curl betasatellite infecting tobacco in Oman. *Virology J.* 16:131. https://doi.org/10.1186/s12985-019-1235-4.

Sharma, A., Sharma, P., Dixit, A. et al. 2017. Karnal bunt of wheat in India and its management: a review. *Plant Pathology and Quarantine*, 7(2): 165–173.

Sharma, M. and Ghosh, R. 2016. An update on genetic resistance of chickpea to Ascochyta blight. *Agronomy*, 6, 18. doi:10.3390/agronomy6010018.

Sharma, M., Ghosh, R., Tarafdar, A., et al. 2019. Exploring the genetic cipher of chickpea (*Cicer arietinum* L.) through identification and multi-environment validation of resistant sources against fusarium wilt (*Fusarium oxysporum* f. sp. *ciceris*). *Front. Sustain. Food Syst.* 3:78. doi: 10.3389/fsufs.2019.00078.

Sharma, O.P., Singh, S.K., Vennila, S., et al. 2014. ICAR–National Centre for Integrated Pest Management, LBS Building, IARI Campus, New Delhi–110012, India. *Technical Bulletin* No. 36, pp. 36.

Shu, Y., Luo, M. and Dong, Z. 2021. Whole genome sequence resource for *Fusarium oxysporum* f. sp. *capsici* 14003, the causative agent of pepper wilt. *Plant Disease*, 105:1183–1186.

Simone, N.D., Pace, B., Grieco, F., et al. 2020. *Botrytis cinerea* and table grapes: A Review of the main physical, chemical, and bio-based control treatments in post-harvest. *Foods*, 9, 1138. doi:10.3390/foods9091138.

Singh, K.P., Kumari, P. and Rai, P.K. 2021. Current Status of the disease-resistant Gene(s)/QTLs, and strategies for improvement in *Brassica juncea*. *Front. Plant Sci.* 12:617405.

Singh, R.P., Hodson, D.P., Huerta-Espino, J., et al. 2011. The emergence of Ug99 races of the stem rust fungus is a threat to world wheat production. *Annual Review of Phytopathology*, 49: 465–81.

Suthaparan, A., Stensvand, A., Solhaug, K.A., et al. 2012. Suppression of powdery mildew (*Podosphaera pannosa*) in greenhouse roses by brief exposure to supplemental UV-B radiation. *Plant Dis.* 96: 1653–1660.

Taliansky, M., Mayo, M.A. and Barker, H. 2003. Potato leaf roll virus: a classic pathogen shows some new tricks. *Molecular Plant Pathology*, 4(2): 81–89.

Thompson, M.G., Moore, W.M., Hummel, N.F.C., et al. 2020. *Agrobacterium tumefaciens*: A bacterium primed for synthetic biology. *BioDesign Research*. https://doi.org/10.34133/2020/8189219

Tipu, M.M.H., Masud, M.M., Jahan, R., et al. 2021. Identification of citrus greening based on visual symptoms: A grower's diagnostic toolkit. *Heliyon* 7, e08387. https://doi.org/10.1016/j.heliyon.2021.e08387.

Vishwakarma, S.K., Kumar, P., Nigam, A., et al. 2013. Pokkah Boeng: An Emerging Disease of Sugarcane. *J. Plant Pathol. Microb.* 4:170. doi:10.4172/2157-7471.1000170.

Viswanathan, R., Chinnaraja, C., Karuppaiah, R., et al. 2011. Genetic diversity of sugarcane grassy shoot (SCGS)-Phytoplasmas causing grassy shoot disease in India. *Sugar Tech, DOI* 10.1007/s12355-011-0084-2.

Viswanathan, R. 2013. Status of sugarcane wilt: one hundred years after its occurrence in India. *Journal of Sugarcane Research*, 3(2): 86–106.

Williamson, B., Tudzynski, B., Tudzynski, P. et al. 2007. *Botrytis cinerea*: the cause of grey mould disease. *Molecular Plant Pathology*, 8(5): 561–580.

Wu, T.Y. and Kirschner, R. 2017. A brief global review on the species of cucurbit powdery mildew fungi and new records in Taiwan. *Mycologia Iranica*, 4(2): 85–91.

Yadav, M.S., Singh, S., Ahmad, N., et al. 2012. Sclerotinia rot: A Serious Problem of Rapeseed-Mustard in Rajasthan and Haryana. ICAR-NCIPM, LBS Building, IARI Campus, New Delhi 110 012, *Technical Bulletin*, pp. 15.

Yao, Z., Zou, C., Peng, N., et al. 2020. Virome identification and characterization of *Fusarium sacchari* and *F. andiyazi*: Causative agents of pokkah boeng disease in sugarcane. *Front. Microbiol.* 11:240. doi: 10.3389/fmicb.2020.00240.

Zhao, Y-qiang, Tian Yan-li, Wang Li-min, et al. 2019. Fire blight disease, a fast-approaching threat to Apple and pear production in China. *Journal of Integrative Agriculture*, 18(4): 815–820.

Index

पिशाच की प्रशंसा में...

'आपके लिए दुआ है मेरी। लिखते रहें, कंसिस्टेंसी रखें।'

—सुरेंद्र मोहन पाठक (लेखक)

'संजीव पालीवाल का नया क्राइम थ्रिलर, *पिशाच*, पढ़ा। मज़ा आ गया। हिंदी क्राइम थ्रिलर को नया सुपर स्टार मिल गया।'

—आशुतोष (वरिष्ठ पत्रकार)

'अपराध कथा लेखन की परंपरा को संजीव पालीवाल ने नए सिरे से खड़ा कर दिया है।' *—प्रभात रंजन (लेखक और मॉडरेटर जानकी पुल)*

'*नैना* से अगर एक लेखक के रूप में संजीव पालीवाल का उदय हुआ, तो *पिशाच* निश्चित ही उन्हें स्थापित करता है।'

—शम्स ताहिर खान (सीनियर एक्ज़ीक्यूटिव एडिटर, आजतक)

'*पिशाच* तो बवाल है।'

—संजय पुगलिया (प्रेसीडेंट और एडिटोरियल डायरेक्टर, दि क्विंट)

'यक़ीनन शब्दों में वो ताक़त होती है जो आपकी सोच को दूर तक ले जाती है। *पिशाच* को पढ़कर अहसास होता है।' *—चित्रा त्रिपाठी (एंकर, आजतक)*

'*पिशाच* एक ऐसा उपन्यास है जिसको पढ़ते समय पाठक स्व को भूल जाता है। लेखक पाठकों को ऐसी दुनिया में ले जाते हैं, जहां जाने का दुस्साहस कम ही लोग कर पाते हैं। इसकी भाषा और कथ्य उत्तर आधुनिक है।'

—अनंत विजय (एसोसिएट एडिटर, दैनिक जागरण)

'*पिशाच* एक ऐसी मर्डर मिस्ट्री है, जिसमें आपको क़ातिल से प्यार हो जाता है।' *—ऋचा अनिरुद्ध (एंकर, ज़िंदगी लाईव, ज़िंदगी विद ऋचा)*

'*पिशाच* सम्मोहित करती है। संभवत: अंग्रेज़ी में इसे ही अनपुटडाउनेबल कहते हैं।'

—सत्य व्यास (लेखक)

'*पिशाच* को पढ़ना शुरू किया तो रुकना मुश्किल हो गया।'

—विनोद कापड़ी (फिल्म निर्देशक, लेखक, पत्रकार)

'*पिशाच* में एक सपना कुतर रहा है दूसरे सपने का जिस्म। *पिशाच* में बहुत कुछ अनकहा भी है और उस अनकहे से उपजे स्पेस में पाठकों की कल्पना ढेर सारी कुलांचे भरती है।'

—देव प्रकाश चौधरी (अमर उजाला)

'संजीव पालीवाल अपने दूसरे उपन्यास *पिशाच* के साथ हाज़िर हैं। इसका भी कथानक उनके पहले उपन्यास की तरह अपराध, हत्या और चैनलों की दुनिया का स्याह सच है। पर *पिशाच* का कैनवस पहले के मुकाबले में बड़ा और ज़्यादा चौंकाने वाला है।'

—दैनिक हिन्दुस्तान

'पहली बार किसी लेखक ने साहित्य के भीतर के अपराध पर ऐसी दहला देने वाली साहसिक कथा लिखी है। उपन्यास पढ़ते हुए झटके खाने को तैयार रहें।'

—गीताश्री (न्यूज़18.कॉम)

'*पिशाच* एक प्रतिशोध, यात्रा के हर पड़ाव पर रिश्तों, चर्चित चेहरों की चमक, ख़बरों की प्रतिद्वंद्विता और कथित विद्रोही तेवरों पर चढ़ी नैतिकता को नोंच कर फेंक देती है।'

—तस्लीम खान (नवजीवन)

'आप पढ़ना शुरू करते हैं, फिर आपको अपने साथ बहा ले जाती है। आप कथानक के ताने-बाने में इस कदर उलझते हैं कि आमतौर पर दो-दो पंक्तियां कूद कर उपन्यास ख़त्म करने वाले भी, दम साधे आगे बढ़ते हैं।'

—विनोद अनुपम (लेखक, फिल्म समीक्षक)

'हिंदी में खो गई पाठकीयता को बढ़ाने के लिए *पिशाच* एक बड़ी किताब है।'

—रत्नेश्वर सिंह (लेखक और आयोजक, पटना पुस्तक मेला)